Agent-Based Social Systems

Volume 14

This series is intended to further the creation of the science of agent-based social systems, a field that is establishing itself as a transdisciplinary and cross-cultural science. The series will cover a broad spectrum of sciences, such as social systems theory, sociology, business administration, management information science, organization science, computational mathematical organization theory, economics, evolutionary economics, international political science, jurisprudence, policy science, socioinformation studies, cognitive science, artificial intelligence, complex adaptive systems theory, philosophy of science, and other related disciplines.

The series will provide a systematic study of the various new cross-cultural arenas of the human sciences. Such an approach has been successfully tried several times in the history of the modern science of humanities and systems and has helped to create such important conceptual frameworks and theories as cybernetics, synergetics, general systems theory, cognitive science, and complex adaptive systems.

We want to create a conceptual framework and design theory for socioeconomic systems of the twenty-first century in a cross-cultural and transdisciplinary context. For this purpose we plan to take an agent-based approach. Developed over the last decade, agent-based modeling is a new trend within the social sciences and is a child of the modern sciences of humanities and systems. In this series the term "agent-based" is used across a broad spectrum that includes not only the classical usage of the normative and rational agent but also an interpretive and subjective agent. We seek the antinomy of the macro and micro, subjective and rational, functional and structural, bottom-up and top-down, global and local, and structure and agency within the social sciences. Agent-based modeling includes both sides of these opposites. "Agent" is our grounding for modeling; simulation, theory, and realworld grounding are also required.

As an approach, agent-based simulation is an important tool for the new experimental fields of the social sciences; it can be used to provide explanations and decision support for real-world problems, and its theories include both conceptual and mathematical ones. A conceptual approach is vital for creating new frameworks of the worldview, and the mathematical approach is essential to clarify the logical structure of any new framework or model. Exploration of several different ways of real-world grounding is required for this approach. Other issues to be considered in the series include the systems design of this century's global and local socioeconomic systems.

More information about this series at http://www.springer.com/series/7188

Aki-Hiro Sato
Editor

Applications of Data-Centric Science to Social Design

Qualitative and Quantitative Understanding of Collective Human Behavior

Springer

Editor
Aki-Hiro Sato
Yokohama City University
Kanazawa-ku, Yokohama-shi,
Kanagawa, Japan

Japan Science and Technology Agency PRESTO
Kawaguchi-shi, Saitama, Japan

Statistical Research and Training Institute,
Ministry of Internal Affairs and Communications
Shinjuku-ku, Tokyo, Japan

ISSN 1861-0803
Agent-Based Social Systems
ISBN 978-981-10-7193-5 ISBN 978-981-10-7194-2 (eBook)
https://doi.org/10.1007/978-981-10-7194-2

This Springer imprint is published by the registered company Springer Nature Singapore Pte Ltd.
The registered company address is: 152 Beach Road, #21-01/04 Gateway East, Singapore 189721, Singapore

Preface

This book aims to understand human behavior and to address methods for social design from both qualitative and quantitative perspectives based on the methodology of data-centric science. Recently, a large amount of data formatted by various expressions are accumulated in different fields of socioeconomic systems. However, at this moment, it seems that the utilization of such data has not been matured, yet. Since data are not answers but questions inspiring people who are working in the fields, many efforts of data analysis are required to obtain insights from data. Furthermore, the outcomes of data analysis should support human decision-making from qualitative and quantitative points of view.

The design process is a sequence of actions with the decision-making. There must be a lot of useful points of data-centric science to design. Primarily, this book has been edited by motivating to collect methodologies and practical examples for contributing to data-driven social design. The intentions behind this book are to develop the methodology to support people who are joining or attempt to join all the activities related to "social design" based on data.

This book contains three main parts, such as:

Part I: Methods for data analysis and design
Part II: A mathematical foundation of collective human behavior
Part III: Applications of data analysis to social design

Part I mentions a conceptual framework of applications of data-centric science to social design and includes four chapters. Readers may find methodological ideas and frameworks that can be used when they consider design for social systems from the chapters in Part I.

The chapter titled "How to Design Society from a Data-Centric Point of View" by Aki-Hiro Sato explains the intentions behind this book including the three types of aspects: human behavior, data-centric science, and social design. The chapter titled "Practical Methods for Data Analysis" by Aki-Hiro Sato explains practical methods used in data utilization. The data analysis is an essential part to understand the phenomena observed in the actual environment. The data analysis forms a workflow consisting of data acquisition, data collection, data visualization, and quantification and data interpretation. The purpose of data analysis is to find insights on phenomena that we have attended to and make decision-makers change their behavior. The chapter titled "An Approach to Product Design Involving Heterogeneous Stakeholders" by Aki-Hiro Sato discusses how to design our product regarding data and communications among multi-stakeholders. Several methods to design our product as group work are addressed. Specifically, eight techniques that can be used in the group work are introduced step by step. The chapter titled "Designing Human-Machine Systems Focusing on Benefits of Inconvenience" by Hiroshi Kawakami proposes benefits of inconvenience in human-machine systems to prevent loss of human ability and joy in too convenient human-machine systems.

Part II introduces collective human behavior and information cascade in social groups from a mathematical physics point of view. Collective human behavior is often seen in socioeconomic systems and should be understood when readers want to deal with socioeconomic systems involving heterogeneous stakeholders. Readers can understand collective human behavior with information interactions from the chapters in this part.

The chapter titled "Information Cascade and Phase Transition" by Masato Hisakado and Shintaro Mori discusses a voting model with two candidates and concludes that the model features a phase transition beyond which a state where most voters make the correct choice coexists with one where most of them are wrong as the fraction of herders increases; the chapter titled "Information Cascade, Kirman's Ant Colony Model, and Kinetic Ising Model" by Masato Hisakado and Shintaro Mori discusses a voting model in which voters can obtain information from a finite number of previous voters. In this case, the phase transition can be observed in infinite-size limit, but no phase transition happens when the reference number is finite. The chapter titled "Information Cascade and Networks" by Masato Hisakado and Shintaro Mori discusses a voting model on networks and concludes that the influence of networks can be seen in the convergence speed only and cannot be seen in the information cascade transition. The chapter titled "The Pitman-Yor Process and Choice Behavior" by Masato Hisakado and Shintaro Mori discusses choice behavior using a voting model in which voters can obtain information from a finite number of some previous voters and concludes that the posting process is described by the proposed model for analog herders for a small reference number. The chapter titled "Domino Effect in Information Cascade" by Shintaro Mori and Masato Hisakado studies several simple models for information cascade and proposes that the memory length in the sequential decisions and the number of stable states (equilibrium) play the key role in the domino effect in the information cascade. The chapter titled "Information Cascade Experiment: General Knowledge Quiz" by Shintaro Mori and Masato Hisakado and the chapter titled "Information Cascade Experiment: Urn Quiz" by Shintaro Mori and Masato Hisakado deal with information cascade experiments. The difficulty of the questions depends on the sequential collective behavior. The chapter titled "Information Cascade and Bayes Formula" by Masato Hisakado and Shintaro Mori considers a voting experiment using two-choice questions and attempts to establish the Bayes formula to correct the wrong decisions.

Part III provides readers with several case studies of social systems based on a data-driven approach. The final four chapters provide useful insights and domain knowledge in each case study. Readers may obtain insights and domain knowledge to deepen understandings of socioeconomic systems from data analysis. These chapters may be useful when they obtain domain knowledge, practical examples, need specification, and design of socioeconomic systems. All chapters are useful when readers attempt to solve social issues and deal with social design by using the methodology of data-centric science.

The chapter titled "How Betters Vote in Horse Race Betting Market" by Shintaro Mori and Masato Hisakado analyzes JRA (Japan Racing Association) win betting data and how the efficiency and the accuracy improve as betting proceeds. The chapter titled "Smart Micro-sensing: Reaching Sustainability in Agriculture via Distributed Sensors in the Food Chain" by Rob Dolci and Laura Boschis proposes a method to detect contaminants during the whole food supply chain by using portable biosensor devices. The proposed method can improve food safety in both agriculture and food industries. The chapter titled "High-Frequency Data Analysis of Foreign Exchange Markets" by Aki-Hiro Sato proposes a method to characterize collective human behavior based on a multivariate Poisson model and shows results obtained from empirical analysis with high-frequency foreign exchange data. The common mode may measure collective human behavior from observable activity data. The chapter titled "On Measuring Extreme Synchrony with Network Entropy of Bipartite Graphs" by Aki-Hiro Sato proposes a method to quantify the structure of a bipartite graph using a network entropy per link. The network entropy on a bipartite network per link was used to quantify the degree of collective behavior in the foreign exchange market. The proposed method is applicable to detecting extreme synchrony in various types of socioeconomic systems.

This book will be devoted to promoting data-driven social design with multi-stakeholders through design activities consisting of data acquisition, data collection, data analysis, data interpretation, reporting, and decision-making by accumulating data, methods, and connections with stakeholders.

Yokohama, Japan — Aki-Hiro Sato

Contents

Part I
Methods for Data Analysis and Design

How to Design Society from a Data-Centric Point of View

Aki-Hiro Sato

1 Introduction

Many literatures are now engaged in terms of data-driven *X*, in which *X* can be replaced with various concepts in different areas of investigation, research, and practice. Examples include the following concepts: data-driven innovation, data-driven decision-making, data-driven investigation, data-driven network analysis, data-driven design, data-driven simulation, data-driven management, data-driven programming, data-driven marketing, data-driven security, and data-driven science. Data-driven innovation concerns product development based on data. Data-driven decision-making is the principle of making decisions based on data. A data-driven investigation is an investigation methodology both based on and driven by data. Data-driven network analysis is a method to analyze networks based on data. Data-driven design is a paradigm for products and service design based on data. Data-driven approaches may have the potential to improve various aspects of knowledge creation in society.

The author published a book entitled *Applied Data-Centric Social Sciences* (Sato 2014), which set out several fundamental ideas concerning data analysis, mathematical expressions, and computational abilities.

Ideally, science can be separated into four faces based on two axes such as deductive-inductive and human-cyber dimensions (Kitagawa 2010). The deductive-inductive axis characterizes scientific methods according to their philosophical basis. The deductive method assumes that all insights can be derived from a few

A.-H. Sato (✉)
Yokohama City University, Kanazawa-ku, Yokohama-shi, Kanagawa, Japan

Japan Science and Technology Agency PRESTO, Kawaguchi-shi, Saitama, Japan

Statistical Research and Training Institute, Ministry of Internal Affairs and Communications, Shinjuku-ku, Tokyo, Japan
e-mail: ahsato@yokohama-cu.ac.jp

A.-H. Sato (ed.), *Applications of Data-Centric Science to Social Design*,
Agent-Based Social Systems 14, https://doi.org/10.1007/978-981-10-7194-2_1

fundamental theoretical principles. In this case, we believe that there are basic equations or a few basic components (models) that can be used to explain the variety of observations that occur in reality. When following a deductive approach, the major task is to construct fundamental theories that will explain actual observations. Meanwhile, the inductive method takes for granted that we do not already possess fundamental models that could explain all empirical data. In this sense, inductive methods aim to uncover patterns and categorize the relationships among observations. Meanwhile, the human-cyber axis assesses scientific methods in a behavioral manner. The human-centric method focuses on scientific operations by manual. Theoretical sciences are located at the intersection of deductive and human-centric science. Experimental science is located where inductive and human-centric science coincide. Meanwhile, computational science is located at the coextent of deductive and cyber-enabled science, and data-centric or data science is located in the quadrant belonging to inductive and cyber-enabled science.

Data-centric science, then, studies data processing techniques, data visualization techniques, and data analytics in relation to a data analysis pipeline consisting of data acquisition, data collection, data analysis, interpretation, and decision-making. Data analysis may start with data acquisition by making connections with potential stakeholders who manage data. It is often necessary to establish contracts with several stakeholders to receive their data. Next, data collection must be undertaken, consisting of available data. Data normalization should be designed to incorporate database servers at the next stage, which may coincide with validation or verification of data. Data quality management and data management should then be considered. Afterward, data can be analyzed to understand the nature of the phenomena that they express. Eventually, this process accumulates knowledge about the events and phenomena that correspond to the data. Consequently, we can construct a picture of the relationship between these phenomena and collected data at the data interpretation stage for decision-making.

2 How to Use Data for Social Design

Three pillars of data analysis applications are nowcasting, forecasting, and doing design. The nowcast is a tool with which to understand past and current situations, including ongoing events, phenomena, and environmental factors. Meanwhile, forecasting is intended to predict future situations based on insights concerning past and current situations. Design, meanwhile, relies on the ability to identify mechanisms of artificial and natural events and phenomena and find a solution to solve issues by using controllable causal and correlational relationships among them. Specifically, data analysis has a potential to drastically improve design from both quantitative and qualitative perspectives since it enables us to improve our understandings of causal and correlational relationships and help us make decisions in design components.

Social systems can be constructed based on several components including rules, organizations, individuals, roles, natural environments, artificial objects, technology, and human relationships (Lazer et al. 2009).

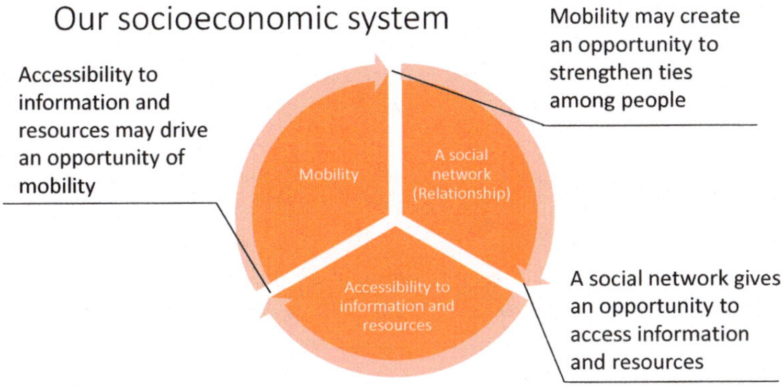

Fig. 1 Socioeconomic system

From a functional perspective, society can be modeled using three attributes: a social network (relationships), access to information and resources, and mobility. These three components establish circular causality as shown in Fig. 1.

Social networks give us opportunities to access information and resources, which may in turn propel an increasing opportunity for mobility. Finally, mobility may create an opportunity to strengthen or weaken ties among individuals.

How can data be used to create the characteristics desired to serve specified social functions? This is the main issue dealt with in this chapter. While many studies use data accumulated within society, few of them have yet used these data to improve society. To verify our interpretation derived from data analysis and available data, we should understand the difference between data collection to problem-solving in a nuanced way. Data are not answers in themselves but do inspire questions for both researchers and practitioners. Data tell us how a phenomenon behaves and at the same time prompt the question why a phenomenon behaves as observed.

3 How to Use Data for Social Storytelling

Data about society reflects the activities of each of its individual members. In fact, each one independently executes their activities while only interacting with part of the rest of the population. Therefore, it is worthwhile to find the patterns of our social activity based on data and understand our society for obtaining insights to construct and improve our social systems.

Researchers have two main types of methodologies available: data-driven and purpose-driven methods. A data-driven method aims to identify a societal issue as a focus of attention, if the researcher does not begin with a specific purpose in mind. A data-driven method is useful to find an issue (problem) that should then be solved. Meanwhile, a purpose-driven method aims to construct procedures to solve the issue or resolve the issue. If researchers begin a study having already identified a problem to be solved, then they can collect data about the problem and attempt to find a solution to solve it. In this case, identifying causality related to the issue is useful if the problem results from a clear, linear relationship of cause and effect.

However, many social problems result from a circular relationship between cause and effect, also referred to as circular causality. This implies that the action of one person or organization results in an effect for other persons or organizations and that at the same time the potential actions of any person or organization themselves result from effects caused by other persons or organizations.

In general, it is difficult to detect time differences between a cause and an effect in circumstances of circular causality. All researchers can do is to measure correlations between the cause (effect) and the effect (cause). To analyze the problem from a causal perspective, a story (narrative) about cause and effect is also required.

To tell a story based on data about society, two types of approaches may be taken. One approach is to construct stories based on available data. Data about human activities are collected by individuals or organizations. Researchers can tell a story about the available data using a journey map technique (Miaskiewicz and Kozar 2011; Richardson 2010; Moon et al. 2016), for example. To examine a story about the data, analysts can draw a journey map related to the data. The overall story about the data can be expressed using several journey maps that draw different Personas (Nielsen 2013). Such a story is also profoundly related to workflow about data collection, data summarization, and data dissemination.

Another approach is to understand human life using data based on a holistic perspective since society consists of many individuals. The lifelog data (Botzheim et al. 2013; Stefan 2016) and survey-based official statistics (Bureau of Statistics in Japan 2016) about everyday life are useful from this perspective.

In the case of Japan, we have official statistics about everyday Japanese life in the *Survey on Time Use and Leisure Activities*, (Bureau of Statistics in Japan 2016). This survey aims to obtain comprehensive data on the daily patterns of time allocation and leisure activities in Japanese society; it has been conducted every 5 years since 1976. This survey provides statistics that are not obtainable from others, which focus almost exclusively on economic aspects of living. These statistics make it possible to observe the lifestyles of various groups and evaluate preferences for certain activities over others, as well as to improve the interpretation and understanding of various social and economic phenomena. This survey also provides essential background information on economic conditions.

The Survey on Time Use and Leisure Activities in 2016 classifies daily activities into 20 categories and surveyed based on 15-minute time slots. The respondents classify and record their activities on the reference date of the survey. When respondents are engaged in more than one activity at the same time, they report activity that they consider to be the main one. The 20 categories of activities are grouped into three broad areas, called primary, secondary, and tertiary activities.

Primary activities are those that are physiologically necessary and consist of "sleep," "personal care," and "meals." Secondary activities comprise those which each person is committed to performing as a member of a family or of the wider society. This category includes "commuting to and from school or work," "work (for pay or profit)," "school-work," "housework," "caring or nursing," "child care," and "shopping." Tertiary activities include all other activities such as "learning, self-education, and training (except for schoolwork)," "hobbies and amusements," "sports," and "volunteer and social activities." Time spent in tertiary activities corresponds to what is usually called "free time." Table 1 shows the weekly average time used in Japan for each kind of activity, grouped by sex, in 2011 and 2016. We can see that in 2016, 10.41 h are used for primary activities, 6.57 h for secondary activities, and 6.22 h for tertiary activities.

By using Personas and the journey map technique (Cooper and Saffo 1999; Nielsen 2013), we can draw typical journey maps of the three types of activities as shown in Fig. 2. The journey map of everyday life is quite meaningful. The utility curve of primary activities is a function that increases with time. Meanwhile, the utility curve of secondary and tertiary activities may increase or decrease as a function of time. The maxima of the tertiary activities are usually higher than the maxima of the secondary activities.

4 How to Use Data to Design Social Systems

Data consists in a collection of observations or facts. Efforts to analyze data and to apply insights obtained from data analysis to solve problems are also important when designing social systems. In such cases, participatory processes (Kita et al. 2008; Kita and Mori 2009) and a CAPD cycle (Check, Action, Plan, and Do) is quite a useful method, representing a version of the Deming cycle (Deming 1986).

The first step of the CAPD cycle is to Check. In the Check phase, researchers attempt to analyze the current situation of the social system that they aim to improve. The next step of the cycle is Action, when they identify what they want to improve in the selected social system more specifically. In the Plan phase, they carefully construct a plan to modify the social system (organization, communication, information, management, and so on). Finally, in the Do phase, they execute the plan constructed in the Plan phase. In terms of this procedure, the C and A phases form the basis of the gap analysis at the initial stage.

So how can the CAPD cycle be applied to solving social problems? A classical approach to solving social problems consists of the following:

1. Specialists analyze social problems using advanced techniques and various types of data and propose several potential solutions using insights about phenomena.
2. Specialists design plans to implement the optimal and most efficient solutions identified.
3. Stakeholders in the actual social world select an optimal solution from various proposals to solve social problems initiated outside of political organizations.
4. Stakeholders implement the optimal solution they have chosen.

Table 1 Weekly average time usage (in hours) in Japan for each kind of activity by sex in 2011 and 2016

	Total			Male			Female		
	2011	2016	Difference	2011	2016	Difference	2011	2016	Difference
Primary activities	10.40	10.41	0.01	10.33	10.34	0.01	10.46	10.49	0.03
Sleep	7.42	7.40	−0.02	7.49	7.45	−0.04	7.36	7.35	−0.01
Personal care	1.19	1.22	0.03	1.09	1.11	0.02	1.29	1.31	0.02
Meals	1.39	1.40	0.01	1.36	1.38	0.02	1.42	1.43	0.01
Secondary activities	6.53	6.57	0.04	6.49	6.50	0.01	6.57	7.03	0.06
Work and work-related activity	4.43	4.49	0.06	6.08	6.08	0.00	3.23	3.35	0.12
Commuting to and from school or work	0.31	0.34	0.03	0.40	0.43	0.03	0.23	0.25	0.02
Work	3.33	3.33	0.00	4.46	4.41	−0.05	2.23	2.29	0.06
School work	0.39	0.42	0.03	042	0.44	0.02	0.37	0.41	0.04
Housework and related work	2.10	2.08	−0.02	0.42	0.44	0.02	3.35	3.28	−0.07
Housework	1.27	1.23	−0.04	0.18	0.19	0.01	2.32	2.24	−0.08
Caring and nursing	0.03	0.04	0.01	0.02	0.02	0.00	0.05	0.06	0.01
Child care	0.14	0.15	0.01	0.05	0.06	0.01	0.23	0.24	0.01
Shopping	0.26	0.26	0.00	0.17	0.17	0.00	0.35	0.34	−0.01
Tertiary activities	6.27	6.22	−0.05	6.38	6.36	−0.02	6.16	6.09	−0.07
Moving (excluding commuting)	0.30	0.29	−0.01	0.29	0.28	−0.01	030	0.30	0.00
Watching TV, listening the radio, reading newspapers, or magazines	2.27	2.15	−0.12	2.31	2.19	−0.12	2.24	2.11	−0.13
Rest and relaxation	1.31	1.37	0.06	1.31	1.37	0.06	1.31	1.36	0.05
Learning, self-education, and training (excluding schoolwork)	0.12	0.13	0.01	0.13	0.13	0.00	0.12	0.12	0.00
Hobbies and amusements	0.44	0.47	0.03	0.53	0.57	0.04	0.37	0.37	0.00
Sports	0.14	0.14	0.00	0.18	0.18	0.00	0.11	0.10	−0.01
Volunteer and social activities	0.04	0.04	0.00	0.04	0.04	0.00	0.04	0.04	0.00
Social life	0.19	0.17	−0.02	0.18	0.15	−0.03	0.20	0.19	−0.01
Medical examination or treatment	0.08	0.08	0.00	0.07	0.07	0.00	0.10	0.09	−0.01
Other activities	0.17	0.19	0.02	0.15	0.17	0.02	0.18	0.20	0.02

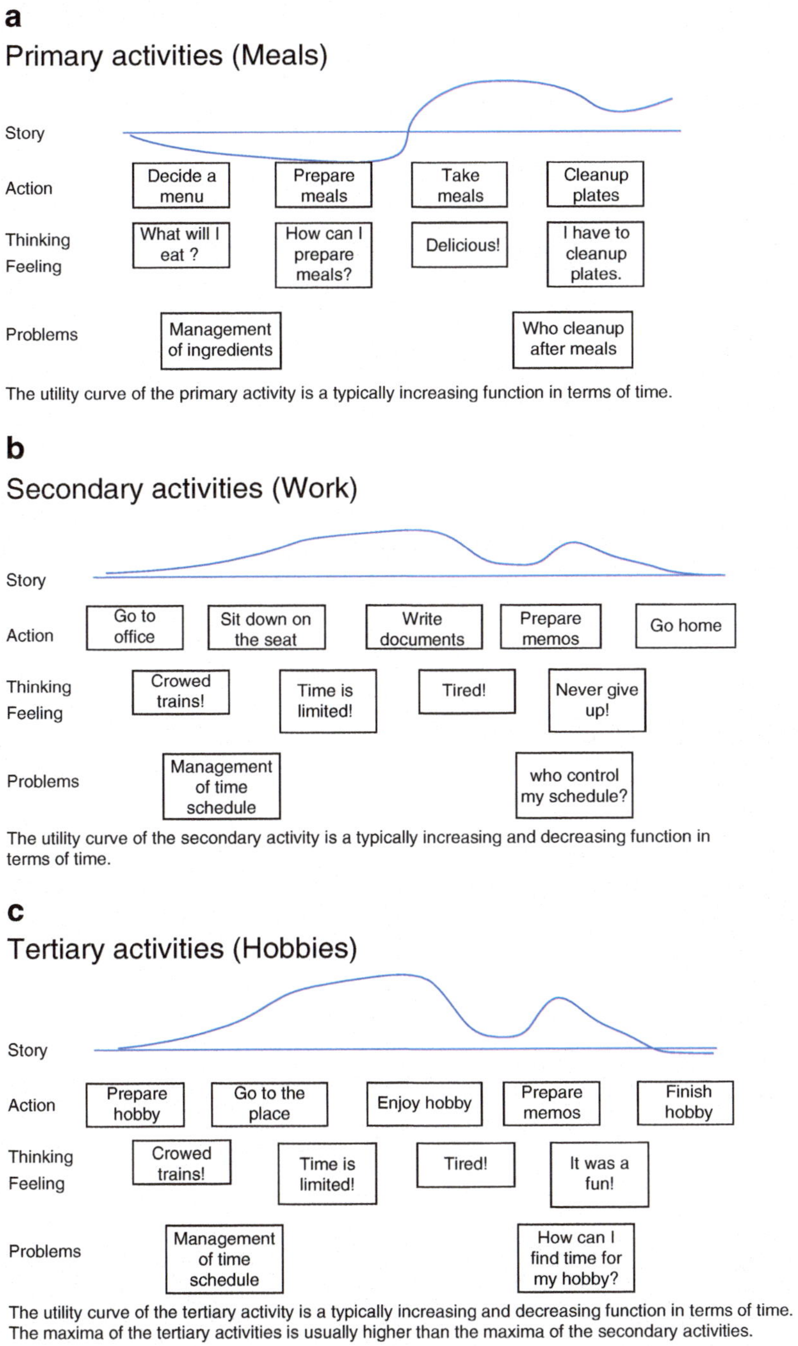

Fig. 2 Examples of journey maps concerning three types of activities. (**a**) Primary activity (meals), (**b**) secondary activity (work), and (**c**) tertiary activity (hobbies)

However, the classical approach to solving social problems only works in principle, since in practice specialists cannot understand all social phenomena due to their complexity, while stakeholders cannot understand the plan designed by specialists either, due to its complexity. Therefore, a modern approach to solving social problems has been proposed as follows:

1. Specialists identify general areas of interest to focus on based on their research activities.
2. Specialists organize groups of varied stakeholders in the areas identified, trying to analyze and discuss social problems through participatory processes.
3. Both stakeholders and specialists analyze data related to social problems and share their understanding of the problems along with visions of future society.
4. Based on shared visions, stakeholders design and implement plans to solve social problems within their areas of influence and control.

This modern approach to solving social problems has also been articulated in the 12 steps used to develop indicators of sustainable development for tourism by the World Tourism Organization (*Indicators of Sustainable Development for Tourism Destinations: A Guidebook* 2004). Good indicators offer some of the benefits needed for better decision-making, illuminating emerging issues, identifying impacts, measuring the performance of monitored activities, and reducing the risk of planning mistakes. Indicators have several levels of resolution in terms of both their spatial and temporal aspects. From a spatial viewpoint, there is a hierarchical structure such as the global, national, regional, subregional, and gridded levels. Moreover, specific places, organizations, and establishments may be used as a unit of measurement for indicators.

Indicators are categorized into six types:

1. Early warning indicators
2. Indicators of stresses on the system
3. Measures of the current state of the industry
4. Measures of the impact of implementation on the environment
5. Measures of management effort
6. Measures of management effect and performance

Measurements are separated into two types such as quantitative and qualitative (or normative) measurements. Typical examples of quantitative measurements include raw quantities, ratios, and percentages, while, typical qualitative and normative measurements are mainly opinion-based indicators.

Moreover, we need to understand collective human behavior often observed in socioeconomic systems. The simplest model of social collective behavior is an Ising model, which was firstly proposed for understanding magnetisms. In the Ising model, interactions among components and noises produce order parameters governing each component. The order parameters appearing in a social group also may coordinate human individual behavior. Key performance indicators (KPIs) often may be introduced in a socioeconomic system to subject order parameters based on data-driven methods.

5 UNWTO Framework

This section explains the 12 steps for sustainable tourism development proposed by the World Tourism Organization (UNWTO) (*Indicators of Sustainable Development for Tourism Destinations: A Guidebook* 2004). The 12 steps are classified into three phases. The three phases of this process consist of research and organization, indicator development, and implementation. KPIs may be used to direct our society toward what we desire. The three phases grouping 12 steps are as follows:

Phase 1. Research and organization

Step 1. Definition/delineation of the destination
Step 2. Use of participatory processes
Step 3. Identification of tourism assets and risks
Step 4. A long-term vision for a destination

Phase 2. Indicator development

Step 5. Selection of priority
Step 6. Identification of desired indicators
Step 7. Inventory of data sources
Step 8. Selection of procedures

Phase 3. Implementation

Step 9. Evaluation of feasibility/implementation
Step 10. Data collection and analysis
Step 11. Accountability, communication, and reporting
Step 12. Monitoring and evaluation of indicators application

As depicted in these 12 steps, data collection and data analysis take place at the implementation phase. These activities are organized in the form of workshops or conferences in which stakeholders participate.

The essential activity across the 12 steps is a gap analysis between the current situation of the community to which participating stakeholders belong and the long-term vision that can be shared by this community. After selecting priority issues in the gap recognized by the community, participants attempt to construct KPIs that can be measured by community members and to implement an organizational reporting activity. KPIs can thus be used to improve the collective behavior of the community, if all community members pay attention to the KPIs and change their behavior based on what they observe.

The most essential components of this approach include the data (information) flow needed to derive the KPIs and an organization willing to maintain them. Specifically, the dominant factor concerns human collective behavior. Data can form human relationships as they pass on information. Thus, the design and implementation of data flow and organization can be the most essential step in developing indicators. Data analysis and deployment of indicators sometimes need experts such as data scientists and ICT engineers due to high technical levels and the complexity of social systems.

5.1 *Phase 1. Research and Organization*

Step 1. Definition/Delineation of the Place or Domain Definition of the place and domain is a necessary first step. Practitioners should find key stakeholders representing the place or domain as part of their research and organizing activities. Moreover, they need to understand the current state of the place or domain. At this step, they understand that significant amounts of information may already exist by which to understand the issues and the scope of the project to be planned. To develop indicators, a participatory process is essential; this is also beneficial for identifying potential sources of information for indicators.

Step 2. Use of Participatory Processes Key factors to consider when building a participatory process are timing, frequency, duration, and consultation techniques. Evaluating the timing for participants to obtain knowledge about the system is essential. The frequency of meeting or events may be adjusted. If they are not frequent enough, participants may forget the processes of the project. Meanwhile, meeting too frequently may result in less participation since participants feel overloaded. Indicator development should be undertaken as a long-term activity. However, the development phase should be done based on a project with a finite duration and budget. The size of the group for any participatory process must also be considered. Managing large groups is quite difficult at the initial stage. A growth model for participatory processes may be used when introducing a CAPD cycle. As a consultation technique, organizing meetings with selected stakeholders may be useful.

Step 3. Identification of Assets and Risks The purpose of this step is to identify the assets within the relevant place or domain and to understand what the key assets are and which of their elements are valued by suppliers as well as current and potential consumers. Assets are defined by the benefits they produce within the place or domain. Moreover, it is necessary to understand the sensitivity of such assets to changes in demand, which may entirely reconfigure their importance. Moreover, it is useful to assess the strengths, weaknesses, opportunities, and threats (SWOT) of the place or domain.

Step 4. The Long-Term Vision for a Place or Domain Indicators used to organize participants should be based on the agreement of stakeholders within the place. To accomplish this, a long-term vision should be developed that can be broadly shared within the place. At a workshop, participants representing each stakeholder may discuss a long-term vision that can be shared broadly within the place or domain. For example, maintaining a community over an extended period of time is typically more essential than high economic performance. In the case of an industrial organization, profit is often more important than environmental issues. Thus, priorities depend strongly on stakeholder perspectives. This results in trade-offs among stakeholders and may turn out to be a multi-objective optimization problem.

5.2 *Phase 2. Indicator Development*

Step 5. Selection of Priority Issues and Policy Questions The purpose of this step is to identify the most important issues while considering the perspectives of all stakeholders. If possible, priority issues and policy questions should be identified using a participatory group approach.

The objective at this stage is to obtain consensus involving a broad range of participants on a list of important issues identified in a participatory workshop.

Step 6. Identification of Desired Indicators Based on the risks and policy questions identified, multi-stakeholders may form a technical group consisting of experts from several domains and can define a list of possible indicators that might be used to understand the risks and issues and to manage the place or domain. In certain cases, some of the most potentially useful indicators may not be feasible due to technical, financial, staff, or other constraints related to data collection and data processing. Such indicators can be set aside for further development. In some cases, the range of processes to be used in indicator identification is highly technical; stakeholders may need help from a technical group to identify the indicators they desire and find most useful.

Step 7. Inventory of Data Sources To calculate indicators, data sources are required. Indicator selection requires information about currently available and potentially obtainable data sources to calculate indicators. Two methods may be used to develop indicators, namely, data-driven and issue-driven approaches. The data-driven approach is to develop indicators that can be calculated based on a list of available data sources. This approach is particularly effective at places or domains where extensive data source are already available. Meanwhile, the issue-driven approach is to develop indicators for what stakeholders want to understand to solve political issues. After identifying the issue and developing an indicator, the feasibility of the indicator must be evaluated. This approach is useful in cases where further development is needed to improve the current situation. Based on desired indicators, a mechanism to collect data that has not previously existed can be installed in the system. The process of indicator development needs to ask questions: what information is needed to calculate indicators, what data can be created or obtained now, and how information sources can be improved in the future.

Step 8. Selection Procedures Indicators should be scored based on five criteria including (1) the relevance of the indicator, (2) the feasibility of obtaining and analyzing data, (3) the credibility of the information and reliability of the data for users, (4) the clarity and understandability of the indicator for users, and (5) comparability over time and across places and domains.

In many cases, the desired indicators will not be easy to produce due to technical, financial, or organizational reasons. This can occur because information sources are widely dispersed, and because the cost of data collection may be unacceptably high, or there is no authority with the capacity to gather the information. However, the selection procedure activates stakeholders throughout the discussion, which may stimulate a discussion among stakeholders and eventual future implementation activity.

5.3 Phase 3. Implementation

Step 9. Evaluation of Feasibility/Implementation Procedures To evaluate feasibility and implementation procedures, each indicator in a list of selected candidates should be specified in terms of:

1. Specific data source(s) to be used to compute the indicator
2. Specific characteristics of the data
3. The frequency of data collection
4. Time lags between collection and availability
5. Considerations of access and confidentiality
6. Reporting units, validity, and accuracy
7. Responsibility for the provision of data, data analysis, and any additional manipulation
8. Costs and technical requirements of data collection and analysis

In practice, it may not always be necessary to calculate indicators based on detailed scientific data. We need to consider what decision-makers need.

Step 10. Data Collection and Analysis For each indicator, it is essential to clarify the specific means to be used to obtain information. Existing data collected by stakeholders, extraction and manipulation of data from existing data sources, the creation of new comprehensive data, and the creation of sample data are all possibilities. It is often useful to consider data sources collected by other administrations or organizations that have not been included in a participatory process. The four types of possible indicators are quantitative, qualitative, normative, and descriptive in nature. Quantitative means variables that may be collected to measure something such as liters of water consumption per person. Qualitative measures include percentages of the population who experience a particular feeling. Normative means adaption to socially defined standards, and it is measurable from data that includes, for example, the ratio of yes/no decisions to the total survey samples.

Step 11. Accountability, Communication, and Reporting The purpose of indicators is to be used in decision-making and communication. Regular reporting to stakeholders, the public, and specific decision-makers is intended to maximize this influence. The key to effective implementation is commitment. Essentially, indicators should become part of a planning process for the place or the domain.

Step 12. Monitoring and Evaluation of the Application of Indicators Indicators should be reviewed regularly. Periodic review of indicator applications can lead to maintaining implemented KPIs from both organizational and technical perspectives or shading light on further local priorities or emerging issues. Decision-makers need to confirm current situations based on indicators and to understand changes in situations by comparing indicators with past data. It is still worthwhile to revisit indicators every few years seeking improvement.

6 Conclusions

This chapter has discussed how to understand society using data and apply insights obtained data analysis to social design. It presented examples such as journey maps about three types of activities (primary, secondary, and tertiary activities) and statistics on time use for these activities based on Japanese official statistics. In such cases, data storytelling is quite meaningful.

Participatory design approach is a helpful tool used to enhance design work with different stakeholders. This chapter also mentioned the advantages of design workshops with heterogeneous participants and how to position data analysis in relation to the design workshop. Data analysis can be used to understand current situations in both qualitative and quantitative terms. This analysis can be used as an input into the design workshop with various types of stakeholders.

A CAPD (check, act, plan, and do) cycle was introduced to consider how to influence social relationships and the behavior of members belonging to the community, from a social design perspective.

Specifically, it is crucial to consider and manage the interaction among multiple stakeholders during the design task based on the purpose of the design. In these cases, understanding of herding behavior based on dynamical models is needed for managing the design processes as well as for shaping social systems that we want to design.

References

Botzheim J et al (2013) Extraction of daily life log measured by smart phone sensors using neural computing. Procedia Comput Sci 22:883–892

Bureau of Statistics in Japan (2016) Survey on time use and leisure activities (Online). Ministry of Internal Affairs and Communications, Tokyo. Available from: http://www.stat.go.jp/english/data/shakai/2016/pdf/timeuse-a2016.pdf. Accessed 1 Aug 2018

Cooper A, Saffo P (1999) The inmates are running the asylum. Macmillan Publishing Co., Indianapolis
Deming WE (1986) Out of the crisis. Massachusetts Institute of Technology, Center for Advanced Engineering Study, Cambridge, MA
Indicators of Sustainable Development for Tourism Destinations: A Guidebook (2004) Madrid: World Tourism Organization
Kita H, Mori M (2009) Projects of participatory production and their management: practice in industrial accumulation in Suwa. In: Ohara S, Asada T (eds) Japanese project management, world scientific, pp. 361–373
Kita H, Mori M, Tsuji T (2008) Toward field informatics for participatory production, international conference on informatics education and research for knowledge-circulating society, pp 68–72
Kitagawa G (2010) Data centric science for information society. In: Takayasu H, Takayasu M, Watanabe T (eds) Econophysics approaches to large-scale business data and financial crisis. Springer, Tokyo
Lazer D et al (2009) Computational social science. Science 323:721–723
Miaskiewicz T, Kozar KA (2011) Personas and user-centered design: how can personas benefit product design processes? Des Stud 32(5):417–430
Moon H, Han SH, Chun J, Hong SW (2016) A design process for a customer journey map: a case study on mobile services. Hum Factors Ergon Manuf 26:501–551. https://doi.org/10.1002/hfm.20673
Nielsen L (2013), 30. Persona in The encyclopedia of human-computer interaction, 2nd ed (Online). Available from: https://www.interaction-design.org/literature/book/the-encyclopedia-of-human-computer-interaction-2nd-ed. Accessed 16 Aug 2018
Richardson A (2010) Series on customer journey: using customer journey maps to improve customer experience. Available at: https://hbr.org/2010/11/using-customer-journey-maps-to?referral=03759&cm_vc=rr_item_page.bottom
Sato AH (2014) Applied data-centric social sciences. Springer, Tokyo
Stefan S (ed) (2016) Lifelogging: digital self-tracking and lifelogging – between disruptive technology and cultural transformation. Springer, Wiesbaden

Practical Methods for Data Analysis

Aki-Hiro Sato

Data analysis is essential for understanding the phenomena observed in the actual environment. Data analysis forms a workflow consisting of data acquisition, collection, visualization and quantification, and interpretation. The purpose of data analysis is to find insights into phenomena that we have attended to and make decision-makers change their behavior. Data utilization should form an improvement cycle with Check, Action, Plan, and Do (CAPD). This chapter shows a fundamental definition of data (four types of data formats, such as time series, network, spatial, and linguistic data). Several methodological frameworks for analyzing data and how to use the results obtained from data analysis are discussed.

1 Introduction

Why do we need data analysis? We have some likely answers to this question. According to some, visualizing and quantifying data by applying statistical and mathematical methods to the given data should be a goal of data analysis. Others think that we should begin with data analysis to produce sufficient data collection. Both answers are partially correct. Each mentions a methodological aspect of data analysis but does not tell us the purpose of data analysis.

An essential answer to this question may be that we want to make decisions in a specific domain from a data-centric point of view. Thus, the final goal of data analysis should be to contribute to someone's decision-making. To do so, we need to consider a method for extracting information (interpretation, implication,

---

A.-H. Sato (✉)
Yokohama City University, Kanazawa-ku, Yokohama-shi, Kanagawa, Japan

Japan Science and Technology Agency PRESTO, Kawaguchi-shi, Saitama, Japan

Statistical Research and Training Institute, Ministry of Internal Affairs and Communications, Shinjuku-ku, Tokyo, Japan
e-mail: ahsato@yokohama-cu.ac.jp

A.-H. Sato (ed.), *Applications of Data-Centric Science to Social Design*, Agent-Based Social Systems 14, https://doi.org/10.1007/978-981-10-7194-2_2

and insights) from the data that we want to tackle. In this sense, data analysis should form a process consisting of four steps: acquiring, collecting, visualizing and quantifying, and interpreting data.

Specifically, in this chapter, we address existing workflows and frameworks that can be useful in practical data analysis and compare those workflows. Moreover, we discuss how to find a connection with problem discovery, problem identification, problem-solving, and design of a system in societal contexts from a data-centric point of view.

The Cross-Industry Standard Process for Data Mining (CRISP-DM) is a data mining process model commonly used in business and management (Chapman et al. 2000). This framework includes six crucial components: understanding the business, understanding the data, preparing the data, modeling, evaluating, and deploying. The business understanding is located in the initial phase of data analysis. This step focuses on understanding the project objectives and requirements from a business perspective. Knowledge about business understanding can be used as a preliminary plan designed to achieve the given project objectives. The data understanding phase starts with the initial data collection to discover the first insights into the data and interesting hypotheses.

Define, Measure, Analyze, Improve, and Control (DMAIC; ISO 13053-1:2011) is another method for using data for decision-making. This approach was developed in Six Sigma projects. This procedure is mainly used in the improvement process to produce products in manufacturing and service industries. The initial phase of DMAIC is Define. The purpose is to clarify business problems, goals, potential resources, the project scope, and hierarchical project timeline from conceptual levels to detailed levels.

The Generic Statistical Business Process Model (GSBPM; European Commission 2014) is a workflow model for producing statistics within and between statistical offices in different ways. This procedure is to be used when we produce a data product. The GSBPM includes nine components: Specify Needs, Design, Build, Collect, Process, Analyze, Disseminate, Archive, and Evaluate. The initial phase of this procedure is the specification of needs, which consists of six steps: determine the need for information, consult and confirm needs, establish output objectives, identify concepts (user needs), check data availability, and prepare a business case.

Specifying needs is an essential part of a project. This is almost equivalent to managing projects to provide users with goods or services. However, we often do not have sufficient project definitions, such as needs, project objectives, and plans. In this case, exploratory data analysis (Tukey 1977) is an effective method. We also introduce a cyclic pipeline based on the Check, Action, Plan, and Do (CAPD) cycle. In this chapter, data analysis tools for decision-making are addressed.

2 What Data Are

What are data? Data are a collection of facts that can be described as values or expressions. Data are expressed as a set of descriptions or numbers and can be qualitative or quantitative. Data are collected from the world through devices or surveys. There are several types of data, such as time series, network, geospatial, and text data. The data types are classified into structured, semi-structured, and unstructured. An example of structured data is tabular data, which are expressed as relational databases or. csv/.tsv data. Each column is represented by a field name, and a set of elements filled in some fields in each row is called a record. Each record contains values and expressions. The values are called "measures," and the expressions are "dimensions" (see Sect. 3.3). A dimension is a set of categorical discretized concepts. Discrimination by categorical discretization is not unique. Moreover, we intuitively use projectable predicates in inductive reasoning only. However, we cannot rigorously discriminate projectable predicates from unprojectable predicates when we apply inductive reasoning to problem-solving. This is referred to as "the grue problem" (Goodman 1955). This problem proposes that categorical data are ambiguous. By using continuous or discrete values recognized as a measure, we normally define categories. For example, children and adults are defined by using ages. A 25-year-old woman is categorized as an adult. An 8-year-old boy is called a child. A boundary separates children and adults. This discrimination boundary is not unique but dependent on societies, nations, and organizations. Thus, the selection of the number of categories and a way to determine the discrimination boundaries are arbitrary.

Moreover, we need to understand errors in values. Three types of errors are important: observation, computational, and discrimination errors. Observation errors are made through observation processes for collecting data. To improve observation errors, we need to improve the physical processes for obtaining the original data. Computational errors are errors made through computational processes. Modern digital computers require discretization to store and manipulate numerical data. In this case, rounding, discretization, numerical, and model errors are inevitable. Discrimination errors dominate categorical data. Type I and type II errors are known in model-based categorization problems. Moreover, ambiguity or arbitrariness in categorical data should be mentioned. This means that categories and categorized elements defined by one person are not always the same as categories defined by other persons.

According to Karl Popper (1978), there are three worlds. World 1 is a world that consists of physical bodies. World 2 is the world of mental or psychological states or processes. World 3 is the world of the products of the human mind. Data are collected from each of the three worlds.

Data are described as symbols associated with objects or concepts in the real world. We assume that there is a one-to-one relationship between a symbol and an object or a concept.

3 Methods

3.1 Concepts

Data analysis forms a process consisting of several steps. Figure 1 shows a schematic illustration of the essential workflow of the data analysis. A data analysis pipeline consists of data acquisition, collection, quantification and visualization, and interpretation. In this section, we explain a generic workflow for data analysis.

Data Acquisition In the data acquisition step, people access data sources and construct a concrete connection with an organization to produce data sources.

Data Collection In the data collection step, people collect data from the detected data sources. This includes storing data, constructing new data, and preparing data that can be used in the next step, data visualization and quantification.

Data Visualization and Quantification In the data visualization and quantification step, people may draw graphs or pictures from data and calculate representative quantities called indicators or indices. Descriptive statistics, such as mean and variance, are useful for characterizing quantitative data.

Data Interpretation In the data interpretation step, people write documents to explain an objective with the graphs and representative quantities obtained in the previous step. This activity is often called reporting. Reports that explain what the data told us should be used to deliver messages or stories about insights obtained from the data analysis to a decision-maker.

3.2 Data Acquisition

If we want to conduct data analysis, we always need data. Therefore, the first step in data analysis is data acquisition. How do we acquire data? There are three ways to acquire data:

1. Access data that were collected in our own organization.
2. Access data that were collected in other organizations.
3. Design experiments, survey, and investigations, and collect data from the actual environment.

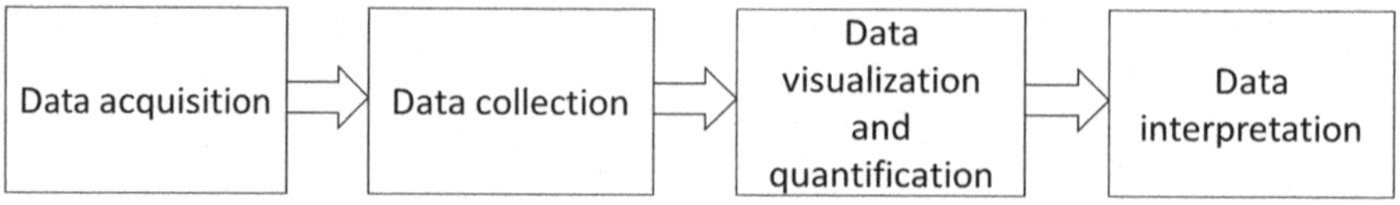

Fig. 1 A schematic illustration of a pipeline for data analysis

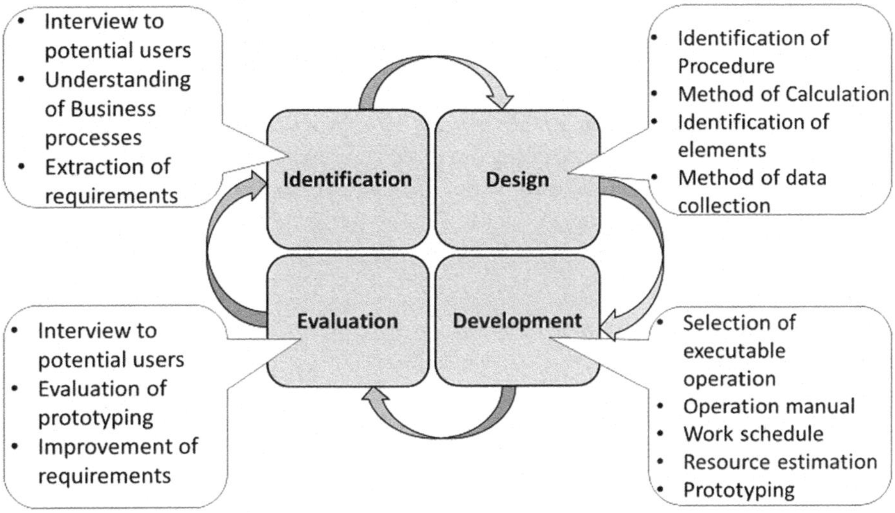

Fig. 2 A schematic illustration of R&D for data and statistics

(1) When we access data that have accumulated in our own organization, we plan a secondary use for the data. In this case, we do not always obtain the data that we want to access because the designers of the data sets had different intentions for the data and had different purposes for collecting the data. Nevertheless, we may find a way to use data that were collected in other projects. Data repositories and data libraries can be used to find useful data sets that we need.

(2) If we want to access data that were accumulated in other organizations, then we may need to purchase data from commercial entities or obtain data based on contracts. In this case, a data catalog and a data company are channels for accessing data sources we need.

(3) If we want to collect original data based on an experiment, survey, or investigation, then we need to develop what we want.

The data acquisition step is related to research and development (R&D) for data analysis. In this sense, we may add a generic workflow about R&D to the data acquisition step. A generic R&D process for data can consist of identification of needs, design, development, and evaluation as shown in Fig. 2.

Identification of Needs We need to understand the needs of the data analysis before we collect data. In this step, we could interview potential users. After we understand the business processes, we can extract the requirements of our data analysis.

Design After identifying the needs for the data analysis, we need to identify procedures, a calculation method, data elements, and the data collection method.

Development After identifying the procedure for collecting data, we need to select an executable operation and describe a series of methods as the operation manual.

In this stage, we need a work schedule and resource estimates for the workforce, finances, and equipment. We can produce a prototype from the developed work schedule.

Evaluation We may evaluate the process for processing data by showing a prototype to potential users and collecting their opinions.

These four steps should form a cycle. We can eventually improve our requirements and work schedules for producing data by using the cycle.

3.3 Data Collection

How do we collect data? What kinds of data are needed? To understand the phenomena observed in the actual environment, we need to understand the fundamental characteristics of the available data in advance.

According to the cube model in RDF Data Cube Vocabulary (MacKay and Oldford 2000; W3C 2014), tabular statistical data include three types of categories:

1. Dimension.
2. Measure.
3. Attribute.

The dimension components are used to identify the observations. Categorical data are dimensional. For example, time stamps and regional codes are dealt with in the dimension components. The measure components represent values observed in the phenomenon. For example, physical quantities, such as temperature, length, weight, speed, and the number of objects, are dealt with in the measure components. The attribute components are used to qualify and interpret the observed values. These components enable us to specify the units of measure, any scaling, and metadata, such as the status of the observations (estimation, census, etc.).

If the data are tabular, then a relational data model is applicable. This model is normally used in tabular data consisting of several cells. Rows and columns are used to characterize the tabular data. A relational database must define the meaning of the rows and columns. The action for defining the database structure is called normalization. Creating a table is related to defining a data structure. In data collection, we always define the data elements and the meaning of the dimension components and measure components. This task sometimes is called understanding the data. Understanding the data is the first important task of data collection. In the first stage, we do not need to identify all the data elements. However, we must identify some types of data elements before we analyze the data.

Figure 3 shows an example of tabular data that express statistical data. The field names are expressed as dimension components, and observations are expressed as measures. In this example, the measures and their values are expressed in observations.

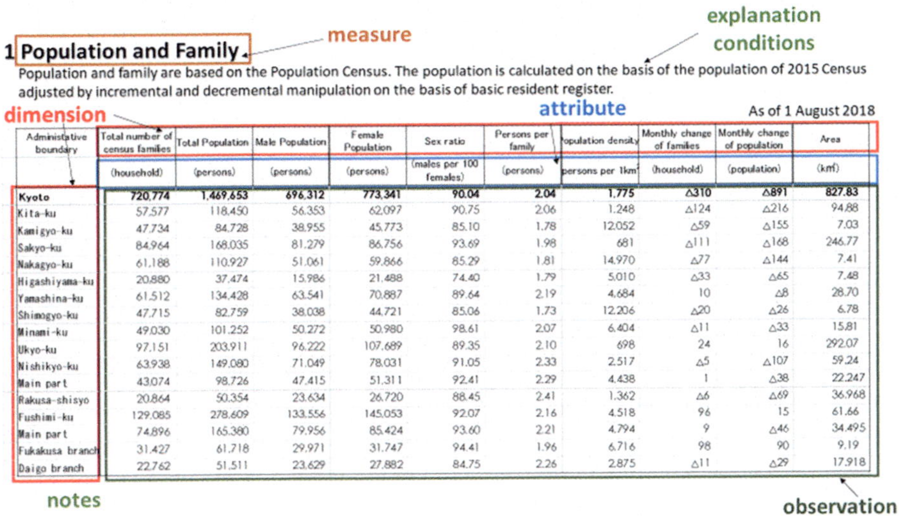

1 Population and Family

Population and family are based on the Population Census. The population is calculated on the basis of the population of 2015 Census adjusted by incremental and decremental manipulation on the basis of basic resident register.

As of 1 August 2018

Administative boundary	Total number of census familes	Total Population	Male Population	Female Population	Sex ratio	Persons per family	Population density	Monthly change of families	Monthly change of population	Area
	(household)	(persons)	(persons)	(persons)	(males per 100 females)	(persons)	persons per 1km²	(household)	(population)	(km²)
Kyoto	**720,774**	**1,469,653**	**696,312**	**773,341**	**90.04**	**2.04**	**1,775**	**△310**	**△891**	**827.83**
Kita-ku	57.577	118.450	56.353	62.097	90.75	2.06	1.248	△124	△216	94.88
Kamigyo-ku	47.734	84.728	38.955	45.773	85.10	1.78	12.052	△59	△155	7.03
Sakyo-ku	84.964	168.035	81.279	86.756	93.69	1.98	681	△111	△168	246.77
Nakagyo-ku	61.188	110.927	51.061	59.866	85.29	1.81	14.970	△77	△144	7.41
Higashiyama-ku	20.880	37.474	15.986	21.488	74.40	1.79	5.010	△33	△65	7.48
Yamashina-ku	61.512	134.428	63.541	70.887	89.64	2.19	4.684	10	△8	28.70
Shimogyo-ku	47.715	82.759	38.038	44.721	85.06	1.73	12.206	△20	△26	6.78
Minami-ku	49.030	101.252	50.272	50.980	98.61	2.07	6.404	△11	△33	15.81
Ukyo-ku	97.151	203.911	96.222	107.689	89.35	2.10	698	24	16	292.07
Nishikyo-ku	63.938	149.080	71.049	78.031	91.05	2.33	2.517	△5	△107	59.24
Main part	43.074	98.726	47.415	51.311	92.41	2.29	4.438	1	△38	22.247
Rakusa-shisyo	20.864	50.354	23.634	26.720	88.45	2.41	1.362	△6	△69	36.968
Fushimi-ku	129.085	278.609	133.556	145.053	92.07	2.16	4.518	96	15	61.66
Main part	74.896	165.380	79.956	85.424	93.60	2.21	4.794	9	△46	34.495
Fukakusa branch	31.427	61.718	29.971	31.747	94.41	1.96	6.716	98	90	9.19
Daigo branch	22.762	51.511	23.629	27.882	84.75	2.26	2.875	△11	△29	17.918

Fig. 3 An example of tabular data in statistics

3.4 *Data Visualization and Data Quantification*

Data visualization and quantification are important tasks for explaining the characteristics of the data to other people. For this purpose, various types of graphs and visualized data representations have been invented by many researchers and practitioner (Tufte 2001; Krum 2014; Kirk 2016; Naparin and Saad 2017). Edward Tufte (2001), the father of data visualization, published *The Visual Display of Quantitative Information*. The purpose of data visualization is to make audiences understand the information from data. Thus, a better method for visualizing data is to convey a strong sense of understanding and credibility.

The appropriate method for data visualization strongly depends on the type of data. Three main types of data expression, such as time series, network, and spatial data, are used. Combinations of the three types of data expressions are also possible. For time series data, we can use a bar graph, a line graph, and a point graph. For network data, we can use nodes (vertexes) and links (edges) to show relationships among elements. For spatial data, we can use geographic mapping into spatial data. We need to understand the characteristics of the data by using graph expressions.

3.5 *Data Interpretation*

From data analysis, we understand the structure of the data and draw the phenomenon described by the data. Such outcomes should be described as a document that includes stories about the phenomenon, graphs, and tables.

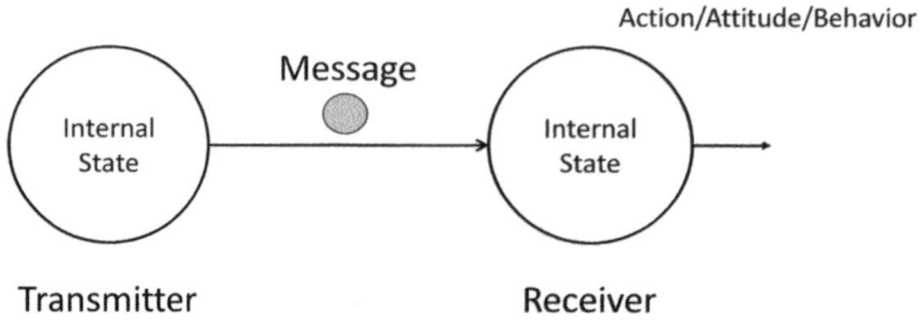

Fig. 4 A conceptual illustration of communication between a transmitter and a receiver

If a decision-maker regards the information provided by the document and changes his/her behavior, then the data analysis turned out to produce value. Otherwise, the data analysis did not produce value. Therefore, the data or data analysis should be measured by how much decision-makers change their behavior by using the data, or an outcome of the data analysis.

This view of the value of data or data analysis is closely related to a general definition of information from a qualitative point of view. The most famous definition of information from the qualitative aspect is Gregory Bateson's. In his book *Steps to an Ecology of Mind*, Bateson (2000) defines information as "a difference which makes a difference."

As shown in Fig. 4, Bateson assumes there are two roles in communication: a transmitter and a receiver. The transmitter and the receiver have their own internal state that is sensitive to their action, attitude, and behavior. The transmitter produces a message and sends it to the receiver. Bateson's statement about the definition of information means that the information in a message can be recognized by the transmitter and the receiver if and only if the receiver's internal state (related to the receiver's attitude and behavior) have been changed by the message that the transmitter sent to the receiver.

In the context of data analysis, the value of data or data analysis should be captured by how much information is extracted from the data through the data analysis. Therefore, we must pay attention to the outcomes of data analysis by measuring how our reporting activities made decision-makers change their internal state.

4 CAPD Cycle

If we follow the conventional data analysis pipeline, then we need to consider data acquisition as the first step. However, normally, we do not have sufficient knowledge about data initially. As a result, we cannot find a good way to access an adequate data source. To avoid deadlock, we should begin with small data related to problems when we start the data analysis. Moreover, we do not sometimes have

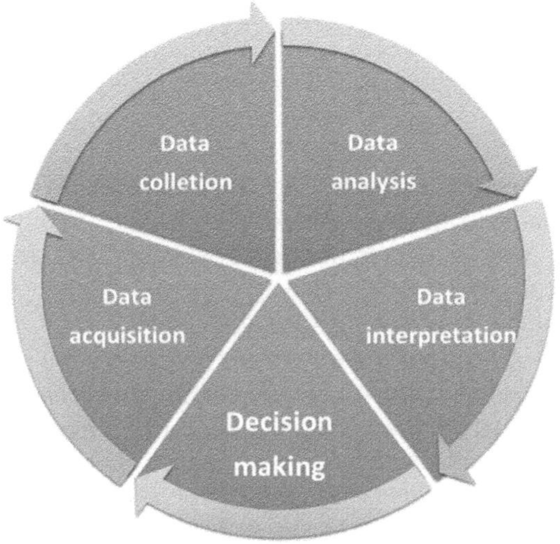

Fig. 5 A conceptual illustration of the data analysis cycle

even fundamental domain knowledge about the data. We often do not know issues and problems that we need to solve. We further may not know the exact meaning of the descriptions included in the data. Thus, if we address the field names included in tabular data, then we may improve our understanding of a specific domain about the given data. In this case, we may use the Check, Action, Plan, and Do (CAPD) cycle with the data analysis pipeline. Then we can form a data analysis cycle. Figure 5 shows a cyclic method for data analysis.

First, we should start with data acquisition. This data analysis pipeline will help us improve the data volume and data variety by deepening the data interpretation. If we employ the data analysis cycle by starting with small data in which we are interested, then we may approach big data and big pictures from small data related to what we want to focus on through an improvement process for data analysis.

Normally, big data exist far beyond the human processing capacity because human data processing speed for long memory through reading, writing, and speaking is estimated at 0.7–2.3 bits per second. The total volume of data that can be expressed by narratives per life per human is estimated as ranging from 300 MB to 500 MB. Therefore, we humans normally need a system to use big data for decision-making. Furthermore, such a system should be connected to several data sources to construct data flows. We can identify three areas of data sources in the social context:

1. Data we have not assessed yet.
2. Data we have accessed but have not extracted sufficient information for interpretation.
3. Data we accessed and extracted sufficient information for interpretation but have not integrated with other data for users.

Big data are thought to be the next frontier for innovation, competition, and productivity. The classical big data characteristics are volume, velocity, and variety. This is abbreviated as the 3Vs.

Big data can be accessed by everybody theoretically; however, big data cannot be analyzed by anybody in practice. This comes from a limitation of our resources in the data analysis. Thus, we need to consider the data analysis workflow and the purposes of data analysis carefully.

As we humans are finite, our capacity is smaller than the currently available big data. Thus, we need to carefully consider how to tackle issues that we need to solve with big data. In this sense, the CAPD cycle with gap analysis is quite helpful. In the Check phase, we start by confirming our current situation, such as our resources, knowledge, and capacity. In the Action phase, we recognize a gap between what we are and what we will be. Our desirable ideal in the future should be designed then. After understanding the gap, in the Plan phase, we can construct a plan to realize our desirable ideal based on the current situation (knowledge, resources, memberships, and organization). In the Do phase, we can conduct our plan step-by-step. If the plan is executable within a reasonable time horizon, we may change our situation and approach our desirable ideals. In the next cycle, we check our current situation and recognize a gap between our current situation and our ideal.

Specifically, big data are characterized by the 3Vs. Volume means a quantitative aspect of data, such as the number of records, the number of elements, and bits included in data. The first, data volume, should be considered in terms of three aspects: units of volume, a time period, and the criterion used to define the volume size. Recently, the data size in some big data use cases reached from terabytes to petabytes. The data-generating speed (velocity), which measured by bit per second (pbs), can be used to characterize big data. This is referred to as the rate of flow at which the data are created, stored, analyzed, and visualized. For example, big data with high velocity means a large quantity of data is being processed within a short amount of time. Velocity has three aspects: the unit, time period, and the criterion used to define the velocity value. Data variety can be used to characterize big data. This can be considered from multiple repositories, domains, or types. The number of categories included in data is one of the indices for measuring various big data. Other aspects of big data can also be characterized. For example, variability refers to changes in the rate and nature of the data. Veracity relates to data quality represented by the completeness and accuracy of the data concerning semantic content, as well as the syntactical quality of the data (missing fields or incorrect values).

5 Previous Studies

We look at existing procedures related to data analysis. Some existing frameworks can be used in practical data analysis. For example, CRISP-DM is a data mining process model commonly used in business and management (Chapman et al. 2000; Wirth and Hipp 2000). The CRISP-DM proposes a cyclic workflow reference

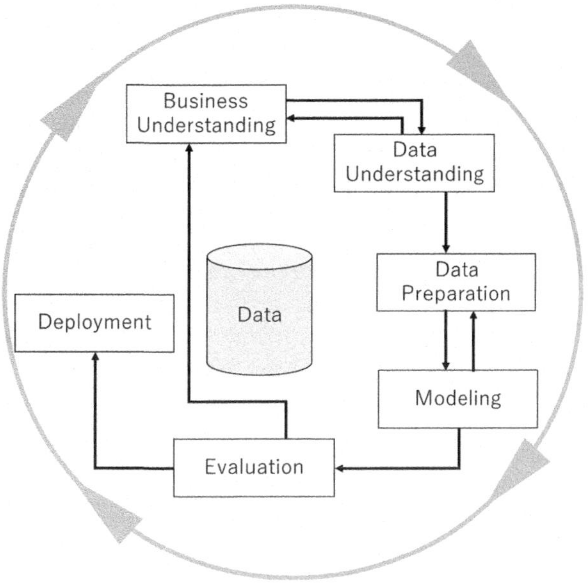

Fig. 6 Phases of the CRISP-DM reference model (Chapman et al. 2000)

model consisting of data understanding, data preparation, data modeling, evaluation, business understanding, and deployment. The CRISP-DM possesses a hierarchical process model, composed of the sets of tasks described at four levels of abstraction: phase, generic task, specialized task, and process instance. Each phase consists of generic tasks related to possible data mining situations. The specialized tasks in a generic task indicate specific situations. The process instance is a record of the actions, decision, and results of an actual data mining engagement. The CRISP-DM reference model forms the life cycle of a data mining project. This life cycle contains six phases, business understanding, data understanding, data preparation, modeling, evaluation, and deployment, as shown in Fig. 6.

Business understanding is the initial phase for understanding the project objectives and requirements from a business perspective. In this phase, we need to find a problem definition and a preliminary plan designed to achieve the objectives. In the data understanding phase, we collect the initial data, and we attempt to become familiar with the data. In this phase, we identify data quality problems, discover first insights into the data, and assume working hypotheses regarding hidden information. The data preparation phase includes all activities needed to construct the final dataset based on the insights obtained during the data understanding phase. In the modeling phase, various modeling techniques are selected and applied. Moreover, model parameters are calibrated to optimal values. Some types of techniques require additional data that are not included in the dataset generated in the previous data preparation phase. Thus, going back to the data preparation phase is often necessary. Evaluation is a phase for checking a model (models) with high-quality data before the final deployment of the model. An essential purpose of this stage is to determine whether there are some critical business issues that have

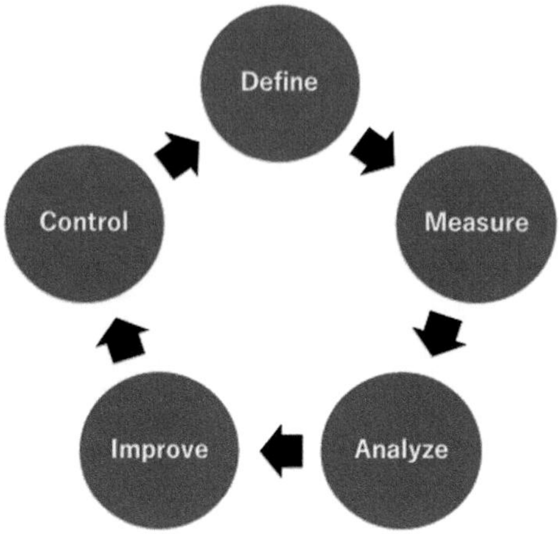

Fig. 7 DMAIC methodology

not been sufficiently considered. If business issues have not sufficiently considered, we need to go back to the business understanding phase. Deploying the model developed through data analysis in the system or decision-making processes is the final goal of the project. In the deployment phase, we generate a report for decision-makers or construct automated systems based on the developed model with calibrated parameters. The deployment phase is one of the end points of the project.

DMAIC (ISO 13053-1:2011) is another approach for using data for decision-making. This approach was developed in Six Sigma projects. This procedure is mainly used the improvement process to produce products in manufacturing and service industries. DMAIC consists of five phases: Define, Measure, Analyze, Improve, and Control. They are shown in Fig. 7. In the Define phase, we understand the problem, project goals, and consumer requirements. We identify methods for improving the system and opportunities for improvement. In the Measure phase, we measure the system. In the Analysis phase, we determine dominant causes of variation and poor performance (bottleneck). In the Improvement phase, we attempt to eliminate dominant causes of the system bottleneck. In the Control phase, we improve the system and consider how to improve the system performance further.

The GSBPM (European Commission 2014) is a workflow model for producing statistics within and between statistical offices in different ways. In around 2006, Statistics New Zealand was the first organization to adopt the GSBPM in official statistics. Then, there was continuous work in joint UNECE/Eurostat/OECD Work Sessions on Statistical Metadata (METIS). The GSBPM has a hierarchical structure consisting of four levels:

Level 0, the statistical business process.
Level 1, the nine phases of the statistical business process.
Level 2, the sub-processes within each phase.
Level 3, a description of those sub-processes.

The nine phases are as follows:

1. Specify needs.
2. Design.
3. Build.
4. Collect.
5. Process.
6. Analyze.
7. Disseminate.
8. Archive.
9. Evaluate.

The preparation is related to phases 1 to 3. The production ranges from phases 4 to 7. Phases 8 and 9 are actions for future improvement of data sets which have been disseminated and archived until the previous phases.

Phase 1 consists of 1.1 "Determine needs for information," 1.2 "Consult and confirm needs," 1.3 "Establish output objectives," 1.4 "Identify concepts," 1.5 "Check data availability," and 1.6 "Prepare the business case."

Phase 2 consists of 2.1 "Design outputs," 2.2 "Design variable descriptions," 2.3 "Design data collection methodology," 2.4 "Design frame and sample methodology," 2.5 "Design statistical processing methodology," and 2.6 "Design production system and workflow."

Phase 3 consists of 3.1 "Build data," 3.2 "Build or enhance process components," 3.3 "Configure workflows," 3.4 "Test production system," 3.5 "Test statistical business process," and 3.6 "Finalize production system."

Phase 4 consists of 4.1 "Select sample," 4.2 "Set up collection," 4.3 "Run collection," and 4.4 "Finalize collection."

Phase 5 consists of 5.1 "Integrate data," 5.2 "Classify and code," 5.3 "Review Variable and edit," 5.4 "Impute," 5.5 "Derive new variables and statistical units," 5.6 "Calculate weights," 5.7 "Calculate aggregates," and 5.8 "Finalize data files."

Phase 6 consists of 6.1 "Prepare draft outputs," 6.2 "Validate outputs," 6.3 "Scrutinize and explain," 6.4 "Apply disclosure control," and 6.5 "Finalize output."

Phase 7 consists of 7.1 "Update output systems," 7.2 "Produce dissemination procedure," 7.3 "Manage release of dissemination products," and 7.5 "Manage user support."

Phase 8 consists of 8.1 "Define archive rules," 8.2 "Manage archive repository," 8.3 "Preserve data associated metadata," and 8.4 "Dispose of data and associated metadata."

Phase 9 consists of 9.1 "Gather evaluation inputs," 9.2 "Conduct evaluation," and 9.3 "Agree action plan." The GSBPM is a framework for producing data and statistics for public and commercial uses through collaborative work by different persons.

The Problem-Plan-Data-Analysis-Conclusion (PPDAC) model is a cyclic methodological framework for approaching an analytical or research question based on the scientific method (MacKay and Oldford 2000). The PPDAC model provides a fundamental workflow for applying scientific methodology to solving problems.

Data storytelling is one of the important issues in design thinking. Specifically, in storytelling in the persona design (Cooper and Saffo 1999), it is essential to use quantitative and qualitative data to understand consumer behavior (Nielsen 2013). According to Miaskiewicz and Kozar (2011), the most significant characteristic of personas introduced in a group task is their ability to let a group focus on the actual goals of the target customers. If participants share real-life characteristics of the target consumers, then the participants can use a common understanding of the customer as a central driver of their design task. In general, to conduct effective group work, we need expressionability, communicability, exchangeability, and decisionability. Personas enable us to share fundamental assumptions about consumers and markets, to express consumer behavior, to communicate with others, and to make decisions under the same framework. Personas are required to draw a customer journey map for design of goods and services. Data analysis about consumers and markets is quite useful to assume the personas.

6 Conclusion

This chapter introduced several methods that are useful in data analysis. The final goal of data analysis is to transmit insights obtained from data analysis and reality expressed by the data. I introduced several existing methodologies related to data analysis. A Check, Action, Plan, and Do (CAPD) cycle consisting of data acquisition, data collection, data quantification and visualization, and data interpretation is applicable to starting data analysis with small data.

We may improve our understanding of data and business models eventually by using the CAPD cycle of the data pipeline. The outcome of data analysis should be a document that reports results obtained from the analysis. If a report that includes messages and insights obtained from the data analysis changes the behavior of decision-makers, then the outcomes of the data analysis are meaningful. The existing frameworks related to data analysis and data production exist depending on the purpose of the applications. It was found that the DMAIC and PPDAC are used for understanding problems and problem-solving based on the improvement mechanism. The CRISP-DM was developed for problem-solving with business modeling. The GSBPM enables different organizations to produce high-quality data and statistics for public and commercial use.

The data analysis should begin with small data. The CAPD cycle is essential to approach to big data analytics by accumulating data, domain knowledge, and connections with stakeholders. Moreover, data storytelling is one of the important issues in both data utilization and design thinking. Specifically, in storytelling the persona design and journey map approach are useful and quantitative, and qualitative data analysis to understand consumer behavior is one of the most exciting applications of data analysis to social design.

References

Bateson G (2000) Steps to an ecology of mind: collected essays in anthropology, psychiatry, evolution, and epistemology. University of Chicago Press, Chicago

Chapman P, Clinton J, Kerber R, Khabaza T, Reinartz T, Shearer C, Wirth R (2000) CRISP-DM 1.0 Step-by-step data mining guides. ftp://ftp.software.ibm.com/software/analytics/spss/support/Modeler/Documentation/14/UserManual/CRISP-DM.pdf. Accessed 29 Apr 2018

Cooper A, Saffo P (1999) The inmates are running the asylum. Macmillan, Indianapolis

European Commission (2014) Collaboration in research and methodology for official statistics, European Commission, General observations–GSBPM. https://ec.europa.eu/eurostat/cros/system/files/General%20Observations-06-T-GSBPM%20v1.0_1.pdf. Accessed 29 Apr 2018

Goodman N (1955) Fact, fiction, and forecast. Harvard University Press, Cambridge, MA

ISO 13053-1:2011 (2011) Quantitative methods in process improvement – Six Sigma—Part 1: DMAIC methodology. https://www.iso.org/standard/52901.html. Accessed 29 Apr 2018

Kirk A (2016) Data visualization: a handbook for data driven design. SAGA Publication Ltd, London

Krum R (2014) Cool infographics effective communication with data visualization and design. John Wiley & Sons, Indianapolis

MacKay RJ, Oldford RW (2000) Scientific method, statistical method and the speed of light. Stat Sci 15(3):254–278

Miaskiewicz T, Kozar KA (2011) Personas and user-centered design: how can personas benefit product design processes. Des Stud 32(5):417–430

Naparin H, Saad AB (2017) Infographics in education: review on infographics design. Int J Multimed Appl. https://doi.org/10.5121/ijma.2017.9602

Nielsen L (2013) Personas. In: The encyclopedia of human-computer interaction, 2nd ed. https://www.interaction-design.org/literature/book/the-encyclopedia-of-human-computer-interaction-2nd-ed. Accessed 16 Aug 2018

Popper K (1978) Three worlds: the tanner lecture on human values. https://tannerlectures.utah.edu/_documents/a-to-z/p/popper80.pdf. Accessed 15 June 2015

Tufte E (2001) The visual display of quantitative information. Graphics Press, Cheshire

Tukey JW (1977) Exploratory data analysis. Addison-Wesley, Reading

W3C (2014) RDF data cube vocabulary. https://www.w3.org/TR/vocab-data-cube/. Accessed 15 Aug 2018

Wirth R, Hipp J (2000) CRISP-DM: towards a standard process model for data mining. In: Proceedings of the fourth international conference on the practical application of knowledge discovery and data mining, pp 29–39

An Approach to Product Design Involving Heterogeneous Stakeholders

Aki-Hiro Sato

1 Introduction

What is the purpose of data and data analysis? In general, we use data to determine something, and data analysis should provide insights that will assist decision-making. While society is sometimes unpredictable, data can be used to shape its future. To create and change that future, we need innovation. Rather than just producing new things, innovation can be understood as a process of finding new combinations of goods and services by creating and strengthening connections between products and stakeholders that did not previously exist or whose significance has not yet been recognized. Design techniques can help us to develop innovations by supporting decisions about structure, function, form, amount, time, place, and occasion. Given the complexity of design decision-making, data can inform the process in multiple ways.

According to Schumpeter (1983), our modern capitalist society creates "boom-and-bust" business cycles as an inevitable consequence of how firms behave to make profits, which are generated only while a given market is developing. Eventually, that market becomes mature and must be destroyed in order to create a new market. This process of innovation is used to attract consumers and generate profits by exploiting a new business opportunity, resulting in the business cycle of boom-and-bust.

Jetzek et al. (2014) proposed a conceptual model of data-driven innovation comprising four multidimensional enabling factors (absorptive capacity, openness,

A.-H. Sato (✉)
Yokohama City University, Kanazawa-ku, Yokohama-shi, Kanagawa, Japan

Japan Science and Technology Agency PRESTO, Kawaguchi-shi, Saitama, Japan

Statistical Research and Training Institute, Ministry of Internal Affairs and Communications, Shinjuku-ku, Tokyo, Japan
e-mail: ahsato@yokohama-cu.ac.jp

A.-H. Sato (ed.), *Applications of Data-Centric Science to Social Design*,
Agent-Based Social Systems 14, https://doi.org/10.1007/978-981-10-7194-2_3

resource governance, and technical connectivity), three innovation mechanisms (idea generation, idea conversion, and idea diffusion), and two forms of impact (on social and economic values). The adjustable factors in organizations allow people to use data to more easily produce, convert, and diffuse ideas (Laursen and Salter 2006), resulting in the generation of macro-level value, and production of economic and social value can further improve these enabling factors and the innovation environment.

In general, design can be said to involve four phases: conceptual design, fundamental design, detailed design, and production design. Conceptual design defines the components of goods or services, addressing fundamental entities, consumers, and stories about goods or services embedded in the actual environment. Fundamental design is about identifying the necessary technologies to deliver fundamental requirements, as well as specifying goods and services in terms of costs and benefits. Detailed design determines the shapes and structures needed to realize the requirements. Finally, production design must consider how to produce and operate the goods or services in question, including fabrication, assembly, and delivery from production line to end users.

Conceptual design requires careful identification of the potential users of goods and services to be produced, as well as estimation of market size based on current demand. This cannot be achieved solely by engineers with technical domain knowledge but must also involve marketing, operations, fabrication, and management specialists. In that sense, heterogeneous group work is more critical than individual design work. The aim of this chapter is to explain how to organize a design workshop for participants from heterogeneous backgrounds.

2 Design Workshops: A General Framework

A design workshop may include multiple elements such as lectures, field investigations, brainstorming, ideation, prototyping, and reviewing. These are participatory processes, in which individuals from different domains of expertise and knowledge come together as a group to share their visions and work collaboratively. The quality of outcomes from a design workshop depends on appropriate task design and on bringing the right people together. Typically, stakeholders participate in groups of four to eight from different backgrounds, along with a few moderators or facilitators to promote effective communication among the participants.

The following procedure is typical of a design workshop.

1. Inputs.
2. Brainstorming.
3. Ideation (from demand and supply sides).
4. Matrix expansion of ideas.
5. Persona design.
6. Data journey mapping.
7. Prototyping.
8. Review and reflection.

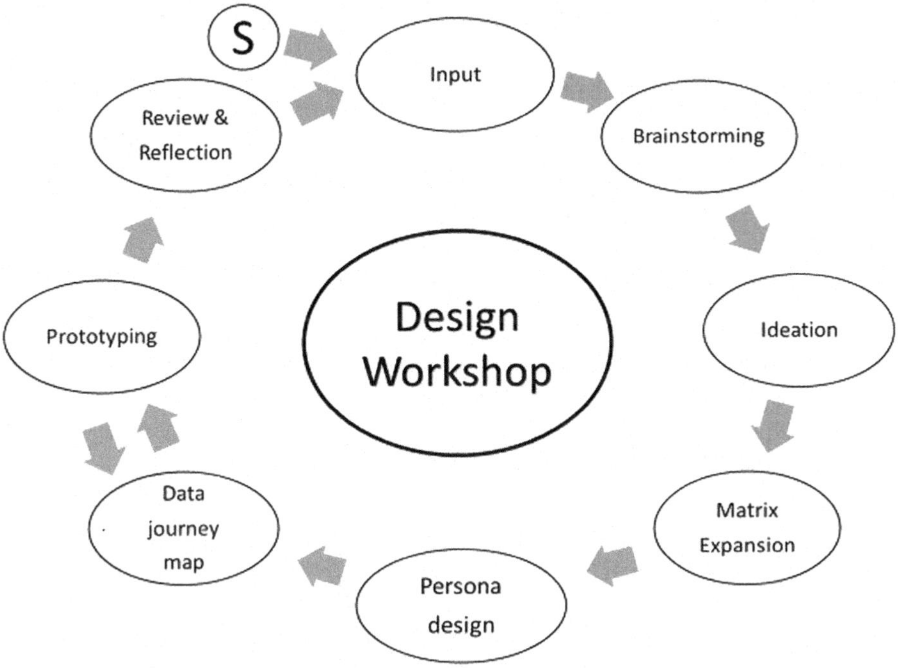

Fig. 1 Cyclic procedure for a design workshop (S represents the starting point)

At the end, review and reflection can inform the inputs for subsequent groups. In general, we can identify a cyclic procedure for organizing a series of design workshops to generate ideas and to design goods and services. Figure 1 shows a typical cycle of eight methodologies.

2.1 Inputs

Inputs are essential activities that provide participants with specific domain knowledge and basic ideas about relevant fields. As they can be drawn from lectures, expert interviews, field investigations, and data analysis, inputs are an adjustable and controllable element of the design workshop.

- Lectures: Participants listen to talks by experts, acquiring previously unknown domain knowledge and identifying fundamental ideas and principles, as well as asking the experts any questions they may have.
- Expert interviews: Expert interviews are denser than lectures, as participants can design their own questions to ask experts and stakeholders, listen to experts' knowledge and experiences, and actively acquire information about a specific field, leading to new ideas and insights.

- Field investigation: Participants go into the field to make observations and to discover issues. As well as new ideas and insights, they may gather samples of materials and data.
- Data analysis and visualization: Participants frame questions based on the results obtained from data analysis, sharing problems and definitions, acquiring new insights and knowledge, and asking further questions.

2.2 Brainstorming

As proposed by Osborn (1953), brainstorming is a method of creating ideas through group work. The purpose of brainstorming is to address issues, produce ideas, and expand concepts by working together. To do so while maintaining the appropriate atmosphere, four fundamental rules should be followed.

1. Motivate sharing of wild and unconventional ideas. As a brainstorming session is also a fun tool to keep the whole team involved, it is sometimes useful to look beyond serious solutions.
2. Do not criticize ideas. This is not a debate or a platform for one person to display their superiority.
3. Build on other people's ideas. As an idea suggested by one person can often trigger a bigger and/or better idea, attempts should be made to expand or modify ideas proposed by others.
4. Reverse the idea of "quality over quantity." Here, quantity is more valuable than quality: the more creative ideas we generate, the better brainstorming becomes.

In a brainstorming session, the aim is to produce a story about a specific topic. One of the simplest ways of doing this is to introduce a modified 4W1H game, which uses five questions (who, how, what, when, and where) to develop a story.

Step 1. Participants attempt to produce a story by combining each word in five categories (4W1H).

Step 2. Participants have to write five words on a sticky note and place it on the board to explain to other participants. The participants then listen to each other's stories.

Step 3. To expand and categorize ideas produced through brainstorming, the following five methods can be applied to existing stories.

- Exchange: Swap two items including two different stories.
- Generalization: Find general patterns from entities.
- Alternatives: Find other possibilities.
- Classification: Find categories and other possible categories.
- Patterns: Find rules and patterns from several ideas.

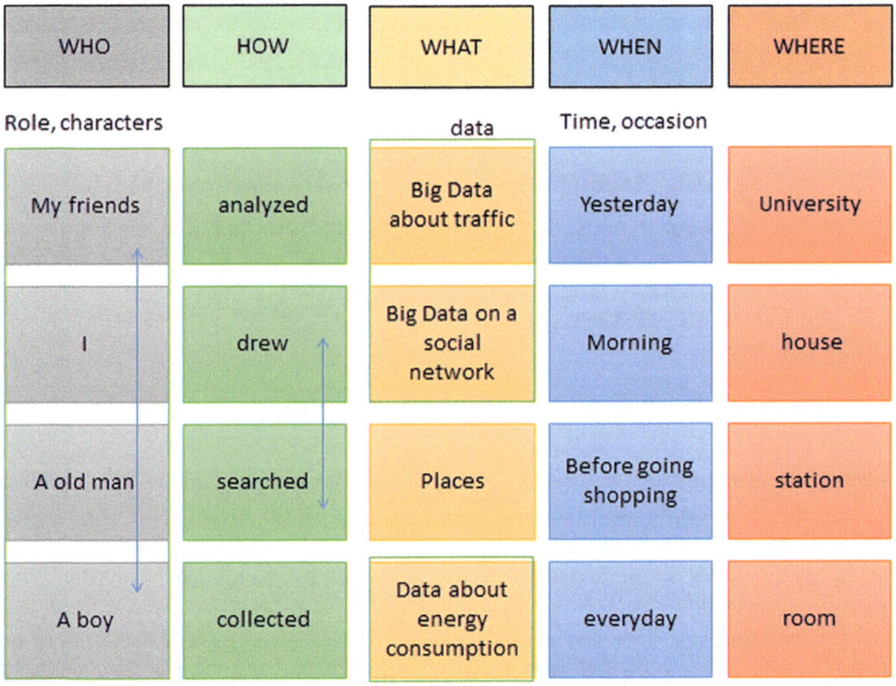

Fig. 2 Using 4W1H words to form four stories about data services

Step 4. Participants propose a new story consisting of five words not yet seen on the board (Step 2) or produce a new story by swapping two words (Step 3) several times.

Figure 2 shows four stories about data services, comprising five words based on the 4W1H approach. The first story in Fig. 2 is about an experience of data analysis in the university; the second story tells about a personal experience of data visualization. Each participant must produce a story. For instance, by swapping "my friends" in the first story and "a boy" in the fourth story, we can create a new story (exchange). We can also produce a new story by swapping "drew" in the second story and "searched" in the third story, and we can categorize the roles of words. For example, *generalization* suggests that experiences of data services may include data analysis, data visualization, data collection, and database searching. Among the methods used to organize a brainstorming session, the scenario graph (Kunifuji 2013) collects various stories on a specific topic, and the KJ method (Kim and Ishii 2007) extracts general patterns.

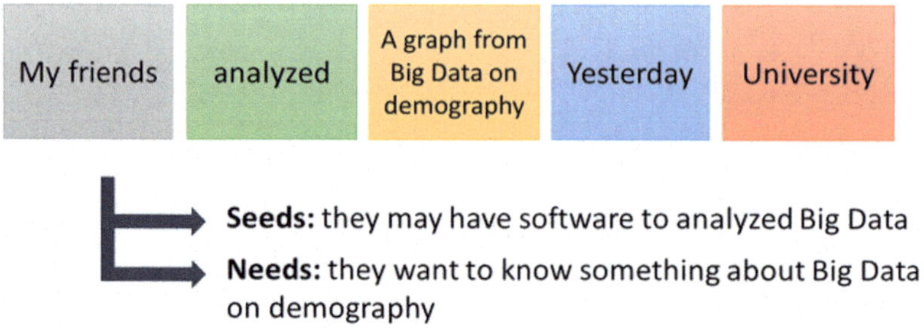

Fig. 3 Example of story separation into technology (seeds) and demand (needs) aspects

2.3 Ideation from Demand and Supply Sides

A story about goods or services can be told from two perspectives: seeds (supplier technology) and needs (user demand). To develop useful ideas, both aspects are included in one story of seeds and needs. The example in Fig. 3 uses seeds and needs to tell a story about data analysis and demography, which can be divided into two separate stories.

From a given story, we can derive a pair of stories of seeds and needs, both of which include the core idea. From the supplier side, the story is "Software to analyze Big Data about demography." From the user side, the story is "Understanding socioeconomic activities." Using this method, each participant selects a useful story and presents a pair of ideas from the supplier and user sides, leading to the collection of a set of ideas produced by participants.

2.4 Matrix Expansion of Ideas

In this step, we can produce new stories about goods or services based on the core ideas of supply and demand. Figure 4 shows a matrix comprising ideas about needs (in the first column) and ideas about seeds (in the first row). We can then produce new stories based on each independent idea about seeds and needs. The diagonal cells of the matrix correspond to stories originally proposed by participants.

Participants rate each new story in terms of possibility, convenience, availability, innovativity, and reliability. Stories are then evaluated on the basis of a sum or a weighted sum of scores. Finally, the group identifies the best story from the set of combinations included in the idea matrix, and that story can be used to draw a data journey map.

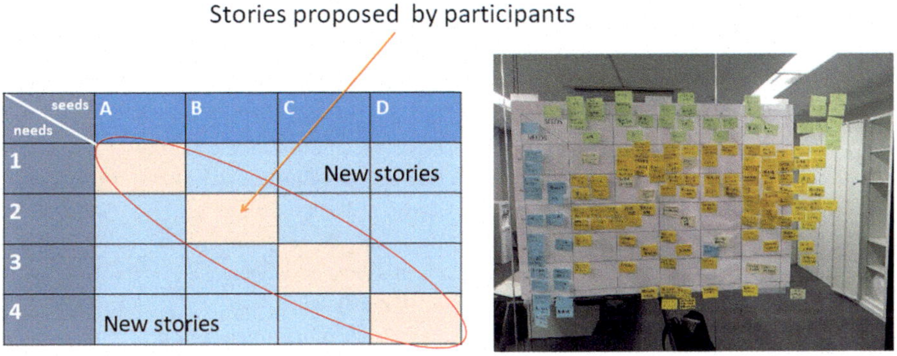

Fig. 4 (left) Matrix of ideas about seeds and needs; from *N* pairs of original stories, *N*(*N*-1) pairs of new stories can be produced. (right) An example of story matrix produced through a design workshop (The photo is provided by Mr. Y. Kenmochi).

2.5 *Persona Design*

To further elaborate the stories about services or goods selected from the matrix expansion, we must envisage an actor who plays the role of the consumer in the story (or stories) and then develop the characteristics of that actor in greater detail. This task is referred to as persona design (Cooper and Saffo 1999); characteristics typically used to express a persona include gender, age, location, place of birth, profession, role, hobbies, preferences, interests, and desires.

According to Miaskiewicz and Kozar (2011), the key reason for introducing a persona is to help the group to focus on the actual goals of the target customer. By sharing characteristics of the target consumer in this way, the group can develop a common understanding of the customer as a central driver of their design task. In general, effective group work depends on expressibility, communicability, exchangeability, and decidability. Personas enable us to share fundamental assumptions about consumers and markets, to express consumer behavior, to communicate with others, and to make decisions based on an agreed framework.

We can also use insights from the analysis of consumption and social activities data to identify typical national characteristics, using official statistics as data sources for developing stories. Lene Nielsen of the Interaction Design Foundation proposes ten steps to create personas and scenarios based on data analysis (Nielsen 2013). After collecting as much data as possible about the users, the insights obtained from data analysis enable formulation of a hypothesis about potential users. With the agreement of all participants, the final number of personas can be established. Group work then involves describing these personas in order to capture the needs and goals of users in the area of interest. The persona's characteristics should include education, lifestyle, interests, values, goals, needs, limitations, desires, attitudes, and patterns of behavior. Fictional personas' names and pictures should be added at the end of each 1–2 page description. The example in Table 1

Table 1 Example of a persona description

Items	Character
Name	Data Raro
Age	32
Gender	M
Profession	Employee of a software company
Hobby	Baseball, traveling
Attitude	Active and enthusiastic
Education	Bachelor's degree (computer science)
Interests	Open-source development and software design
Lifestyle	Rhythmical
Needs	Knowledge of information design
Desires	To launch his own company

describes the persona of a 32-year-old male working in a software company. His hobbies are playing baseball and traveling. He has an interest in open-source development and wants to launch his own company.

2.6 Data Journey Map

The data journey map (DJM) is a useful method of conveying a user's experience and data behavior and may help to clarify issues regarding the story selected in the matrix expansion. The DJM is a modified version of the customer journey map (CJM) (Miaskiewicz and Kozar 2011), which is often used as a service design tool. Among the various types of CJM (Richardson 2010; Moon et al. 2016), no standards exist. Richardson proposed four lanes: actions, motivations, questions, and barriers. Actions describe what the customer is doing (doing); motivations explain why the customer is motivated to take actions to the next stage (feeling); questions refer to uncertainties and issues that prevent customer engagement (problems); and barriers include concerns about the service in relation to structure, process, cost-benefit, and implementation. The CJM can also be used as a process mining tool to describe business workflows (Bernard and Periklis 2017).

The critical difference between a DJM and a CJM is the addition of a lane describing data behavior (e.g., status, touchpoint, how to collect), enabling the DJM to specify how to manage additional data collected from user actions. To capture the user experience, we propose the use of six DJM lanes: story, doing, data, thinking, feeling, and problems (Table 2).

In Fig. 5, the first lane of the DJM is the story of the user's experience. A utility curve displays the temporal development of the user's feeling. If the utility curve goes up, the user's satisfaction may increase; if the utility curve goes down, it may decrease. The second lane shows actions assumed in the story. Typically, there are three types of experience: before the use of services or goods, during use, and after use. The third lane refers to two types of data opportunity related to the user's

Table 2 Steps in constructing a data journey map

Step 1	Extract the persona's possible actions and display on a whiteboard
Step 2	Categorize and sort these actions
Step 3	Describe the persona's feeling and thinking for each action
Step 4	Extract data opportunities related to consumer actions
Step 5	Discuss problems and exchange ideas; identify likely problems for each action
Step 6	Define the temporal development of utility using a utility curve
Step 7	Summarize all individual actions in the form of a story

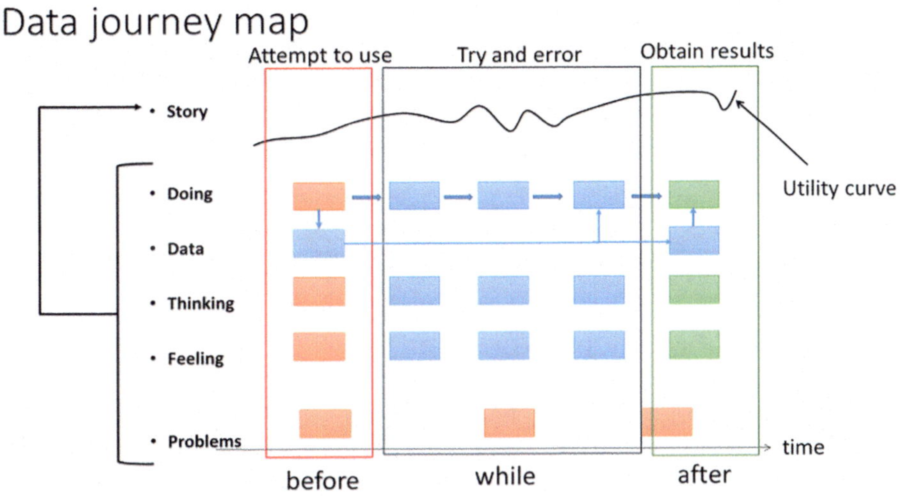

Fig. 5 Schematic of the data journey map

actions: data collection and data usage. The fourth lane refers to what the user thinks during the action, and the fifth lane refers to the user's feelings (good or bad). The sixth lane addresses problems extracted from the user's thinking and feeling. Based on a DJM constructed during group work, participants can discuss user experiences and data behavior (opportunities). To address the problems in the sixth lane, the user's actions and data behavior should be added and modified. In this way, new actions and experiences for services and goods related to the story can be designed by modifying the DJM. Specifically, to address the issues extracted from a set of actions and data opportunities, we can place a new sequence of actions and data opportunities as a part of design process.

2.7 *Prototyping Goods and Services*

In the prototyping stage, we refine the content of the DJM. To implement the developed story discussed in the DJM, we need to detail how consumers behave in the DJM and what kinds of system functions support consumer experiences. Unified Modeling Language (UML 2.0) [13] uses four *compliance levels* related to capability, which are partitioned into horizontal layers.

Level 0: class-based structures used in object-oriented programming languages and providing entry-level modeling capacity
Level 1: adds use cases, interactions, structures, and activities to Level 0.
Level 2: adds deployment, state machine modeling, and profiles to Level 1.
Level 3: extends Level 2, adding information flows, templates, and detailed modeling packages.

This approach to prototyping refines an implementation through trial and error by sharing structures and behaviors to be deployed at an initial stage to repeatedly produce and evaluate system descriptions. At this stage, UML 2.0 Level 1 description may be sufficient to share a system described in group work. Some useful diagrams for product prototyping are described below.

2.7.1 Use Cases

A use case specifies a set of actions performed by one or more actors or system stakeholders.

1. When drawing a use case diagram, the first step is to identify the actors in the system. An actor can be represented by a "stick man" icon bearing the name of the actor or by a class rectangle that includes the word <<actor> > and the actor's name. Figure 6 shows an administrator represented as an actor.
2. The next step addresses essential actions performed by all actors referred to in the first step.
3. A rectangle is drawn to refer to the system and actions in each oval within the box, specifying the relationship between actions such as inclusions and extensions (see Fig. 7).
4. Connections are drawn as a solid line between an actor and an action.
5. Inheritance is drawn as a solid arrow between actors (see Fig. 8).

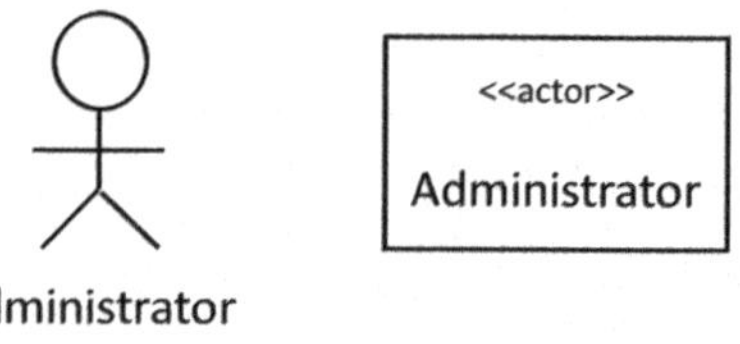

Fig. 6 Representations of an administrator as an actor: (left) "stick man" icon and (right) class rectangle with the keyword <<actor>>

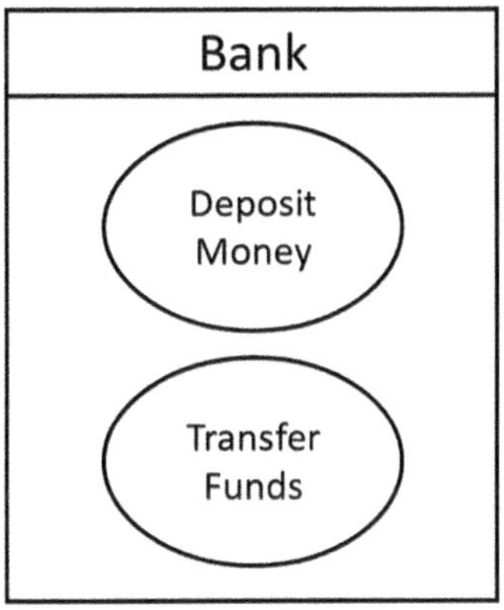

Fig. 7 Example of actions performed in a bank

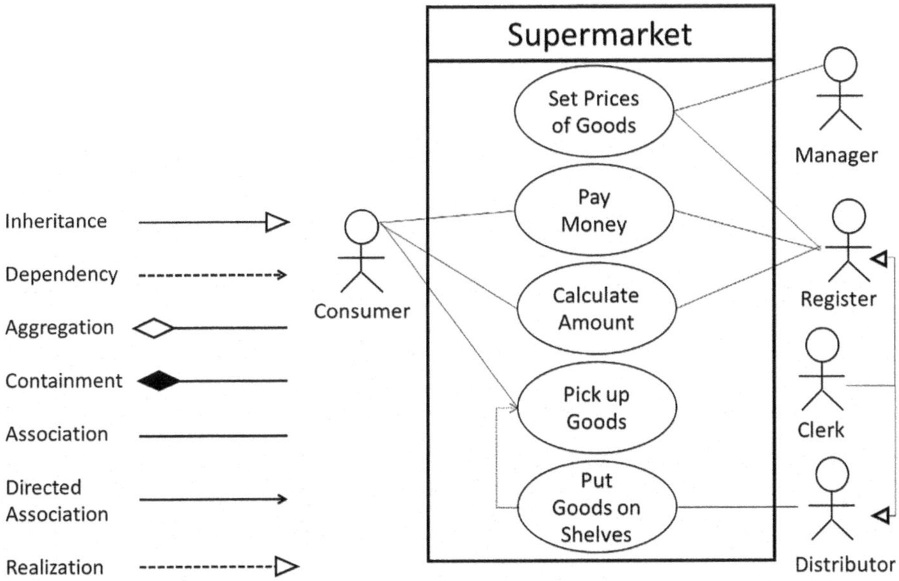

Fig. 8 (left) Relationship representation and (right) use case diagram of a supermarket

2.7.2 Interactions

Interactions are used to archive a common understanding of the situation, and the DJM is one of the diagrams used to express interactions. Other types of diagram can also be used to represent interactions, including sequence diagrams, interaction overview diagrams, and communication diagrams. For example, sequence diagrams express sequences of messages between lifelines. The data that messages convey and lifelines store may be important for making system interactions visible. Sequence diagrams are drawn as follows. Objects are placed from left to right and are expressed by a rectangle containing the object's name. A dotted line is drawn at

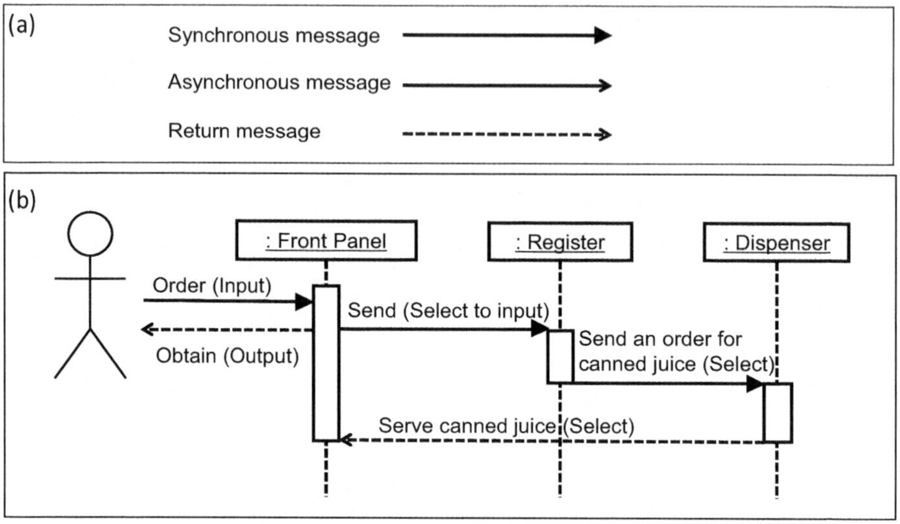

Fig. 9 (**a**) Three types of message used in sequence diagrams. (**b**) Example of an instance sequence diagram from the use case "buying a canned juice"

the center of an object from top to bottom to represent the period during which the object is alive (the "lifeline"). A rectangular object represents "activation," referring to a situation in which the object is executing an action. As shown in Fig. 9a, messages can be classified as one of three types: synchronous, asynchronous, or return. A synchronous message is responded to after waiting for a sender's response; an asynchronous message is a message without any sender's response; and a return message is a direct reply from a receiver. The temporal development of the system is shown on the y-axis. Figure 9b shows an example of an instance sequence diagram, which is so called because it shows only one of several scenarios related to the use case "buying a canned juice from a vending machine." The simple messages shown in the diagram are control instructions from one object to another.

2.7.3 Structures

Structure refers to the relationships among objects and can be expressed by a collaboration diagram, which shows relationships as messages between objects. In such diagrams, multilayer descriptions cannot be used to reduce complexity. In collaboration diagrams, messages are shown with arrows beside the lines connecting two objects; the arrow points to the object that receives the message. The action to be taken by the receiver object is shown beside the arrow. Action parameters can be added at the end of actions with brackets. As messages indicated in interactions can be expressed in collaboration diagrams, it is useful to include orders as numbers in front of each action. Colons divide message numbers and message names. In the

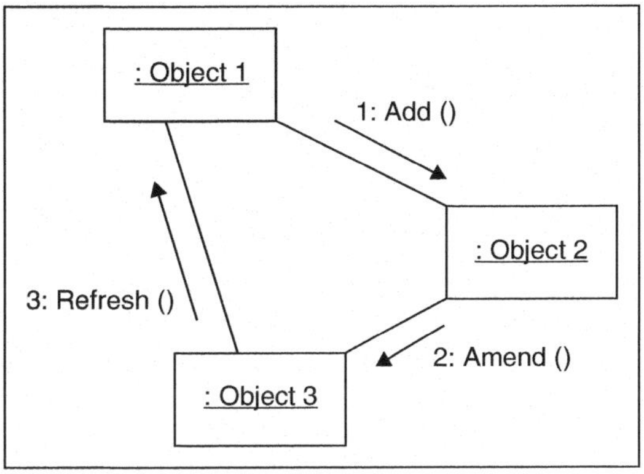

Fig. 10 Example of a collaboration diagram involving three objects sending messages from one to another

example in Fig. 10, three objects send messages such as add, amend, and refresh from one to another.

2.7.4 Activities

Activities consist of a series of actions, each of which may execute zero, one, or more times for each activity. Implementation of a sequence of actions is based on a story about the system. In the prototyping stage, the simplest approach is to use pictogram objects made of paper, with simple illustrations that explain states and processes for each action.

More directly, consumer experiences can be role-played by participants. The goal of prototyping is to make a conceptual design that can be shared by participants in the same group and communicated to other persons who have no knowledge of the ideas produced by the group.

2.7.5 Design Patterns

Design patterns help to reduce the resources needed for prototyping in terms of the time required to assign workers to design tasks. However, some common knowledge is needed to use a design pattern effectively. Pattern languages have been invented to express various design patterns. According to Meszaros and Doble (1997), these patterns can be classified into five types.

Context-Setting Patterns This includes some working definitions that use patterns and pattern language. A pattern is a solution to a problem in a context, and a pattern language is the collection of patterns that solve the same problems or provide parts of a solution to a more significant partitioned problem. The pattern language itself presents the solution as several patterns, each of which describes the solution to a specific smaller problem.

Pattern Structuring Patterns These describe the desired content and structure of individual patterns, explaining the rationale for using the solution as well as describing that solution. Examples include patterns ensuring that all necessary information is addressed and patterns that specify ways of ignoring or focusing on particular elements in order to discuss an issue.

Pattern Naming and Referencing Patterns These describe techniques for labeling a pattern and linking one pattern to another. The names or labels representing a pattern are sometimes referred to as *jargons*. Using evocative pattern names makes them easier to refer to and reduces the need for users to follow by connecting with other pattern names, using meaningful metaphors, using a phrase that refers to the solution, and using understandable names.

Patterns for Making Patterns Understandable These make a pattern or pattern language easier to read, understand, and apply. For example, we can categorize specific patterns by user or employ terminologies and diagram notations that the target users can understand. Sometimes, patterns can be described using examples (such as code samples in software architectures) that only the target users can understand.

Pattern Language Structuring Patterns These describe the desired content and structure of a pattern language—for example, a set of patterns that solves a complex problem in a specific domain, provides a problem/solution summary to help users find the patterns used in specific problems, identifies several probable solutions to a problem, or provides a single example for all patterns in the language. A sequential diagram including such patterns can also be used to solve a problem.

Pattern languages are useful for constructing and sharing common design patterns for problem-solving if all group members share fundamental knowledge of the design patterns. The power of pattern languages lies in enabling members sharing a common language to reduce the resources needed to communicate with one another.

2.8 *Review and Reflection*

This usually involves several groups sharing outcomes. The purpose of this process is to produce and disseminate group work outcomes clearly, using presentation slides or posters to explain prototypes of services and products to other groups. This enables participants to discuss the details of services and goods designed by each group. Slides and/or posters should include (at least) a title, the outcome structure, a prototype, and the DJM, enabling other groups to ask questions and offer comments to indicate what might be improved and/or expanded in the outcome.

Reflection is feedback to explain roles in group work and share their own feeling to other participants. Relationships among participants may be improved by discussion about the group tasks.

3 Adjustable Parameters of Design Workshops

The generalized design work procedure described in Sect. 2 can be applied to various types of design work to move from conceptual design to production design by means of a participatory framework. The most general parameters of the design work procedure are focus and total duration (typically 1 day or 3 successive days). The design work should be defined on the basis of a problem definition and domain. For example, to produce a conceptual design for goods and services, it is necessary to identify related fields and organizational divisions. The main issue for workshop design is the duration of each task.

Each task in the proposed design work procedure involves several parameters. Table 3 shows the adjustable parameters of group work. Selection of inputs is a significant determinant of quality of outcome. To strengthen connections with the environment in question, expert interviews and field investigations are useful as inputs; to collect various types of domain knowledge, use expert lectures as inputs; and to obtain quantitative outcomes, use data analysis.

Typically, inputs and brainstorming should account for 30% of all design work. Ideation, matrix expansion, and DJM should account for a further 30%, and the remaining 40% should be used for prototyping and review and reflection.

Table 3 Adjustable parameters of a design workshop

Inputs	Input types (lectures, expert interviews, field investigations, data analysis)
Lectures	Domain, number, and duration of lectures
Expert interviews	Domain, number of interviewees, duration of interviews
Field investigations	Domain, place, duration of field investigation
Data analysis	Domain, data sources, duration of data analysis
Brainstorming	Duration of brainstorming and enhanced methods to produce, extend, and select ideas
Ideation	Number of seeds-needs pairing ideas per participant, duration of task
Matrix expansion	Types of evaluation function (possibility, convenience, availability, innovativity innovation plus creativity, reliability)
Persona design	Number of personas, characteristics of each
Data journey map	Task duration
Prototyping	Types of prototyping (pictogram, performance, video, presentation slides)
Review and reflection	Duration of the task, types of review (presentation, posters, report), method of reflection

4 Using Insights from Data Analysis in a Design Workshop

As shown in Sect. 3, there are three opportunities across eight methods for exploiting quantitative insights from data analysis in a design workshop. (1) As inputs, insights from data analysis can be used directly. In this case, domains and data sources underpinning domain knowledge must also be determined. Potential data sources include official statistics, open data provided by government offices, and commercial data supplied by commercial vendors. Before commencing a data analysis session, the duration of data analysis should be determined, and potential outputs should be explained. When using data analysis tools, these should be explained in a hands-on lecture. As participants in a design workshop are likely to constitute a heterogeneous group, many may find it difficult to engage with complex techniques for analyzing and visualizing data. To avoid this problem, introduce the simplest tools for data analysis, or include data analysis experts among the participants.

A second opportunity is ideation. Using insights obtained from data analysis, participants may produce a story about needs and seeds based on both quantitative and qualitative evidence. This task is sometimes called data storytelling. For example, by using demographic data to assess market size in a particular area, demand can also be estimated to inform a story about needs.

The third and final opportunity is persona design. By collecting and analyzing data about consumers, participants can more fully understand their likely behavior, so improving persona quality and facilitating development of a DJM.

5 Conclusion

This chapter discussed the use of participatory design methods to perform a conceptual design task from a data-centric perspective. The three essential components of group design work are domain knowledge inputs, heterogeneous membership, and a facilitator. The proposed design workshop procedure was based on eight methodologies: inputs, brainstorming, ideation, idea expansion, persona design, data journey map, prototyping, and review and reflection. The key objectives of conceptual design are to produce a concept that can be communicated and explained to others who have no knowledge of it. Personas enable participants to share fundamental assumptions, to clarify interactions to construct consumer experiences, and to make everything explainable. In proposing an approach to data utilization in a design workshop, three opportunities were identified at the input, ideation, and persona design stages. Cyclical use of the eight methodologies can improve design outcomes.

References

Bernard G, Periklis A (2017) A process mining based model for customer journey mapping. CaiSE-Forum-DC

Cooper A, Saffo P (1999) The inmates are running the asylum. Macmillan, Indianapolis

Jetzek T, Avital M, Bjorn-Andresen N (2014) Data-driven innovation through open government data. J Theor Appl Electron Commer Res 9(2):100–120

Kim SK, Ishii K (2007) Scenario graph: discovering new business opportunities and failure mode. Technical paper, Stanford University, pp 1–8

Kunifuji S (2013) A Japanese problem solving approach: the KJ-Ho method. In: Proceedings of KICSS, pp 333–338

Laursen K, Salter A (2006) Open for innovation: the role of openness in explaining innovation performance among U.K. manufacturing firms. Strateg Manag J 27:131–150

Meszaros G, Doble J (1997) A pattern language for pattern writing. In: Martin RC, Riehle D, Buschmann F (eds) Pattern languages of program design, vol 3. Addison-Wesley Longman, Boston, pp 529–574

Miaskiewicz T, Kozar KA (2011) Personas and user-centered design: how can personas benefit product design processes? Des Stud 32(5):417–430

Moon H, Han SH, Chun J, Hong SW (2016) A design process for a customer journey map: a case study on mobile services. Hum Factors Ergon Manuf 26:501–551. https://doi.org/10.1002/hfm.20673

Nielsen L (2013). 30. Persona Available at: https://www.interaction-design.org/literature/book/the-encyclopedia-of-human-computer-interaction-2nd-ed. Accessed 16 Aug 2018

Osborn AF (1953) Applied imagination. Scribner's, Oxford

Richardson A (2010) Series on customer journey: using customer journey maps to improve customer experience. Available at: https://hbr.org/2010/11/using-customer-journey-maps-to?referral=03759&cm_vc=rr_item_page.bottom

Schumpeter JA (1983) The theory of economic development. Transaction Publishers, New Brunswick

Unified Modeling Language: Superstructure, version 2.0 formal/05–07-04. Available at: https://www.omg.org/spec/UML/2.0/Superstructure/PDF, Accessed 25 Aug 2018

Designing Human-Machine Systems Focusing on Benefits of Inconvenience

Hiroshi Kawakami

1 Introduction

Fuben-eki denotes the further benefits of a kind of inconvenience (Kawakami 2009). It does not reflect nostalgia but a standpoint of reviewing existing things and designing new ones.

In general, designers tend to automatically pursue convenient systems. Although convenience may enrich our lives in some ways, it is not always optimal for users. Convenient systems may inflate such harms as excluding users, depriving the pleasure of using systems, and eroding human motivation and skills. For example, the US Federal Aviation Administration reported that the continual use of convenient auto-flight systems might degrade a pilot's ability to recover an aircraft quickly from undesired states (Federal Aviation Administration 2013).

Some other convenience-derived harms exist for human-machine systems (Norman 2005). Contemplating the downside of convenience prompts us to pursue a new design principle other than convenience. Fuben-eki is one of the promising bases of a new principle for designing human-machine systems.

H. Kawakami (✉)
Kyoto University, Sakyo, Kyoto, Japan
e-mail: kawakami.hiroshi.3s@kyoto-u.jp

A.-H. Sato (ed.), *Applications of Data-Centric Science to Social Design*,
Agent-Based Social Systems 14, https://doi.org/10.1007/978-981-10-7194-2_4

2 Benefits of Inconvenience

2.1 *Definition of Inconvenience*

Generally, inconvenience and benefit are thought to be contradictory terms. Benefits are associated with positive images; inconvenience seems to be negative. Although the meaning of inconvenience is slippery, the word is used for multiple purposes. The Internet provided sentences that included the term convenience and such notions as save labor. The word labor has the following meanings:

physical labor: requiring time and effort,
mental labor: requiring special skills, including the consumption of such cognitive resources as paying attention, memorization, and conception.

In this section, we define convenience/inconvenience:

convenience: saving labor to attain a specific task,
inconvenience: being comparatively not convenient, i.e., requiring more labor than convenience.

Compared with the definition in the Longman Dictionary, which argues that "convenience does not spoil plans or cause problems," the above definition is superficial and may be insufficient for structuralists; but it satisfies most of our understanding of convenience.

Following this definition, inconvenience does not contradict benefit. Labor-derived benefits exist.

2.2 *Unilinear Pursuit of Convenience*

2.2.1 Escalator for Hiking

Hiking is an example of a healthy pursuit. If we follow the above definition of inconvenience, hiking (especially up a mountain) should be called an inconvenient activity. If we are solely pursuing convenience, we have to construct an escalator between the mountaintop and its foot, as shown in Fig. 1. Such convenience strikes the wrong note and hiking is no longer leisure.

2.2.2 Ultimate Magic Bat

Sporting goods should be inconvenient. If there were an ultimate magic bat that always got a hit, baseball games would become meaningless. LZR Racer is a line of swimsuits, that was worn by 94% of the swimmers who won races in the 2008 Summer Olympics. Such dominance ruins the meaning of competition, even though it may be convenient for reducing race times.

Fig. 1 An Escalator for mountaineering (Kawakami 2017)

2.3 Toward a Convenient Near Future

The fictional examples described in the previous section demonstrate that solely pursuing convenience is not the ultimate strategy for designing artifacts. Not only fictional but also the actual trend of recent technical development also provides another discussion point.

With the tertiary AI boom, useful AI applications in various fields have been developed. When it becomes impossible to read every mass-produced scientific paper, an AI application that reads papers and accumulates knowledge will be really convenient.

Good themes elicit arguments that blur the differences between means and ends. Implementing an AI that defeats professional Go players must only be a means for measuring the AI ability. However, if this AI is misused as a convenient system that frees players from thinking, it ruins the meaning of Go.

One of my students accepted a job offer from a car manufacturer because he loves driving. Unfortunately, he was assigned to a department that is developing automatic driving systems that deprive people of the pleasure of driving. Automatic systems are convenient, but they might foment problems when they are applied to a field where the absence of human beings is antithetical.

3 Fuben-eki Systems

Systems that provide users with the benefits of inconvenience are called fuben-eki systems. A conventional system design that produces convenient systems is evaluated in terms of one axis: the amount of labor. Fuben-eki system design, on the other hand, adds another axis that evaluates benefits. In this case, we get the four quadrants shown in Table 1, where the horizontal axis denotes conventional convenience and the vertical axis denotes further benefits. The opposite side of convenience is inconvenience, and the opposite of benefit is harm.

Table 1 Quadrants derived from production of convenience/inconvenience and benefit/harm

Benefits of inconvenience	Benefits of convenience
Harm of inconvenience	Harm of convenience

Overlooking the independence of the two axes restricts one's vision to two quadrants: the upper right and the lower left. This section examines examples from the upper left quadrant, most of which are inextricably linked to the examples in the lower right quadrant that shows the harms of convenience.

3.1 Negative Examples of Fuben-eki

Of course, not every inconvenience provides beneficial aspects. The following three types of inconvenience were eliminated from the examples for investigating beneficial inconveniences:

Retro styles: Fuben-eki is different from nostalgia which tends to design retro-looking objects. From a fuben-eki viewpoint, old-fashioned things must be analyzed not from structural aspects but from functional aspects.

Your inconvenience yields to my benefits: If you get benefits by causing trouble to someone else, that situation is not fuben-eki. The person who gets the benefits of inconvenience should also be the person who suffered from it.

Compromises: Inconvenience is inevitable to get fuben-eki. In other words, inconvenience is not a compromise with benefits. For example, the time and effort to type in a password are only compromised with security. In this case, security is not fuben-eki. Typing in a password is not obligatory and can be replaced by other security measures. However, if a person experiences subjective benefits that are caused by the typing itself, such feelings are fuben-eki.

3.2 Positive Example of Fuben-eki

By eliminating such examples as those that are classified into the three types shown in the previous section, we get positive examples of the benefits of inconvenience.

Figure 2 overviews a part of the relationship among benefits and inconvenience. In the figure, each arc denotes the contribution relation and each bidirectional broken arc denotes the mutual contribution relation. Symbol $\Diamond$ is a modal operator. In the alethic mode, $\Diamond p$ means that "proposition p is possible." In the deontic mode, $\Diamond p$ means that "proposition p is permitted." In either mode, even if the value of p varies subjectively depending on a human attitude, the value of $\Diamond p$ is objectively determined.

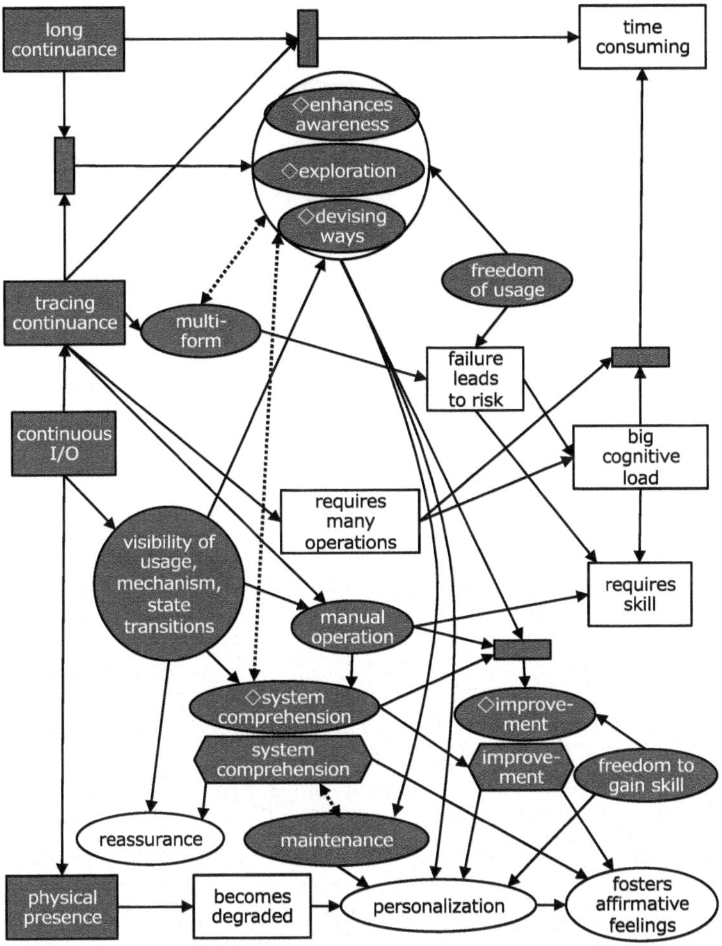

Fig. 2 Contribution relations among benefits and inconveniences. Rectangle nodes denote situations derived from systems, and white rectangles are called inconvenient situations. Ellipsoidal nodes denote benefits, and white ellipsoidal nodes are user-dependent subjective benefits. Rectangle nodes without labels mean conjunctions that work just like Petri-Net transitions

For example, when p denotes that "the user becomes skillful to operate the system," the value of p depends on the user, but the value of $\Diamond p$ is objectively determined depending on the system. When the system allows (permits) users to become skilled, $\Diamond p$ is true. Otherwise, it is false.

4 Analyzing Fuben-eki Examples

4.1 Conditions for Approving Fuben-eki

We call examples that provide users with the benefits of inconvenience fuben-eki examples. In Sect. 3.1, three types of negative examples are shown, which can be rewritten as conditions for approving fuben-eki systems:

- Benefits and inconvenience belong to the user.
- Benefits are directly derived from the inconveniences.

"Your inconvenience is my benefit" is the worst case.

Another viewpoint of fuben-eki approval is the product of subjective/objective and inconvenience/benefit shown in Table 2. The next section discusses fuben-eki from these four viewpoints.

4.2 Objectivity and Subjectivity of Inconvenience

4.2.1 Objective Inconvenience (Labor)

Feelings of convenience/inconvenience are dependent on contexts and on individuals. They vary by the person. Nevertheless, if we need to objectively discuss convenience, individuals cannot be taken into account: they must be treated as a group. In this case, saving labor is often substituted for convenience. Saving labor is an objectively observable phenomenon.

Strictly speaking, when discussing convenience/inconvenience, a targeted task is necessary. In addition, the measure of convenience/inconvenience is at most on a semi-ordinal scale. In other words, they cannot be quantified like interval scales or ratio scales. It is impossible to place a numerical value on the inconvenience degree to a specific event.

Stemming from these facts, this section defines objective convenience as saving labor to attain a specific task, and inconvenience as being comparatively not convenient, as shown in Section 2.1.

Fuben-eki systems have to be objectively inconvenient in this sense.

Table 2 Subjective or objective inconvenience and benefits

Subjective	Inconvenience	Benefit
Objective	Labor	Function

4.2.2 Subjective Inconvenience

Feelings of convenience/inconvenience are intrinsically subjective. When a person is not aware of more convenient tools or methods, she/he does not subjectively feel inconvenience even if she/he is using objectively inconvenient tools.

Furthermore, even if a person objectively knows about more convenient tools or methods, she/he may prefer inconvenient tools. For example, some people prefer manual transmissions to automatic transmissions for driving without feeling inconvenience.

They are examples of these special cases where objective inconvenience is observed but subjective inconvenience is not. It is desirable without being required that subjective inconveniences are observed in each fuben-eki system.

4.3 Objectivity and Subjectivity of Benefits

This subsection discusses the objectivity and subjectivity of benefits.

In general, convenient tools are oriented to be usable by everyone, and users do not have to be skilled with such tools. Inconvenient tools often allow users to become skillful and to gradually master these tools. However, even if such acquisition is allowed, the user does not always become skilled, and some users may not appreciate the benefit of allowance.

Here, objectively "the system is allowing", while subjectively "the user feels it is beneficial." In general, the objective benefits offered by inconvenience can be regarded as a **function** provided by the system. The functions are expressed using $\Diamond$, which is a modal logic symbol (Fig. 2). In this section, we classify benefits into objective ones with $\Diamond$ and subjective ones.

Positive examples provide users with the opportunity to get the following inconvenience-oriented benefits (Hasebe et al. 2015).

4.3.1 Objective Benefits (Function)

When a system requires users to take time and consume cognitive resources, i.e., being objectively inconvenient, the system often performs the following functions:

$\Diamond$devising ways: optimality to a specific task leads convenience, but not allows users to devise ways.

$\Diamond$enhancing awareness: inconvenience offers the opportunity of discovery in the operation process.

$\Diamond$system comprehension: interactions and physical feedbacks enable users to understand systems.

$\Diamond$improvement: inconvenience allows users to become skillful at exploiting the systems.

◊positive contribution to tasks: inconvenience requires users to do a work that in turn enables users to contribute to tasks.
◊preventing skill erosion: to do a work can be a OJT for users.

4.3.2 Subjective Benefits

When the user subjectively regards the objective functions (provided by the system as summarized above) as beneficial, the benefits are classified as follows:

- reassurance,
- personalization,
- motivating tasks,
- fostering affirmative feelings.

5 Synthesizing Fuben-eki Systems

Fuben-eki systems are designed so that users obtain benefits from inconvenience. The time and effort of using fuben-eki systems provide benefits as the by-products of accomplishing the main task.

Section 4 discussed the analysis of fuben-eki systems. This section discusses the synthesis of such systems.

Conventional ideation support methods are based on the combination of divergence and convergence. At the divergence stage of the first half, quantity is required rather than quality, and criticism is forbidden. In this case, inconvenience is passively accepted. On the other hand, this section positively utilizes inconvenience as a key of ideations.

Figure 3 shows the quadrants that reflect Table 1. Ideation styles for fuben-eki systems can be classified with the transitions of quadrants.

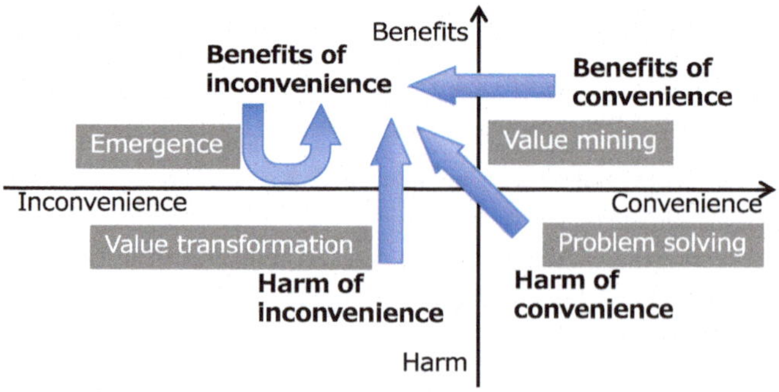

Fig. 3 Four types of ideation for fuben-eki systems

5.1 Ideation Through Problem-Solving

The normal way of engineering is the sequence of two processes: formulating a problem and solving it by developing a new system. This process can be applied to developing fuben-eki systems, i.e., formulating a harm of convenience as a problem and solving it by making target systems inconvenient. The arrow in Fig. 3 from the right lower quadrant to the left upper quadrant illustrates this style of ideation.

Formulating the harm of convenience can be supported by the **objective benefits** shown in Sect. 4.3.1. Checking whether a convenient system prevents users from devising solutions, becoming skillful, understanding systems, contributing positively to tasks, and so on helps identify problems.

As one example of this ideation style, we employ navigation systems, which are convenient because they show clear and accurate information. Unfortunately, the convenience deflates our motivation to remember the environment we are walking through. We only need to follow the fragments of direction to reach our destination, without exploring our surroundings or beginning to proactively understand the area we are moving through. Therefore, we do not remember the roads. Pleasurable explorations are reduced to merely idle transportation.

One way to solve this problem is to introduce inconvenience. We transformed a navigation system by introducing a novel inconvenience: map degradation (Kitagawa et al. 2010). In this new system, the trails followed by users gradually disappear. Therefore, they need to recall the surroundings of the trails when they use the system again; the system encourages users to remember landmarks more precisely. Figure 4 shows the interface of degradation navi.

We developed two types of walking navigation systems: normal and degradation navi. Using these systems, we conducted experiments where examinees walked around a city and answered questions about pictures of the scenes around the pathway of their walk. The answers of the examinees who used degradation navi were significantly more correct than those who used a normal navigation system.

5.2 Ideation via Value Mining

The second type of ideation is value mining, which is different from problem-solving because it does not anticipate problems in the target system. The upper right quadrant in Fig. 3 is for happy states where the target system is convenient and users feel that it is beneficial. That is, there are no problems. We can design fuben-eki systems by transforming such convenient target systems into inconvenient ones that occasionally mine latent benefits. Conceptually, such ideation is a shift from the upper right to the upper left quadrant in Fig. 3.

When the course to transform a system into an inconvenient one reverts to its actual historical development, the lost benefits of the old style system may be revived. On the other hand, when the course introduces novel inconveniences, new benefits are expected.

Fig. 4 Screenshot of degradation navi interface

Figure 5 shows an example of the results of such ideation: a prime number ruler, which is an original item of Kyoto University. A conventional ruler is only a convenient tool for measuring length, but an inconvenient prime number ruler allows users to devise their own ways to measure length.

The process of making target systems inconvenient can be supported by **objective inconvenience**, i.e., taking time and effort, or consuming the cognitive resources of users.

For example, consider a ruler as the target system. Even if a normal ruler has no problems, the first step of this ideation makes it inconvenient. That is, let the ruler consume users' time and cognitive resources. As one practical way to make rulers inconvenient, an idea was provided that restricted the tick marks to just prime numbers.

The second step of this ideation mines benefits from this inconvenient ruler. One benefit is allowing users to make positive contribution to their tasks. A simple subtraction is required to measure a length that is not marked on the scale. As a result, the task of measuring the length became a task that must be positively grappled with in the form of subjective subtraction.

Fig. 5 Prime number ruler

5.3 Ideation by Emergence

There is an emergent type of ideation that is different from both problem-solving and value-mining types. This method produces ideas what pops into a person's mind by shoving fuben-eki examples into the brain and extracting their essence without converting them into linguistic thoughts.

Because emergence is the most non-systematic method, expert skill is required. Professional designers call this method one hundred knocks. This style of thought requires accumulated knowledge and the ability to change knowledge flexibly. Analogies are frequently used.

For example, consider a method for unlocking a smartphone. We conducted a battle-styled brainstorming and got a hundred ideas. Among them, a gesture-unlocking mechanism was proposed. To unlock her smartphone, the user has to shake it with almost the same movement as she registered in advance. This is an inconvenient system because unlocking is almost impossible when an arbitrary gesture is registered. Muscle memory is required to register gestures. That is, this system provides users with a subjective benefit called personalization.

5.4 Ideation by Value Transformation

The fourth type of ideation first identifies the potential fuben-eki features of the target systems without changing the level of convenience. It utilizes the existing inconvenience by transforming annoying labor into a worthy activity. For example, we might increase user fulfillment and affirmative feelings by presenting explicit feedback of actions. Based on psychological theories with respect to motivation, Hiraoka et al. proposed an eco-driving system that shows drivers the evaluations and target scores of fuel saving (Hiraoka 2012). Although this system does not change the amount of labor to drive a car, it increases motivation toward eco-driving by informing drivers of their achievements in terms of the degree of eco-driving.

The arrow in Fig. 3 from the left lower quadrant illustrates such ideation.

6 Conclusions

As one different direction from uncritical pursuits to convenience, we described a direction to utilize the benefits of inconvenience and discussed the system design of that direction. Of course, this direction does not insist on the uncritical pursuit of inconvenience. This chapter organized what kind of inconvenience yields what kind of benefits. In addition, we showed four ways to ideate systems that incorporate the benefits of inconvenience.

References

Federal Aviation Administration (2013) Safety alert for operators, vol 13002. Flight Standards Service, Washington, DC

Hasebe Y et al (2015) Card-type tool to support divergent thinking for embodying benefits of inconvenience. Web Intelligence 13(2):93–102

Hiraoka T et al (2012) Eco-driving support system to encourage spontaneous fuel-efficient driving behavior. J Soc Instrum Control Eng 48(11):754–763

Kawakami H (2009) Toward systems design based on benefit of inconvenience. J Human Interface 11(1):125–134 (in Japanese)

Kawakami H (2017) Standpoint of benefits of inconvenience. Impress, Tokyo ISBN:978-4-295-00092-1 (in Japanese)

Kitagawa H et al (2010) Degrading navigation system as an explanatory example of "benefits of inconvenience." In: Proceedings of the SICE Annaul Confirmed, pp 1738–1742

Norman DA (2005) Human centered design considered harmful. Interactions 12(4):14–19

Part II
Mathematical Foundation of Human Collective Behavior

Information Cascade and Phase Transition

Masato Hisakado and Shintaro Mori

1 Introduction

In general collective herding poses interesting problems in several fields. To cite a few examples in statistical physics, anomalous fluctuations in financial markets (Cont and Bouchaud 2000; Eguíluz and Zimmermann 2000) and opinion dynamics (Stauffer 2002; Galam 1990) have been related to percolation and the random-field Ising model. To estimate public perception, people observe the actions of other individuals; then, they make a choice similar to that of others. Recently, these behaviors have been referred to as Kuki-wo-yomu (follow an atmosphere) in Japanese. Because it is usually sensible to do what other people are doing, the phenomenon is assumed to be the result of a rational choice. Nevertheless, this approach can sometimes lead to arbitrary or even erroneous decisions. This phenomenon is known as an information cascade (Bikhchndani et al. 1992).

A recent agent-based model proposed by Curty and Marsili (2006) focused on the limitations imposed by herding on the efficiency of information aggregation. Specifically, it was shown that when the fraction of herders in a population of agents increases, the probability that herding yields the correct forecast (i.e., individual information bits are correctly aggregated) undergoes a transition to a state in which either all herders forecast rightly or no herder does.

We introduced a voting model that is similar to a Keynesian beauty contest (Keynes 1936; Hisakado and Mori 2010). There are two types of voters – herders and independents – and two candidates. Herders vote for each candidate with the

M. Hisakado (✉)
Nomura Holdings, Inc., Chiyoda-ku, Tokyo, Japan

S. Mori
Faculty of Science and Technology, Department of Mathematics and Physics, Hirosaki University, Hirosaki, Aomori, Japan

A.-H. Sato (ed.), *Applications of Data-Centric Science to Social Design*, Agent-Based Social Systems 14, https://doi.org/10.1007/978-981-10-7194-2_5

probabilities that are proportional to the candidates' votes. In the previous paper, they were known as analog herders. We investigated a case wherein all the voters are herders (Mori and Hisakado 2010). In such a case, the process is a Pólya process, and the voting rate converges to a beta distribution in a large time limit (Hisakado et al. 2006). Next, we doped independent voters in herders. They are a kind of noise which plays the important role for the system. The proposed voting model is a binomial distribution doped in a beta binomial distribution mathematically. In the upper limit of t, the independent voters make the distribution of votes converge to Dirac measure against herders. This model consists of three phases. If herders constitute the majority or even half of the total voters, the voting rate converges more slowly than it would in a binomial distribution. If independents constitute the majority of the voters, the voting rate converges at the same rate as it would in a binomial distribution. The phases differ in terms of the velocity of the convergence. If the independent voters vote for the correct candidate rather than for the wrong candidate, the model consists of no case wherein the majority of the voters select the wrong answer. The herders affect only the speed of the convergence; they do not affect the voting rates for the correct candidate.

In this chapter we consider the digital herder case (Hisakado and Mori 2011). Because herders always select the majority of the votes, which are visible to them, the behavior becomes digital (discontinuous). Digital herders have a stronger herding power than analog herders, and we can find the phase transition like Ising model.

Here, we discuss a voting model with two candidates, C_0 and C_1. We set two types of voters – independent and herders. In this chapter, the herders are digital herders. The voting of independent voters is based on their fundamental values. On the other hand, the voting of herders is based on the number of votes. Herders always select the majority of the previous r votes, which is visible to them.

The remainder of this chapter is organized as follows. In Sect. 2, we introduce our voting model, and we mathematically define the two types of voters – independents and herders. The voters can see the previous r votes of the voters. In Sect. 3, we calculate the exact distribution functions of the votes for the case wherein the voters can see the votes of all the voters. We discuss the phase transition using the exact solutions. In Sect. 4, we discuss the special case, $r = 1$. In this case, we calculate the exact distribution function; however, there is no phase transition. Finally, the conclusions are presented in Sect. 5.

2 Model

We model the voting of two candidates, C_0 and C_1; at time t, they have $c_0(t)$ and $c_1(t)$ votes, respectively. At each time step, one voter votes for one candidate; the voting is sequential. Voters are allowed to see r previous votes for each candidate when they vote so that they are aware of public perception. If $r > t$, voters can see t previous votes for each candidate. At time t, the number of votes for C_0 and C_1 is

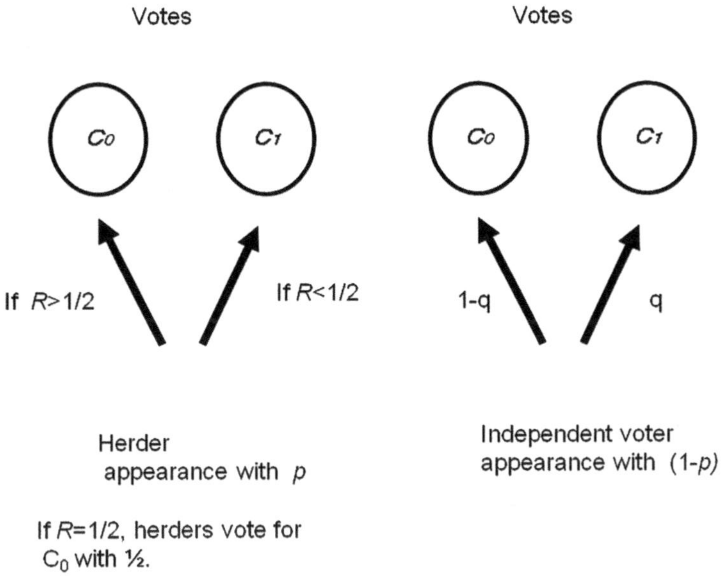

Fig. 1 Demonstration of model. $R = c_0^r/\{c_0^r + c_1^r\}$

$c_0^r(t)$ and $c_1^r(t)$, respectively. In the limit $r \to \infty$, voters can see all previous votes. Therefore, $c_0^\infty(t) = c_0(t)$ and $c_1^\infty(t) = c_1(t)$.

There are two types of voters – independents and herders; we assume an infinite number of voters. Independent voters vote for C_0 and C_1 with probabilities $1 - q$ and q, respectively. Their votes are independent of others' votes, i.e., their votes are based on their fundamental values. Here, we set C_0 as the wrong candidate and C_1 as the correct candidate to validate the performance of the herders. We can set $q \geq 0.5$ because we believe that independent voters vote for the correct candidate C_1 rather than for the wrong candidate C_0. In other words, we assume that the intelligence of the independent voters is virtually correct. It also plays the role of noise.

On the other hand, herders vote for a majority candidate; if $c_0^r(t) > c_1^r(t)$, herders vote for the candidate C_0. If $c_0^r(t) < c_1^r(t)$, herders vote for the candidate C_1. If $c_0^r(t) = c_1^r(t)$, herders vote for C_0 and C_1 with the same probability, i.e.,1/2. It is the response function. The conclusion of the votes depends on this and we will discuss in other chapters. In the previous paper, the herders voted for each candidate with probabilities that were proportional to the candidates' votes (Hisakado and Mori 2010); they were known as analog herders. On the other hand, the herders in this paper are known as digital herders (Fig. 1).

The independent voters and herders appear randomly and vote. We set the ratio of independent voters to herders as $(1 - p)/p$. In this chapter we mainly pay attention in large t limit. It means the voting of infinite voters.

3 Exact Solutions for $r = \infty$

In this section, we study the exact solution of the case $r = \infty$ by using combinatorics.

Here, we map the model to correlated Brownian motion along the edges of a grid with square cells, and we count the number of paths. Let m and n be the horizontal axis and the vertical axis, respectively. The coordinates of the lower left corner are $(0, 0)$; this is the starting point. m is the number of voters who vote for C_1, and n is the number of voters who vote for C_0. A path shows the history of the votes. If a voter votes for C_1, the path moves rightward. If a voter votes for C_0, the path moves upward.

We define $P_i(m, n)$ as the probability that the $(n + m + 1)$th voter votes for the candidate C_i, where $i = 0, 1$. The probability of moving upward is as follows:

$$P_0(m,n) = \begin{Bmatrix} p + (1-p)(1-q) \equiv A & m < n; \\ \frac{1}{2}p + (1-p)(1-q) \equiv B & m = n; \\ (1-p)(1-q) \equiv C & m > n. \end{Bmatrix} \tag{1}$$

The probability of moving rightward is $P_1(m, n) = 1 - P_0(m, n)$ for each case. Here, we introduce $X(m, n)$ as the probability that the path passes through the point (m, n). The master equation is

$$X(m,n) = P_1(m-1,n)X(m-1,n) + P_0(m,n-1)X(m,n-1), \tag{2}$$

for $m \geq 0$ and $n \geq 0$, with the initial condition $X(0, 0) = 1$. This defines $X(m, n)$ uniquely. Hereafter, we refer to the region $m < n$ as I, $m > n$ as II, and $m = n$ as III (Fig. 2).

First, we consider the case $q = 1$. At this limit, independent voters always vote for only one candidate, C_1 (if we set $q = 0$, independent voters vote only for C_0). The probability is reduced from (1) to

$$P_0(m,n) = \begin{Bmatrix} p & m < n; \\ \frac{1}{2}p & m = n; \\ 0 & m > n. \end{Bmatrix} \tag{3}$$

In this case, if the path enters II $(m > n)$, it can only move rightward. Hence, $n = m - 1$ becomes the absorption wall, where $m, n \geq 0$. There is a difference between the probability (3) in I $(m < n)$ and that in III $(m = n)$. Then, we have to count the number of times the path touches the diagonal.

Using (25), we can calculate the distribution for $m \leq n$. (See Appendix A.)

$$X(m,n) = \begin{Bmatrix} \sum_{k=0}^{m} A_{m,n,k} \frac{p^n(1-p)^m}{2^{k+1}} & m < n; \\ \sum_{k=0}^{m} A_{m,m,k} \frac{p^m(1-p)^m}{2^k} & m = n. \end{Bmatrix}, \tag{4}$$

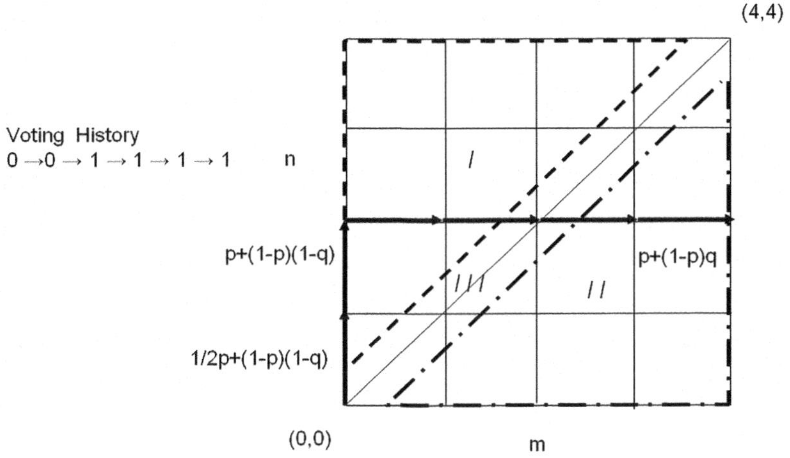

Fig. 2 Voting and path. A path shows the history of the votes. If a voter votes for C_1, the path moves rightward. If a voter votes for C_0, the path moves upward. The sample directed line shows the voting of six voters, 0, 0, 1, 1, 1, 1. We refer to the regions $m < n$, $m > n$, and $m = n$ as I, II, and III, respectively

where $A_{m,n,k}$ is given by (25) and k is the number of times the path touches the diagonal.

The distribution for $m > n$ can be easily calculated for the absorption wall $n = m - 1$, where $m, n \geq 0$. The distribution for $m > n$ is given by

$$X(m,n) = \sum_{k=0}^{n} A_{n,n,k} \frac{p^n(1-p)^n(1-\frac{1}{2}p)}{2^{k+1}} \qquad m > n. \tag{5}$$

We investigate the limit $t \to \infty$. Here, we consider m as a variable; it is the distribution function of the vote for C_1. For large t, we can assume that only the first terms of the summation of (4) and (5) are non-negligible. The first term becomes the difference of the binomial distributions using (25). For $m/t < 1/2$, the peak of the binomial distribution is $1 - p$, and for $m/t > 1/2$, it is 1. Then, we can obtain the distribution in the scaling limit $t = m + n \to \infty$.

$$\frac{m}{t} \Longrightarrow Z. \tag{6}$$

The probability measure of Z is

$$\mu = \alpha\delta_{1-p} + \beta\delta_1, \tag{7}$$

where δ_x is Dirac measure. Z is the ratio of voters who vote to C_1 from (6). The distribution has two peaks, one at $Z = 1$ and the other at $Z = 1 - p$. Now, we calculate α and β, where $\alpha + \beta = 1$. The probability that the path touches the

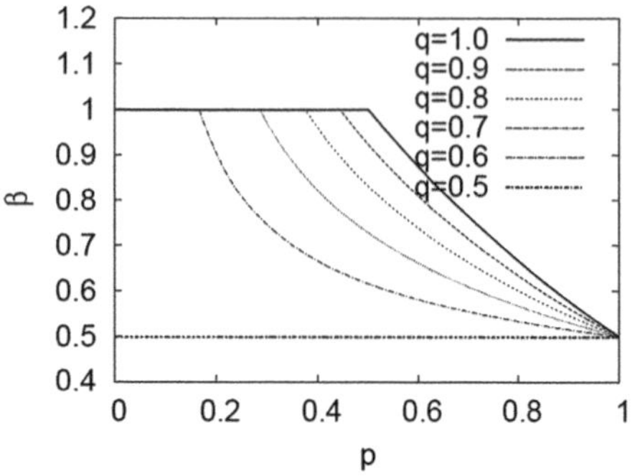

Fig. 3 β or $\bar{s}$, i.e., average votes for candidate C_1 by herders

absorption wall $n = m - 1$ is given by

$$
\begin{aligned}
\beta &= \sum_{m=0}^{\infty} X(m,m)(1-\frac{p}{2}) = \sum_{m=0}^{\infty}\sum_{k=0}^{m} A_{m,m,k}\frac{p^m(1-p)^m}{2^k}(1-\frac{p}{2}) \\
&= (1-\frac{p}{2})[1+\frac{x}{2}C_0(x)+(\frac{x}{2})^2C_1(x)+(\frac{x}{2})^3C_2(x)+\cdots] \\
&= (1-\frac{p}{2})[1+\frac{x}{2}C_0(x)+(\frac{x}{2})^2\{C_0(x)\}^2+(\frac{x}{2})^3\{C_0(x)\}^3+\cdots] \\
&= (1-\frac{p}{2})[\sum_{k=0}^{\infty}\frac{\{xC_0(x)\}^k}{2^k}] = (1-\frac{p}{2})\frac{1}{1-\frac{xC_0(x)}{2}} \\
&= \frac{4-2p}{3+\sqrt{1-4p(1-p)}} = \frac{4-2p}{3+|1-2p|}.
\end{aligned}
\tag{8}
$$

$C_k(x)$ is the generating function of the generalized Catalan number (30), and $x = p(1-p)$. Here, we use the relations (29) and (32). (See Appendix B.)

In Fig. 3, we plot β for the case $q = 1$. We are interested in the average votes for C_1 by the herders to validate the performance of the herders. We define s as the average votes for the correct candidate C_1 by the herders.

$$
s = \frac{Z-(1-p)}{p}. \tag{9}
$$

Here, we take expected values of (9) about several sequences of voting.

$$
\bar{Z} = p\bar{s} + (1-p) = (1-p)\alpha + \beta, \tag{10}
$$

where $\bar{x}$ means the expected value of x. The second equality can be obtained from (7). Using the relation $\alpha + \beta = 1$, we can obtain $\bar{s} = \beta$.

When p is less than 0.5, herding is a highly efficient strategy. The distribution of votes peaks when $Z = 1$. A majority of votes is necessary to select the correct

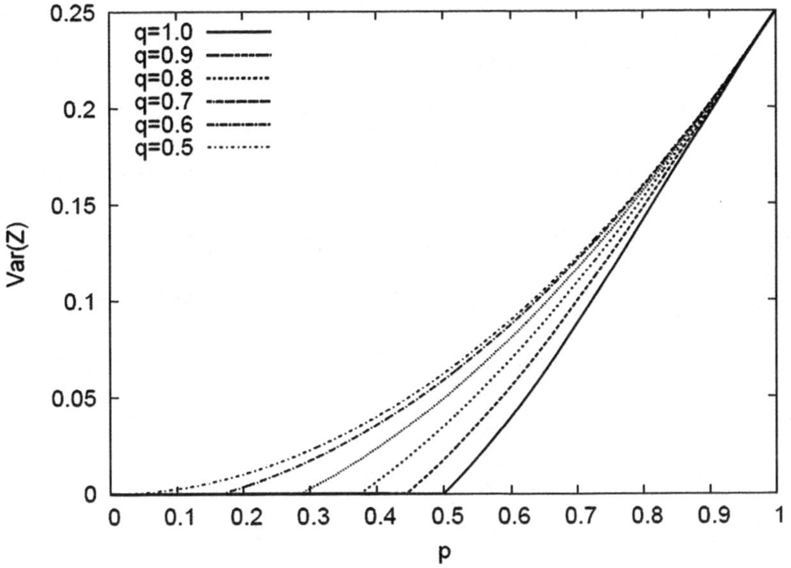

Fig. 4 $Var(Z)$, the variance of Z, is the order parameter. $Var(Z)$ is not differentiable at p_c. Z is the ratio of voters who vote to C_1

candidate C_1. At $p = p_c = 0.5$, there is a phase transition. When p exceeds $p_c = 0.5$, the distribution of votes has two peaks. In this case, a majority may select the wrong candidate C_0. In the language of game theory, this is a bad equilibrium. The probability of falling into bad equilibrium is $1-\beta$. $Var(Z)$, the variance of Z in the large t limit, is the order parameter. It is observed that $Var(Z)$ is not differentiable at $p = 1/2$ (Fig. 4). Hence, the phase transition is of the second order.

When $p \leq p_c$, the distribution has one peak, and it does not depend on $P_0(m, m)$, which is the probability of the vote when the number of votes for C_0 is the same as that for C_1. We can confirm that $Var(Z)$ is 0 in Fig. 4. On the other hand, when $p > p_c$, the limit distribution depends on $P_0(m, m)$; $P_0(m, m)$ is given by (3). We can confirm that $Var(Z)$ is not 0 in Fig. 4. If herders are analog, $Var(Z)$ is 0 in all region of p.

Next, we consider the general q case. In this case, the path goes across the diagonal several times. $n = m - 1$ is no longer the absorption wall, where $m, n \geq 0$. Hence, it is difficult to calculate the exact solution for general q. However, as in the discussion of (7), we can obtain the limit shape of the distribution of the votes for C_1.

$$\frac{m}{t} \Longrightarrow Z. \tag{11}$$

The probability measure of Z is

$$\mu = \alpha\delta_{(1-p)q} + \beta\delta_{p+(1-p)q}, \tag{12}$$

where $\alpha + \beta = 1$. When $q = 1$, (12) becomes (7). We can calculate β as

$$\beta = \tilde{R_1}(1 - R_2)(1 + R_1R_2 + R_1^2R_2^2 + \cdots) = \frac{\tilde{R_1}(1 - R_2)}{1 - R_1R_2}. \tag{13}$$

$\tilde{R_1}$ is the probability that the path starts from (0, 0), goes across the diagonal only once, and reaches the wall $n = m - 1$ in II $(m > n)$. R_1 is the probability that the path starts from the wall $n = m + 1$ in I $(m < n)$, goes across the diagonal only once, and reaches the wall $n = m - 1$ in II $(m > n)$. R_2 is the probability that the path starts from the wall $n = m - 1$ in II $(m > n)$, goes across the diagonal only once, and reaches the wall $n = m + 1$ in I $(m < n)$.

For example, the first term of (13) is the path that starts from (0, 0) and passes through I $(m < n)$ or directly enters II $(m > n)$. The path goes across the diagonal III $(m = n)$ only once; the first step is rightward. The second term is the path that starts from (0, 0), goes across the diagonal III $(m = n)$ three times, and enters II $(m > n)$. We can calculate $\tilde{R_1}$, R_1, and R_2 similarly to (8) (see Appendix B).

$$\begin{aligned}
\tilde{R_1} &= \frac{2(1 - B)}{2 - \gamma_1(1 - \sqrt{1 - 4A(1 - A)})}, \\
R_1 &= \frac{(1 - B)\gamma_1(1 - \sqrt{1 - 4A(1 - A)})}{B\{2 - \gamma_1(1 - \sqrt{1 - 4A(1 - A)})\}}, \\
R_2 &= \frac{B\gamma_2(1 - \sqrt{1 - 4C(1 - C)})}{(1 - B)\{2 - \gamma_2(1 - \sqrt{1 - 4C(1 - C)})\}},
\end{aligned} \tag{14}$$

where A, B, and C are given by (1), $\gamma_1 = B/A$, and $\gamma_2 = (1 - B)/(1 - C)$.

In general, we can calculate the exact value of p_c.

$$p_c = 1 - \frac{1}{2q}. \tag{15}$$

As q increases, p_c increases. At p_c, the model features a phase transition beyond which a state where most agents make the correct forecasts coexists with one where most of them are wrong. Thus, the effectiveness of herding decreases as q decreases. In the limit $q = 0.5$, the phase transition disappears. The distribution becomes symmetric in this case.

If p is greater than p_c, the vote ratios deviate considerably from the fundamental value q. Thus, digital herders account for greater effects than analog herders. Analog herders affect only the speed of convergence to the fundamental value (Hisakado and Mori 2010). Independent voters cannot oppose digital herders.

4 Exact Solutions for $r = 1$

Here, we discuss the cases $r = 1$ besides $p \neq 1$. When $p = 1$, all voters are herders, and the distribution becomes the limit shape of beta distribution, as discussed in Mori and Hisakado (2010). Herders can see only a vote of the previous voter. We define $P_i(t)$ as the probability that the $(t + 1)$th voter votes for C_i, where $i = 0, 1$. Here, t denotes the time.

$$P_0(t) = \begin{Bmatrix} p + (1-p)(1-q) \equiv F \; ; Y_0(t-1) = 1; \\ (1-p)(1-q) \equiv G \quad\quad : Y_0(t-1) = 0. \end{Bmatrix} \tag{16}$$

$Y_i(t) = 1$ indicates that at t, the voter votes for C_i. $Y_0(t-1) = 1$ indicates that the previous voter votes for C_0. On the other hand, $Y_i(t) = 0$ indicates that at t, the voter does not vote for C_i. $Y_0(t-1) = 0$ indicates that the previous voter votes for C_1. Thus, $\sum_{l=1}^{t} Y_i(l)$ is the total number of votes for C_i until t. Here, the relation $P_1(t) = 1 - P_0(t)$ holds. The initial distribution is

$$P_0(0) = \frac{1}{2}p + (1-p)(1-q). \tag{17}$$

The model was studied as a one-dimensional correlated random walk (Bohm 2000; Konno 2002). Here, we introduce $X(m, n)$ as the probability distribution. m is the number of the voters who vote for C_1, and n is the number of the voters who vote for C_0. The master equation is

$$X(m, n) = P_1(t-1)X(m-1, n) + P_0(t-1)X(m, n-1), \tag{18}$$

for $m \geq 0$ and $n \geq 0$, with the initial condition $X(0, 0) = 1$.

In the limit $t \to \infty$:

$$X(m, t-m) \Longrightarrow N(qt, \sqrt{\frac{F(1-G)}{(1-F)G}t}), \tag{19}$$

where $N(\mu, \sigma^2)$ is the normal distribution with mean μ and variance σ^2. (See Theorem 3.1. in Bohm 2000.) F and G are given in (16). Hence, we can obtain the limit shape of the distribution as

$$\frac{m}{t} \Longrightarrow Z. \tag{20}$$

The probability measure of Z is

$$\mu = \delta_q. \tag{21}$$

Then, $r = 1$ involves no phase transition, and the majority of the voters do not select the wrong candidate C_0.

The limit distribution is independent of the initial condition $P_0(0)$ because the distribution Z has only one peak.

5 Concluding Remarks

We investigated a voting model that is similar to a Keynesian beauty contest. We calculated the exact solutions for the special cases $r = 1$ and $r = \infty$. When $r = \infty$, we can obtain the exact solutions and obtain the critical point p_c of phase transition.

We consider herders who always choose the candidate with a majority of the previous votes, which is visible to them. We refer to these herders as digital herders. Digital herders exhibit stronger herd behavior than analog herders. We obtained exact solutions when the voters comprised a mix of digital herders and independents in $r = \infty$ and $r = 1$ cases. Here r is the referred number. $r = \infty$ case, the voter refers to all previous voters. $r = 1$ case, the voter refers to the latest voter. $r = \infty$ case, as the fraction of herders increases, the model features a phase transition beyond which a state where most voters make the correct choice coexists with one where most of them are wrong. This phase transition is referred to as information cascade transition. It is "spontaneous symmetry braking." We can obtain the critical point p_c and the distribution. The phase transition is the non-equilibrium process. On the other hand, the $r = 1$ case, there is no phase transition.

Ants use chemical signals called pheromones, and they behave as herders. The pheromones evaporate quickly. As an analogy, in our model, r is small. Thus, pheromones may amplify the limited intelligence of individual ants into something more powerful to avoid phase transition. In this chapter we discussed extreme cases, $r = 1$ and $r = \infty$ cases. $r = 1$ case corresponds to that where the pheromones evaporate quickly. On the other hand, $r = \infty$ case corresponds to the no evaporation case. We discuss $r \neq 1, \infty$ cases in the next chapter.

The herding system has an important role in the society of ants. Using the system, they can gather and unify the information. It is well known as an example of "emergence." The system which called ant colony optimization (ACO) is applied to several problems such as traversing salesman problem (Dorigo and Stützle 2004; Hisakado and Hino 2016). The herding strength, whether the herder is digital or analog, and the number of references r are the important parameters. As we discussed in this chapter, with the strong herding strength and long memory, there is the phase transition, and the system may be trapped by the bad equilibrium. It seems that ants adjust parameters to avoid the phase transition. There will be many useful things which we can learn from ants.

Appendix A Catalan Number and Extended

Here, we consider the number of monotonic paths along the edges of a grid with square cells, which do not pass lower the diagonal. Let m and n be the horizontal axis and the vertical axis, respectively. The coordinates of the lower left corner are $(0, 0)$. A monotonic path is one which starts in the lower left corner; finishes in the upper triangle (m, n), where $0 \leq m \leq n$; and consists entirely of edges pointing rightward or upward.

The number of paths from $(0, 0)$ to (m, n) can be calculated as

$$C_{m,n} = \frac{(n-m+1)(n+m)!}{m!(n+1)!} = \binom{n+m}{n} - \binom{n+m}{n+1}. \tag{22}$$

These numbers are known as generalized Catalan number.

If the finish point is (m, m), the number of paths becomes the Catalan number.

$$C_{m,m} = c_m = \frac{2m!}{m!(m+1)!} = \binom{2m}{m} - \binom{2m}{m+1}. \tag{23}$$

Next, we compute the distribution of the number of the paths that starts in the lower left corner, finishes in the upper triangle (m, n), and touches the diagonal k times (Di Francesco et al. 1997; Lang 2000). Let $A_{m,n,k}$ denote the number of paths that touches the diagonal k times. We get a simple recursion relation about $A_{m,n,k}$.

$$A_{m,n,k} = \sum_{j=0}^{m-1} c_j A_{m-j-1,n-j-1,k-1}, \tag{24}$$

for $k \geq 0$, $n, m \geq 0$, and $m \geq k$, with the initial condition $A_{0,0,0} = 1$. This defines the numbers $A_{m,n,k}$ uniquely, and it is easy to prove that

$$\begin{aligned} A_{m,n,k} &= \frac{(n-m+k)(n+m-k-1)!}{n!(m-k)!} \\ &= \binom{n+m-k-1}{n-1} - \binom{n+m-k-1}{n}. \end{aligned} \tag{25}$$

From (22) and (25), we can obtain the relation

$$A_{m,m,k} = C_{m-k,m-1}. \tag{26}$$

The well-known generating function $C_0(x)$ of Catalan numbers is given by

$$C_0(x) = \sum_{n=0}^{\infty} C_{m,m}x^n$$
$$= 1 + x + 2x^2 + 5x^3 + 14x^4 + 42x^5 + 132x^6 + \cdots, \quad (27)$$

subject to the algebraic relation

$$xC_0(x)^2 = C_0(x) - 1, \quad (28)$$

and we can obtain

$$C_0(x) = \frac{1 - \sqrt{1-4x}}{2x}. \quad (29)$$

Here, we obtain the generating function $A_{m,k}(x)$ of $A_{m,m,k}$.

$$A_{m,k}(x) = \sum_{m-k=0}^{\infty} A_{m,m,k}x^{m-k} = \sum_{m-k=0}^{\infty} C_{m-k,m-1}x^{m-k}$$
$$= \sum_{l=0}^{\infty} C_{l,l+k-1}x^l = C_{k-1}(x). \quad (30)$$

We use (26) for the second equality. $C_j(x) = \sum_{l=1}^{\infty} C_{l,l+j}x^l$ is the generating function of the generalized Catalan number (22). The generating function of the generalized Catalan number is given by

$$C_1(x) = \sum_{n=0}^{\infty} C_{m,m+1}x^n$$
$$= 1 + 2x + 5x^2 + 14x^3 + 42x^4 + 132x^5 + 429x^6 + \cdots,$$
$$C_2(x) = \sum_{n=0}^{\infty} C_{m,m+2}x^n$$
$$= 1 + 3x + 9x^2 + 28x^3 + 90x^4 + 297x^5 + 1001x^6 + \cdots,$$
$$C_3(x) = \sum_{n=0}^{\infty} C_{m,m+3}x^n$$
$$= 1 + 4x + 14x^2 + 48x^3 + 165x^4 + 572x^5 \cdots.$$

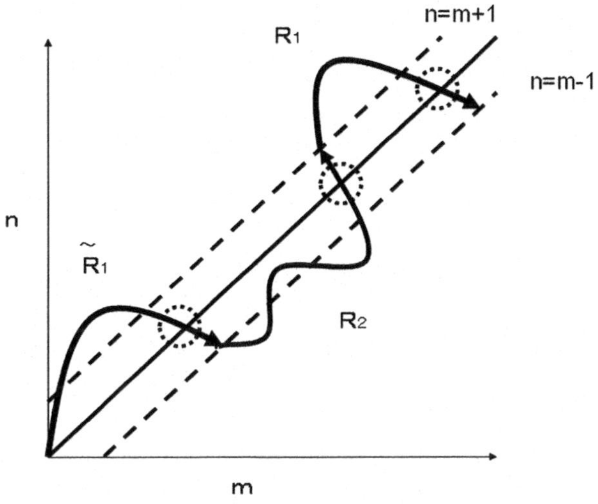

Fig. 5 $\tilde{R}_1$, R_1, and R_2. $\tilde{R}_1$ is the probability that the path starts from (0, 0), goes across the diagonal only once, and reaches the wall $n = m - 1$ in II $(m > n)$. R_1 is the probability that the path starts from the wall $n = m + 1$ in I $(m < n)$, goes across the diagonal only once, and reaches the wall $n = m - 1$ in II $(m > n)$. R_2 is the probability that the path starts from the wall $n = m - 1$ in II $(m > n)$, goes across the diagonal only once, and reaches the wall $n = m + 1$ in I $(m < n)$

From (24), we can obtain

$$C_j(x) = C_{j-1}(x)C_0(x). \tag{31}$$

Thus, the simple relation between the generating functions is given by

$$C_j(x) = \{C_0(x)\}^{j+1}. \tag{32}$$

Appendix B Derivation of $\tilde{R}_1$, R_1, and R_2

$\tilde{R}_1$ is the probability that the path starts from (0, 0), goes across the diagonal only once, and reaches the wall $n = m - 1$ in II $(m > n)$ (Fig. 5).

$$\begin{aligned}
\tilde{R}_1 &= (1-B)[1 + y\gamma_1 C_0(y) + (y\gamma_1)^2 C_1(y) + (y\gamma)^3 C_2(y) + \cdots] \\
&= (1-B)[1 + y\gamma_1 C_0(y) + (y\gamma_1)^2\{C_0(y)\}^2 + (y\gamma_1)^3\{C_0(y)\}^3 + \cdots] \\
&= (1-B)[\sum_{k=0}^{\infty}\{\gamma_1 y C_0(y)\}^k] = \frac{1-B}{1-\gamma_1 A C_0(A)} \\
&= \frac{2(1-B)}{2-\gamma_1(1-\sqrt{1-4A(1-A)})},
\end{aligned} \tag{33}$$

where A and B are given by (1), $\gamma_1 = B/A$, and $y = A(1-A)$. $C_k(y)$ is the generation function of the generalized Catalan number (30). Here, we use the relations (32) and (29). When $q = 1$, (33) reduces to (8).

R_1 is the probability that the path starts from the wall $n = m+1$ in I $(m < n)$, goes across the diagonal only once, and reaches the wall $n = m-1$ in II $(m > n)$.

$$\begin{aligned} R_1 &= \frac{1-B}{B}[y\gamma_1 C_0(y) + (y\gamma_1)^2 C_1(y) + (y\gamma_1)^3 C_2(y) + \cdots] \\ &= \frac{1-B}{B}[y\gamma_1 C_0(y) + (y\gamma_1)^2\{C_0(y)\}^2 + (y\gamma_1)^3\{C_0(y)\}^3 + \cdots] \\ &= \frac{1-B}{B}[\frac{\tilde{R_1}}{1-B} - 1] = \frac{(1-B)\gamma_1(1-\sqrt{1-4A(1-A)})}{B\{2-\gamma_1(1-\sqrt{1-4A(1-A)})\}}. \end{aligned} \quad (34)$$

R_2 is the probability that the path starts from the wall $n = m-1$ in II $(m > n)$, goes across the diagonal only once, and reaches the wall $n = m+1$ in I $(m < n)$.

$$\begin{aligned} R_2 &= \frac{B}{1-B}[z\gamma_2 C_0(z) + (z\gamma_2)^2 C_1(z) + (z\gamma_2)^3 C_2(z) + \cdots] \\ &= \frac{B}{1-B}[z\gamma_2 C_0(z) + (z\gamma_2)^2\{C_0(z)\}^2 + (z\gamma_2)^3\{C_0(z)\}^3 + \cdots] \\ &= \frac{B\gamma_2(1-\sqrt{1-4C(1-C)})}{(1-B)\{2-\gamma_2(1-\sqrt{1-4C(1-C)})\}}, \end{aligned} \quad (35)$$

where C is given by (1), $\gamma_2 = (1-B)/(1-C)$, and $z = C(1-C)$.

References

Bikhchandani S, Hirshleifer D, Welch I (1992) A theory of Fads, fashion, custom, and cultural changes as information cascades. J Polit Econ 100:992–1026

Böhm W (2000) The correlated random walk with boundaries: a combinatorial solution. J Appl Prob 37:470–479

Curty P, Marsili M (2006) Phase coexistence in a forecasting game. JSTAT P03013

Cont R, Bouchaud J (2000) Herd behavior and aggregate fluctuations in financial markets. Macroecon Dyn 4:170–196

Di Francesco P, Golinelli O, Guitter E (1997) Meander, folding, and arch statistics. Math Comput Model 26(8):97–147

Dorigo M, Stützle T (2004) Ant colony optimization. MIT Press, Cambridge

Eguíluz V, Zimmermann M (2000) Transmission of information and herd behavior: an application to financial markets. Phys Rev Lett 85:5659–5662

Galam G (1990) Social paradoxes of majority rule voting and renormalization group. Stat Phys 61:943–951

Hisakado M, Hino M (2016) Between ant colony optimization and generic algorithm. J IPS Jpn 9(3):8–14 (in Japanese)

Hisakado M, Mori S (2010) Phase transition and information cascade in a voting model. J Phys A 43:315207
Hisakado M, Mori S (2011) Digital herders and phase transition in a voting model. J Phys A 22:275204
Hisakado M, Kitsukawa K, Mori S (2006) Correlated binomial models and correlated structure. J. Phys. A. 39:15365–15378
Keynes JM (1936) General theory of employment interest and money. Palgrave Macmillan, London
Konno N (2002) Limit theorems and absorption problems for quantum random walks in one dimension. Quantum Inf Comput 2:578–595
Lang W (2000) On polynominals related to power of the generating function of Catalan's numbers. Fib Quart 38:408
Mori S, Hisakado M (2010) Exact scale invariance in mixing of binary candidates in voting. J Phys Soc Jpn 79:034001–034008
Stauffer D (2002) Sociophysics: the Sznajd model and its applications. Comput Phys Commun 146(1):93–98

Information Cascade, Kirman's Ant Colony Model, and Kinetic Ising Model

Masato Hisakado and Shintaro Mori

1 Introduction

While collective herding behaviour is popularly studied among animals, it can also be observed in human beings. In this regard, there are interesting problems across the fields of sociology (Tarde 1890), social psychology (Milgram et al. 1969), ethnology (Partridge 1982; Couzin et al. 2002), and economics. Statistical physics has been an effective tool to analyse these macro phenomena among human beings and has led to the development of an associated field, sociophysics (Galam 2008; Castellano et al. 2009). For example, in statistical physics, anomalous fluctuations in financial markets (Cont and Bouchaud 2000; Egíluz and Zimmermann 2000) and opinion dynamics (Stauffer 2002; Curty and Marsili 2006) have been discussed.

Most individuals observe the actions of other individuals to estimate public perception and then make a choice similar to that of the others; this is called social learning. Because it is usually sensible to do what other people are doing, collective herding behaviour is assumed to be the result of a rational choice that is based on public perception. While this approach could be viable in some ordinary cases, as a macro phenomenon, it can sometimes lead to arbitrary or even erroneous decisions. This phenomenon is known as an information cascade (Bikhchandani et al. 1992). Here we show that an information cascade is described by the Ising model.

We introduced a sequential voting model (Hisakado and Mori 2010). At each time step t, one voter opts for either of two candidates. As public perception, the tth voter can see all previous votes, that is, $(t-1)$ votes. To identify the

M. Hisakado (✉)
Nomura Holdings, Inc., Chiyoda-ku, Tokyo, Japan

S. Mori
Faculty of Science and Technology, Department of Mathematics and Physics, Hirosaki University, Hirosaki, Aomori, Japan

A.-H. Sato (ed.), *Applications of Data-Centric Science to Social Design*, Agent-Based Social Systems 14, https://doi.org/10.1007/978-981-10-7194-2_6

relationship between information cascade and phase transition, we introduce two types of voters—herders and independents. We also introduce two candidates.

The herders' behaviour is known as the influence response function, and threshold rules have been derived for a variety of relevant theoretical scenarios representing this function. Some empirical and experimental evidence supports the assumption that individuals follow threshold rules when making decisions in the presence of social influence (Watts and Dodds 2007). This rule posits that individuals will switch from one decision to another only when sufficiently many others have adopted the other decision. Such individuals are called digital herders (Hisakado and Mori 2011). From our experiments, we observed that human beings exhibit a behaviour between that of digital and analog herders, that is, the tanh-type herder (Hisakado and Mori 2012). We obtained the probability that a herder makes a choice under the influence of his/her prior voters' votes. This probability can be fitted by a tanh function (Mori et al. 2012).

Here, we discuss a voting model with two candidates. We set two types of voters: independents and herders. As their name suggests, the independents collect information independently, that is, their voting depends on their fundamental values and rationality. In contrast, the voting of herders is based on public perception, which is visible to them in the form of previous votes. In this study, we consider the case wherein a voter can see the latest r previous votes.

When $r \rightarrow \infty$ is the upper limit of t, we can observe several phenomena (Hisakado and Mori 2011). In the case where there are independent voters and digital herders, the independents cause the distribution of votes to converge to one-peak distribution, a Dirac measure when the ratio of herders is small. However, if the ratio of herders increases above the transition point, we can observe the information cascade transition. As the fraction of herders increases, the model features a phase transition beyond which a state where most voters make the correct choice coexists with one where most of them are wrong. Further, the distribution of votes changes from one peak to two peaks.

Here we show that there is no phase transition when r is finite and the solution oscillates between two equilibria. Furthermore, we show relations among our voting model, Kirman's ant colony model, and kinetic Ising model.

The remainder of this chapter is organised as follows. In Sect. 2, we introduce our voting model and mathematically define the two types of voters—independents and herders. In Sect. 3, we discuss the case where there are digital herders and independents. In Sect. 4, we verify the transitions between voting choices through numerical simulations. In Sect. 5, we discuss the case where there are analog herders and independents and show the relation between our voting model and Kirman's ant colony model. In Sect. 6, we discuss the relation between our voting model and the kinetic Ising model. In the final section, we present the conclusions.

2 Model

We model the voting of two candidates, C_0 and C_1. The voting is sequential, and at time t, C_0 and C_1 have $c_0(t)$ and $c_1(t)$ votes, respectively. In each time step, one voter votes for one candidate. Hence, at time t, the tth voter votes, after which the total number of votes is t. Voters are allowed to see just r previous votes for each candidate; thus, they are aware of public perception. r is a constant number.

We assume an infinite number of two types of voters—independents and herders. The independents vote for C_0 and C_1 with probabilities $1 - q$ and q, respectively. Their votes are independent of others' votes, that is, their votes are based on their own fundamental values.

Here, we set C_0 as the wrong candidate and C_1 as the correct one in order to validate the performance of the herders. We can set $q \geq 0.5$, because we believe that independents vote for C_1 rather than for C_0. In other words, we assume that the intelligence of the independents is virtually accurate.

In contrast, the herders' votes are based on the number of previous r votes. At time t, the information of r previous votes are the number of votes for C_0 and C_1:$c_0^r(t)$ and $c_1^r(t)$, respectively. Hence, $c_0^r(t) + c_1^r(t) = r$ holds. If $r > t$, voters can see t previous votes for each candidate. For the limit $r \to \infty$, voters can see all previous votes. We define the number of all previous votes for C_0 and C_1 as $c_0^\infty(t) \equiv c_0(t)$ and $c_1^\infty(t) \equiv c_1(t)$.

Now we define the majority's correct decision. If the ratio to the candidate C_1 who is correct is $c_1/t > 1/2(< 1/2)$, we define the majority as correct (wrong). This ratio is important to evaluate the performance of the herders. In this chapter, we consider three kinds of herders, namely, digital, analog, and tanh-type herders. We define z_r as $z_r = c_1^r/r$. The probability that a herder who refers to z_r votes to the candidate C_1 is defined as $f(z_r)$. Digital herders always choose the candidate with a majority of the previous r votes, which are visible to them (Hisakado and Mori 2011). In this case, $f(z_r) = \theta(z_r - 1/2)$, where θ is a Heaviside function. Analog herders vote for each candidate with probabilities that are proportional to the candidates' votes (Hisakado and Mori 2010). Thus, $f(z_r) = z_r$. The other herder is the tanh-type herder, who is an intermediate between analog and digital herders (Hisakado and Mori 2012). In this case, $f(z_r) = 1/2(\tanh(z_r - 1/2) + 1)$.

The independents and herders appear randomly and vote. We set the ratio of independents to herders as $(1 - p)/p$. In this study, we mainly focus on the upper limit of t. This refers to the voting of infinite voters.

In this model we introduce the two types of noise (Young 2011). One is independents. They appear with the probability $1 - p$ and vote to the correct candidates with the probability q. If q is high, the error becomes small. They choose an action uniformly at random. The model is called "uniform error process".

The other is "log-normal". In this case the noise is included in the herders' response function. The reason of the name is that log probability of choosing the candidate C_1 is linear. In the limit of the no noise case, the herder becomes the digital. In the limit of the noisy case, he/she becomes the analog.

3 Digital Herder

In this section, the herder is a digital herder (Hisakado and Mori 2011) and, hence, votes for the majority candidate; if $c_0^r(t) > c_1^r(t)$, herders vote for the candidate C_0. If $c_0^r(t) < c_1^r(t)$, herders vote for the candidate C_1. If $c_0^r(t) = c_1^r(t)$, herders vote for C_0 and C_1 with the same probability, that is, 1/2.

In the previous paper, we discussed this case by using mean field approximations (Hisakado and Mori 2011). In this chapter, we discuss this using stochastic partial differential equations. First, we consider an approximation to introduce the partial differential equations. The voter selects r votes randomly from the latest previous r voters with overlapping. We can write the process as

$$\begin{aligned} c_1(t) = k \rightarrow k+1 &: P_{k,t:l,t-r} = p\pi((k-l)/r) + (1-p)q, \\ c_1(t) = k \rightarrow k &: Q_{k,t:lt-r} = 1 - P_{k,t}, \end{aligned} \tag{1}$$

where $c_1(t-r) = l$. $P_{k,t}$ and $Q_{k,t}$ are the probabilities of the process. The sum of $P_{k,t}$ and $Q_{k,t}$ is 1. This means that at time $(t-r)$, the number of votes for C_1 is $c_1(t-r) = l$. Here, we define $\pi(Z)$ as the majority probability of binomial distributions of Z. $\pi(Z)$ can be calculated as follows:

$$\pi(Z) = \frac{(2n+1)!}{(n!)^2}\int_0^Z x^n(1-x)^n dx = \frac{1}{B(n+1,n+1)}\int_0^Z x^n(1-x)^n dx, \tag{2}$$

where $r = 2n + 1$.

For convenience, we define a new variable Δ_t such that

$$\Delta_t = 2c_1(t) - t = c_1(t) - c_0(t). \tag{3}$$

We change the notation from k and l to Δ_t and Δ_{t-r} for convenience. Given $\Delta_t = u$ and $\Delta_{t-r} = s$, we obtain a random walk model:

$$\begin{aligned} \Delta_t = u \rightarrow u+1 &: P_{u,t:s,t-r} = \pi(\frac{1}{2} + \frac{u-s}{2r})p + (1-p)q, \\ \Delta_t = u \rightarrow u-1 &: Q_{u,t:s,t-r} = 1 - P_{u,t}. \end{aligned}$$

We now consider the continuous limit $\epsilon \rightarrow 0$:

$$\begin{aligned} X_{\hat{\tau}} &= \epsilon\Delta_{[t/\epsilon]}, \\ P(x,\hat{\tau}) &= \epsilon P(\Delta_t/\epsilon, t/\epsilon), \end{aligned} \tag{4}$$

where $\hat{\tau} = t/\epsilon, \hat{r} = r/\epsilon$ and $x = \Delta_t/\epsilon$. On approaching the continuous limit, we can obtain the stochastic partial differential equation:

$$dX_{\hat{\tau}} = \left[(1-p)(2q-1) - p + 2p\frac{(2n+1)!}{(n!)^2}\int_0^{\frac{1}{2}+\frac{X_{\hat{\tau}}-X_{\hat{\tau}-\hat{r}}}{2r}} x^n(1-x)^n dx\right]d\hat{\tau} + \sqrt{\epsilon}, \tag{5}$$

where we used (2). Equation (5) depends on $X_{\hat{\tau}-\hat{r}}$ and is the feedback system.

We are interested in the behaviour of $X_{\hat{\tau}}$ in the limit $\hat{\tau} \to \infty$. The relation between X_∞ and the voting ratio to C_1,Z is $2Z - 1 = X_\infty/\hat{\tau}$.

We can assume the stationary solution to be

$$X_\infty = \bar{v}\hat{\tau} + (1-p)(2q-1)\hat{\tau}, \tag{6}$$

where $\bar{v}$ is constant. Substituting (6) into (5), we can obtain

$$\bar{v} = -p + \frac{2p \cdot (2n+1)!}{(n!)^2}\int_0^{\frac{1}{2}+\frac{(1-p)(2q-1)}{2}+\frac{\bar{v}}{2}} x^n(1-x)^n dx. \tag{7}$$

Equation (7) is self-consistent with a permission of overlapping. It is the same as the mean field approximation in Hisakado and Mori (2011). If we set $n = 2r+1 \to \infty$, which is the second term of RHS, the herders behave as digital herders and (7) becomes the strict form for them.

When $r = 1, 2$, there is only one solution of (7) in the range p. However, when $r \geq 3$, there is only one solution in $p < p_c$ and three solutions in $p > p_c$, where p_c is critical p (Hisakado and Mori 2011). The middle solution is unstable, while the other two solutions are stable. As there exist both good and bad equilibria, there may be phase transitions.

In the case where $r \to \infty$, the description is correct. In fact, there is phase transition when r is infinite. The voting rates converge to one of the two stable points. However, when r is finite, the solution oscillates between good and bad equilibria. Hence, there is no phase transition. We elaborate this below.

Here, we consider a random walk between the two states, $c_1^r(t)/r > 1/2$ (good equilibrium) and $c_0^r(t)/r > 1/2$ (bad equilibrium). It is coarse-grained votes and corresponds to the block spin transformation (Kadanoff 1966). We define the hopping probability from the state $c_0^r(t)/r > 1/2$ to $c_1^r(t)/r > 1/2$ as a and that from the state $c_1^r(t)/r > 1/2$ to $c_0^r(t)/r > 1/2$ as b. Note that a and b are not the function of t. When $t > r$, the transition matrix A of this random walk is

$$A = \begin{pmatrix} 1-a & a \\ b & 1-b \end{pmatrix}. \tag{8}$$

The random walk of the two states, $\tilde{X}_n$, is defined as the transition matrix A when $t > r$ and the initial condition $\tilde{X}_0 = 0$. If $r > t$, voters can see t previous votes for each candidate.

The model was studied as a one-dimensional correlated random walk (Böhm 2000; Konno 2002). For the limit $t \to \infty$,

$$\tilde{X}_\infty \Longrightarrow N(\frac{a-b}{a+b}t, \frac{4ab(2-(a+b))}{(a+b)^3}t), \tag{9}$$

where $N(\mu, \sigma^2)$ is the normal distribution with mean μ and variance σ^2. (See Theorem 3.1, in Böhm 2000.) a and b are given in (8).

If consecutive independent voters choose the candidate $C_1(C_0)$ when $c_0^r(t)/r > 1/2(c_1^r(t)/r > 1/2)$, the state changes from $c_0^r(t)/r > 1/2(c_1^r(t)/r > 1/2)$ to $c_1^r(t)/r > 1/2(c_0^r(t)/r > 1/2)$. Thus, independent voters behave as a switch for hopping. When the independent voters who vote $C_1(C_0)$ are the majority, the state hops from $c_0^r(t)/r > 1/2(c_1^r(t)/r > 1/2)$ to $c_1^r(t)/r > 1/2(c_0^r(t)/r > 1/2)$. Hence, the hopping rates a and b are estimated to be

$$a = \pi[(1-p)q] = \frac{(2n+1)!}{(n!)^2}\int_0^{(1-p)q} x^n(1-x)^n dx \sim (1-p)^{\frac{r+1}{2}} q^{\frac{r+1}{2}},$$

$$b = \pi[(1-p)(1-q)] \sim (1-p)^{\frac{r+1}{2}}(1-q)^{\frac{r+1}{2}}, \tag{10}$$

where the approximations are in $p \sim 1$. In the case where $r = 1$, $a = (1-p)q$ and $b = (1-p)(1-q)$. We obtained an identical solution in Hisakado and Mori (2010).

In the finite r case, the hopping rates a and b do not decrease as t increases, and the state oscillates between good and bad equilibria. Hence, the distribution of $X_{\hat{t}}$ becomes normal, and there is no phase transition. The voting rates converge to $(1-p)q + pa/(a+b) \sim (1-p)q + pq^{\frac{r+1}{2}}/(q^{\frac{r+1}{2}} + (1-q)^{\frac{r+1}{2}})$. The first term is the number of votes by independent voters, and the second term is the number of votes by the digital herders. The herders' votes oscillate between good and bad equilibria in (7). As r increases, the stay in the good equilibrium becomes longer. The ratio of stay in the good equilibrium to that in the bad equilibrium is $a/b \sim (\frac{q}{1-q})^{r+1/2}$.

When $r = \infty$, we consider two cases. One is the case where r increases with t. Also, in this case, r increases rapidly, that is, the annealing schedule is not adequately slow (Geman and Geman 1984). The voters can refer to all historical votes. The hopping rates a and b decrease exponentially as t increases, and $\bar{v}$ converges to the solutions of the self-consistent equation (7). Hence, above p_c, which is the critical p, the voting rate could converge to the bad equilibrium. Thus, this is the information cascade transition.

The other case is where r increases slowly, that is, the annealing schedule is adequately slow. At first, we set a finite r. After an adequate number of votes, the state frequently oscillates between good and bad equilibria. We increase r after several oscillations of the states. We continue this process until r reaches ∞. Thus, we deem the annealing schedule to be adequately slow. (See Sect. 4, where the schedule $r \sim \log t$.) The ratio of stay in the good equilibrium to the bad equilibrium is $a/b \sim (\frac{q}{1-q})^{r+1/2} \rightarrow \infty$. It means that the state is always in the good equilibrium, and the voting rate converges only to the good equilibrium $(1-p)q + pa/(a+b) \rightarrow (1-p)q + p$ when r is large.

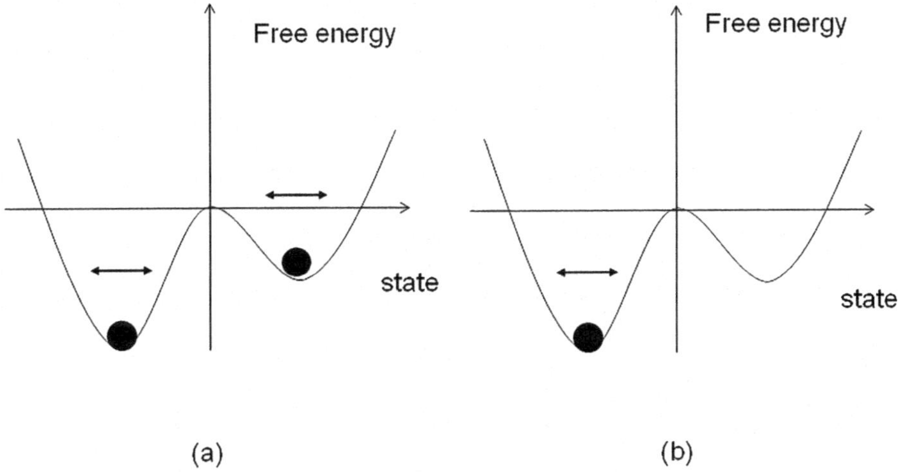

Fig. 1 Illustration of voting rate convergence using the analogy of physical potential. The ball stops at the bottom of the potential, corresponding to the convergence of the voting rate. The deeper (shallower) potential corresponds to good (bad) equilibrium. (**a**) is the case where $r = \infty$. As the annealing schedule is not adequately slow, sometimes, the voting rate converges to the bad equilibrium, and there is information cascade. (**b**) is the case where the annealing schedule is adequate slow. If we increase r slowly, the voting rate converges only to the good equilibrium

In Fig. 1, we depict the convergence as r increases. (a) is the case where r increase rapidly, or the annealing schedule is not adequately slow. $\bar{v}$ converges to one of the solutions of the self-consistent equation (7). Hence, the average correct ratio $E(c_1(t)/t)$ decreases above p_c, as p increases. The average correct ratio is the ratio of the number of votes for the candidate C_1 to all votes. Further, (b) is the case where r and p increase slowly from finite r. In this case, the voting rate converges only to the good equilibrium.

4 Numerical Simulations

To confirm the analytic results of Sect. 3, we performed numerical and Monte Carlo (MC) integration of the master equation for the digital herder case. For the Monte Carlo study, we solve the master equation for 10^6 times and calculate the average value of ratios.

In Fig. 2, we show the average votes ratio for the correct candidate vs p for $q = 0.6$ and $r = 5, 101$ and $r = \infty$. For the time horizon t, we choose $t = 10^2, 10^4$, and 10^6 in order to observe the limit behaviour $t \to \infty$. We also plot the theoretical results for the limit value as

$$\lim_{t\to\infty} E(c_1/t) = (1-p)q + p\frac{a}{a+b}, \tag{11}$$

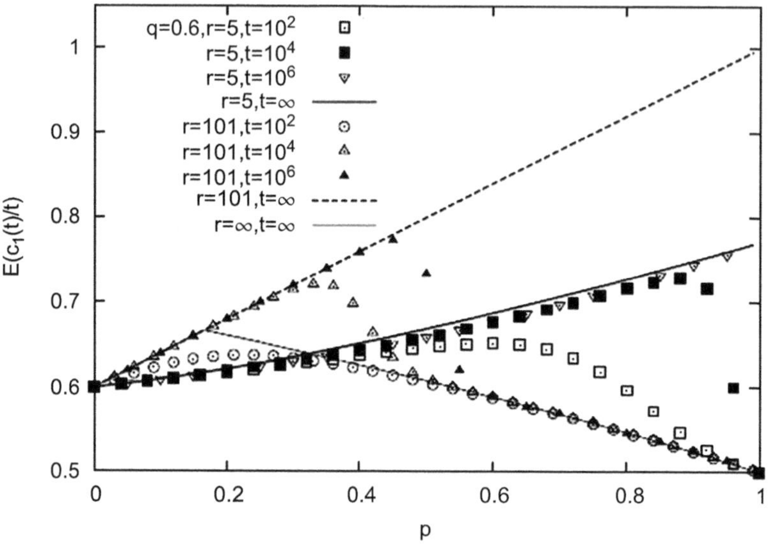

Fig. 2 $E(c_1(t)/t)$ vs p for $r = 5, 101, \infty$ and $t = 10^2, 10^4, 10^6, \infty$. The symbols show the results from numerical studies. The lines represent the theoretical results. The theoretical results for finite r (11) are plotted with solid ($r = 5$) and broken lines ($r = 101$). For $r = \infty$, the result is given in Hisakado and Mori (2011) and is plotted with thin solid line. We set $q = 0.6$

where a, b are given in (10). The thin solid line plots the exact result of the limit value. For $p < p_c = 1/6$, the herders make the correct choice, and it behaves as $(1-p)q + p \cdot 1$. Above p_c, the probability that all herders choose the wrong candidate becomes finite, and it becomes smaller than $(1-p)q + p \cdot 1$.

When $r = 5$ and $t = 10^2$, $E(c_1(t)/t)$ slightly increases with p for small p. For large p, the probability that the system is in the bad equilibrium becomes large, and $E(c_1(t)/t)$ becomes small. As p increases further, the probability that the system escapes from the bad equilibrium becomes negligibly small, and it approaches the limit value for $r = \infty$. As t increases from 10^2 to 10^4, 10^6, the probability that the system is in the bad equilibrium decreases, and $E(c_1(t)/t)$ becomes larger. As p approaches 1, like in the $t = 10^2$ case, it approaches the limit value for $r = \infty$. We also see that $E(c_1(t)/t)$ approaches the theoretically estimated limit value (11) as t becomes large. For the limit $t \to \infty$, the probability that herders make the correct choice can be estimated as $a/(a+b)$. For $r = 101$, we also see the same feature with $r = 5$. Compared with $r = 5$, for $r = 101$, the peak position of $E(c_1(t)/t)$ in p-axis becomes smaller, and it rapidly approaches the limit value for $r = \infty$. As r becomes large, the mean oscillation time between good and bad equilibria becomes large and the probability that the system escapes from the bad equilibrium becomes small. Despite this, the system finally escapes from the bad equilibrium for the limit $t \to \infty$, and $E(c_1(t)/t)$ approaches the theoretically estimated result. The probability that herders make the correct choice $a/(a+b)$ is an increasing function of r, and the slope of $E(c_1(t)/t)$ vs p becomes larger for large r. At $r = 100$ and

$q = 0.6$, $a/(a+b)$ is almost one and $E(c_1(t)/t)$ goes to one for the limit $p \to 1$. For better correct ratio $E(c_1(t)/t)$, r should be large. However, in order to realise high value of $E(c_1(t)/t)$, t also should be large. Otherwise, one cannot necessarily realise a high correct ratio. Therefore, there needs to be a trade-off between time and correct ratio.

In order to realise the high value for $E(c_1(t)/t)$, r should be large. For large r, for the probability in the bad equilibrium to be small, t should be large, and the system should oscillate several times between the two equilibria. As the mean cycle of the oscillation behaves as $1/[(1-p)^{(r+1)/2}(1-q)^{(r+1)/2}] = e^{Cr}$ with $C = -\frac{1}{2}\log(1-p)(1-q)$, a good annealing schedule $r(t)$ is estimated as

$$r(t) = \log t / C. \tag{12}$$

The annealing schedule $r(t)$ depends on t logarithmically and is very slow.

In Fig. 3, we plot $E(c_1(t)/t)$ vs t for $r = 5, 11, 21, 51, 101, \infty$ and $r = r(t)$ in Eq. (12). We set $q = 0.6$ and $p = 0.5 > p_c = 1/6$. For $r = \infty$, the bad equilibrium is stable, and $E(c_1(t)/t)$ saturates at low value ~ 0.6 for $t \geq 10^2$. For $r = 5$, the herder's probability of correct choice is low, and $c_1(t)/t$ soon reaches the maximum value ~ 0.65 at about $t = 2 \times 10^2$. For $r = 11$, $E(c_1(t)/t)$ reaches the maximum value for about $t = 10^3$. As r increases, the time t needed for the high value of $E(c_1(t)/t)$ increases. For $r = 51$, $c_1(t)/t$ reaches about 0.8 at $t = 10^5$. If the probability that herders choose the correct candidate C_1 is one, $c_1(t)/t$ takes $0.8 = (1-p)q + 1 \cdot p$ for $q = 0.6$ and $p = 0.5$. $r = 51$ is large enough to maximise the limit value $\lim_{t\to\infty} c_1(t)/t$. If one adopts $r(t) = 3\log t$ as the

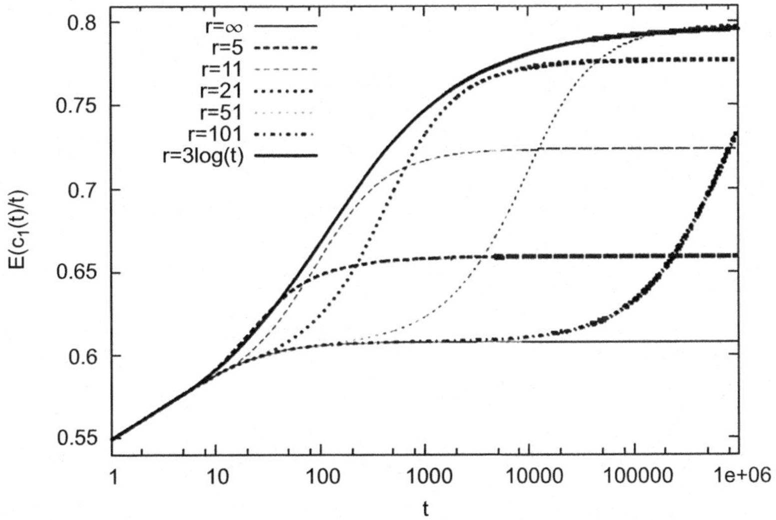

Fig. 3 Plots of $E(c_1(t)/t)$ vs t for $r = 5, 11, 21, 51, 101, \infty$ and $r = 3\log t$. We set $q = 0.6$ and $p = 0.5 > p_c = 1/6$

annealing schedule for $r(t)$, $E(c_1(t)/t)$ increases smoothly to the maximum value 0.8 without saturation at some value between $(1-p)q + p \cdot 0.5 = 0.55$ and the maximum value. If we adopt $r = 51$, at $t = 10^6$, $c_1(t)/t$ reaches the maximum value. However, for $t < 10^5$, $E(c_1(t)/t)$ with $r = r(t)$ is higher than that for $r = 51$. If we set $r = 51$, the probability to escape from the bad equilibrium is small, and it is necessary to wait for a long time t for $c_1(t)$ to reach the maximum value. In contrast, for $r = r(t)$, r slowly increases, and the probability to stay in the bad equilibrium is minimised. Furthermore, unnecessary stay in good equilibrium with medium $E(c_1(t)/t)$ is also avoided by increasing r and smoothly increasing $c_1(t)$.

5 Analog Herder and Kirman's Ant Colony Model

In this section, we consider the case of the analog herder (Hisakado and Mori 2010). As mentioned earlier, the herders vote for the candidate with the probability that is proportional to the previous votes ratio that can be referred. The voter can see the latest previous r voters. The transition is

$$\begin{aligned} c_1(t) = k \to k+1 : P_{k,t:l,t-r} &= (1-p)q + p\frac{k-l}{r} = \frac{q(1-\rho) + \rho(k-l)}{(1-\rho)+\rho r}, \\ c_1(t) = k \to k : Q_{k,t:lt-r} &= 1 - P_{k,t:l,t-r} = \frac{(1-q)(1-\rho) + \rho(r-(k-l))}{(1-\rho)+\rho r}, \end{aligned} \tag{13}$$

where $c_1(t-r) = l$. $P_{k,t:l,t-r}$ and $Q_{k,t:l,t-r}$ are the probabilities of the process. The voting ratio for C_1 is $c_1(t-r) = l$. We changed the parameters from p to ρ, which is the correlation of rth beta-binomial model (Hisakado et al. 2006). The relation between p and ρ is

$$p = \frac{\rho r}{(1-\rho)+\rho r}. \tag{14}$$

Hence, we can map independent voters and herders to beta-binomial distribution. As r increases, $1-p$ decreases as $1/r$, and the independent voters' ratio decreases.

Here, we consider the hopping rate among $r+1$ states $\hat{k} = k - l = 0, 1, \cdots, r$. The dynamic evolution of the process is given by

$$\begin{aligned} \hat{k} \to \hat{k}+1 : P_{\hat{k},\hat{k}+1,t} &= \frac{r-\hat{k}}{r}\frac{q(1-\rho)+\rho\hat{k}}{(1-\rho)+\rho r}, \\ \hat{k} \to \hat{k}-1 : P_{\hat{k},\hat{k}-1,t} &= \frac{\hat{k}}{r}\frac{(1-q)(1-\rho)+\rho(r-\hat{k})}{(1-\rho)+\rho r}, \end{aligned}$$

$$\hat{k} \rightarrow \hat{k} : P_{\hat{k},\hat{k},t} = 1 - P_{\hat{k},\hat{k}-1,t} - P_{\hat{k},\hat{k}+1,t}. \tag{15}$$

This process means that a new vote is added by using the r references, where the oldest vote of the r references exists. If we set $\epsilon = \frac{1}{2}(1-\rho)/[(1-\rho)+\rho r]$, $1-\delta = (r-1)\rho/[(1-\rho)+\rho r]$, and $q = 1/2$, we obtain an equivalent of Kirman's ant colony model (Kirman 1993; Alfarano et al. 2005). The ant model has been reasonably successful in replicating the statistical features of financial returns, like volatility clustering and power low tails of the return distribution. They are famous characteristic of financial return in econophysics. Kirman's colony model corresponds to $r \geq 2$ case. In our model, there exists a constraint between the parameters:

$$2\epsilon + \frac{r}{r-1}(1-\delta) = 1. \tag{16}$$

Here, we define $\mu_r(\hat{k}, t)$ as a distribution function of the state $\hat{k}$ at time t. The number of all states is $r+1$. Using the fact that the process is reversible, we have

$$\frac{\mu_r(\hat{k}+1,t)}{\mu_r(\hat{k},t)} = \frac{P_{\hat{k},\hat{k}+1,t}}{P_{\hat{k}+1,\hat{k},t}} = \frac{r-\hat{k}}{\hat{k}+1}\frac{q(1-\rho)+\rho\hat{k}}{(1-q)(1-\rho)+(r-\hat{k}-1)\rho}. \tag{17}$$

For the limit $t \rightarrow \infty$, we can obtain the equilibrium distribution

$$\mu_r(\hat{k},\infty) = {}_rC_{\hat{k}} \frac{\Pi_{j=0}^{\hat{k}-1}(q(1-\rho)+(j\rho))\Pi_{j'=0}^{r-\hat{k}-1}((1-q)(1-\rho)+(j'\rho))}{\Pi_{i=0}^{r-1}((1-\rho)+r\rho)}. \tag{18}$$

It is the $(r-1)$th beta-binomial distribution that is constructed on the lattice (Hisakado et al. 2006). It is the same distribution as that in Kirman's colony model.

From the central limit theorem, as the distribution of states converges to the beta-binomial distribution, the voting rates for the candidate C_1 converge:

$$\lim_{t\rightarrow\infty} \frac{c_1}{t} = \delta_q, \tag{19}$$

where the reference number r is finite and δ_q is a Dirac measure. Further, when $r \rightarrow \infty$, the distribution of states is the same as the voting rates for the candidate C_1, that is, the beta distribution.

Now, we discuss the case where r increases slowly, that is, annealing schedule is adequately slow. At first, we set a finite r. After an adequate number of votes, the distribution of states converges to the rth beta-binomial distribution. We increase r after the state converges. We continue this process until r reaches ∞. We call it the case where the annealing schedule is adequately slow. In this case, the voting rates for the candidate C_1 converge to the same point, that is, independent voter's correct ratio q, as the finite r case.

6 Tanh-Type Herder and Kinetic Ising Model

In this section, we identify the relation between our voting model and the infinite-range kinetic Ising model. (See Appendix A.)

The state of the Ising model is denoted by the vector $\boldsymbol{\sigma} = (\sigma_1, \cdots, \sigma_{r+1})$ with $\sigma_j = \pm 1$.

The Hamiltonian is defined as

$$H(\boldsymbol{\sigma}) = -J \sum_{i \neq j}^{r+1} \sigma_i \sigma_j, \tag{20}$$

where J is the exchange interaction.

We define F_j as a spin flip operator on jth site: $F_j\boldsymbol{\sigma}$ is the state in which jth spin is flipped from $\boldsymbol{\sigma}$ with the other spins fixed. The Markov chain is characterised by a transition probability $w_j(\boldsymbol{\sigma})$per unit time from $\boldsymbol{\sigma}$ to $F_j\boldsymbol{\sigma}$. Let $p(\boldsymbol{\sigma}, t_n)$ be the probability distribution for finding the spin sate $\boldsymbol{\sigma}$ at time t_n. Then, the discrete master equation is written as follows:

$$p(\boldsymbol{\sigma}, t_{n+1}) = p(\boldsymbol{\sigma}, t_n) - [\sum_j w_j(\boldsymbol{\sigma})]p(\boldsymbol{\sigma}, t_n)\Delta t + [\sum_j w_j(F_j\boldsymbol{\sigma})]p(F_j\boldsymbol{\sigma}, t_n)\Delta t, \tag{21}$$

where the second and third terms in RHS are ongoing and incoming probabilities, respectively.

For the master equation (21), we consider the sufficient condition that $p(\boldsymbol{\sigma}, t_n)$ converges to the equilibrium distribution $\pi(\boldsymbol{\sigma})$ as $t_n \to \infty$ is that the transition probability $w_j(\boldsymbol{\sigma})$ satisfies a detail balance condition:

$$w_j(\boldsymbol{\sigma})\pi(\boldsymbol{\sigma}) = w_j(F_j\boldsymbol{\sigma})\pi(F_j\boldsymbol{\sigma}). \tag{22}$$

This condition is known as reversibility. From this condition, the transition probability is given as

$$\frac{w_j(\boldsymbol{\sigma})}{F_j w_j(\boldsymbol{\sigma})} = \frac{\pi(F_j\boldsymbol{\sigma})}{\pi(\boldsymbol{\sigma})} = \frac{\exp[-\sigma_j \beta h_j]}{\exp[\sigma_j \beta h_j]}, \tag{23}$$

where $h_j = J\sum_{i \neq j}^{r+1} \sigma_i$ and an inverse temperature β, in the units where the Boltzmann constant is 1.

Here, we set

$$w_j(\boldsymbol{\sigma}) = \frac{1}{2}(1 - \sigma_j \tanh \beta h_j). \tag{24}$$

We define the total number of $\sigma_i = 1$, where $i \neq j$, is $\hat{c}_1$ and the total number of $\sigma_i = -1$, where $i \neq j$, is $\hat{c}_{-1}$. Hence, $\hat{c}_1 + \hat{c}_{-1} = r$. The transition is given by

$$\sigma_j = 1 \rightarrow -1 : w_j(\boldsymbol{\sigma}) = \frac{1}{2}(1 - \tanh \beta J(\hat{c}_1 - \hat{c}_{-1})),$$
$$\sigma_j = -1 \rightarrow 1 : F_j w_j(\boldsymbol{\sigma}) = \frac{1}{2}(1 + \tanh \beta J(\hat{c}_1 - \hat{c}_{-1})). \quad (25)$$

Note that as $w_j(\boldsymbol{\sigma}) + F_j w_j(\boldsymbol{\sigma}) = 1$, the transition does not depend on the previous state of σ_j and depends on the other spins σ_i where $i \neq j$.

In an ordinary case, an updated spin is chosen randomly. Here, we consider the case where the updated spin is chosen by the rules. The ordering of update is from σ_1 to σ_{r+1}. After the update of σ_{r+1}, we update σ_1 and so on. We repeat this process. Hereafter, we define the updated spin σ_j after n as $\sigma_j^{(n)}$. The initial condition is $\sigma_j^{(0)}$, where $j = 1, 2, \cdots, r$.

Here, we consider a voting model where all herders are tanh-type herders. The voter can see the latest previous r voters. The transition is given by

$$c_1(t) = k \rightarrow k+1 : P_{k,t:l,t-r} = \frac{1}{2}[\tanh \lambda(\frac{k-l}{r} - \frac{1}{2}) + 1]$$
$$c_1(t) = k \rightarrow k : Q_{k,t:lt-r} = 1 - P_{k,t:l,t-r}, \quad (26)$$

where $c_1(t-r) = l$ and λ is a parameter. (Please see in Hisakado and Mori 2012 for details.) The number of votes for the candidate C_1 at time $(t-r)$ is $c_1(t-r) = l$.

We define a new variable Δ_t such that

$$\Delta_t = 2c_1(t) - t = c_1(t) - c_0(t). \quad (27)$$

For convenience, we change the notation from k to Δ_t as in Sect. 3. Given $\Delta_t = u$ and $\Delta_{t-r} = s$, we obtain a random walk model:

$$\Delta_t = u \rightarrow u+1 : P_{u,t:s,t-r,t} = \frac{1}{2}(1 + \tanh \frac{\lambda(u-s)}{2r}),$$
$$\Delta_t = u \rightarrow u-1 : Q_{u,t:s,t-r,t} = \frac{1}{2}(1 - \tanh \frac{\lambda(u-s)}{2r}). \quad (28)$$

Given that the voter who voted at time $(t-r-1)$ chose the candidate $C_0(C_1)$, the probability that another voter at time t votes to the candidate $C_1(C_0)$ is $P_{u,t:s,t-r,t}(Q_{u,t:s,t-r,t})$. Hence, (28) means that the hopping rate is $P_{u,t:s,t-r,t}$ and $Q_{u,t:s,t-r,t}$. This is equivalent to (25), where $\beta J = \lambda/2r$, $\hat{c}_1 - \hat{c}_{-1} = u - s$. We use the relations $\hat{c}_1 = k - l$, $\hat{c}_{-1} = r - (k-l)$, $u = 2k - t$, and $s = 2l - (t-r)$.

If we consider the row of spins $\sigma_1^{(0)} \cdots \sigma_{r+1}^{(0)} \sigma_1^{(1)} \cdots \sigma_{r+1}^{(n)} \sigma_1^{(n+1)} \cdots$ where $n = 0, 1, 2 \cdots$ as the voting row. Hence, the voting model is equivalent to the infinite-range kinetic Ising model. This is why we can find the same mean field approximation equation as that in the Ising model in the limit $r \to \infty$ case (Hisakado and Mori 2012). Thus, for the finite r case, we confirm that there is no phase transition (Hisakado and Mori 2011) and that it is not appropriate to use the mean field approximation as well.

When $r \to \infty$, the kinetic Ising model is the voting model when the annealing schedule is adequately slow, and the state is always in the good equilibrium. Hence, the Ising model with external field has no phase transition, as discussed in Sects. 3 and 4. In fact, we can confirm as r increases, the maximum point of $E(c_1(t)/t)$ approaches the maximum point of $r = 3\log(t)$, the slow annealing, in Fig. 3. On the other hand, when the annealing schedule is not adequately slow, the voting model has the phase transition. It means that the voting rate could converge to the bad equilibrium. The phase transition is observed when the size of r is infinite.

In addition, the voting model becomes the Potts model, if there are several candidates above two. This is a simple extension of this section. While in this section, we have discussed the Ising model without external fields, we discuss how prior distribution and independent voters correspond to the outer fields of the Ising model in the appendix of Hisakado and Mori (2015).

7 Concluding Remarks

In this chapter, we investigated a voting model that involves collective herding behaviours. We investigated the case where voters can obtain the information from previous r voters. We observed the states of reference votes and considered three cases of voters:

(i) digital herders and independent voters,
(ii) analog herders and independent voters, and
(iii) tanh-type herders.

We investigated the case where there were only digital herders and independent voters. Through numerical simulations and analysis, we show that for finite r, no phase transitions occur and that there is only one-peak phase.

From the viewpoint of annealing, if the annealing schedule is adequately slow, that is, $r \sim \log t$, the voting rate converges to the good equilibrium only. The correct ratio $E(c_1(t)/t)$ increases as r increases. On the other hand, if the annealing schedule is not adequately slow, that is, $r \sim t$, the phase transitions occur. Further, above p_c, there are good and bad equilibria. The voting rates converge to one of the two equilibria, but we cannot estimate which of these equilibria is selected. The correct ratio decreases when phase transition exists. Therefore, there exists a trade-off between time and the correct ratio.

In case (ii), we considered analog herders and independent voters. We show that the states of reference votes in this model are equivalent to Kirman's ant colony model. In this case, the states follow beta-binomial distribution, and for large r, this becomes beta distribution.

In case (iii), we show that the model is equivalent to the finite-size kinetic Ising model. This model follows a simple herding experiment if the voters are rational (Bikhchandani et al. 1992; Anderson and Holt 1997). When $r \to \infty$, the voting model behaves as the infinite-range kinetic Ising model when the annealing schedule is adequately slow. Hence, the Ising model with external field has no phase transition. On the other hand, when the annealing schedule is not adequately slow, the voting model with external fields has the phase transition. While there are many studies in which the herding behaviour has been analysed by the Ising model, only the current study explains this behaviour by using the kinetic Ising model.

In beginning of spring, in March, groups of swans return to Siberia from Japan. When does travelling commence? At first, a few swans fly back and forth over a lake. The remaining swans from the group begin to herd the initial few, and the crowd becomes larger. This is an example of a quorum response, which we discussed as the response function.

From the viewpoint of our model, each swan has the threshold. The threshold goes down in spring. A few swans play the role of a lead and begin cascade. From the observation that the large part of groups flies, we can estimate that the threshold is low. The threshold depends on the day light. The research about the birds concludes that birds recognise less than ten other birds. In this chapter we discussed the case, the threshold is symmetric. We will discuss the asymmetric case in other place.

Appendix A Ising Model

Here we introduce the infinite-range model. It is one of the most popular model in statistical physics which explains the phase transition. In the model spins interact all other spins. Hamiltonian is

$$H(\sigma) = -\frac{J}{N}\sum_{i>j}\sigma_i\sigma_j - h\sum_{i=1}^{N}\sigma_i, \tag{29}$$

where N is the number of spins, σ is the spin that has the value ± 1, J is the parameter of interaction, and h is the outer field. Here we define average of spins, as order parameter, $m = 1/N\sum\sigma_i$.

In large $N \to \infty$ limit, the self-consistent equation is

$$m = \tanh\beta(Jm + h), \tag{30}$$

where $\beta = 1/k_B T$. k_B is Boltzmann constant and T is temperature. When infinite-range model, we can obtain the strict solution from the self-consistent equation. When the symmetric case, under the transition temperature T_C, there are two solutions. One of the solution is selected, and it is called spontaneous symmetry braking. On the other hand, above T_c, there is only one solution. It is the simple

Table 1 Response functions of animalsl

No.	Animal	Linear or nonlinear	Reference
1	Honey bee	Linear	Camazine and Sbeyd (1991)
2	*Temnothorax albipennis* (ant)	Linear	Pratt et al. (2002)
3	*Lasius niger* (ant)	Nonlinear	Bekers et al. (1992)
4	Three-spined sticklebacks (fish)	Nonlinear	Ward et al. (2008)
5	*Gasterosteus aculeatus* (fish)	Nonlinear	Ward et al. (2008)
6	Capuchin monkeys (Cebus capuccinus)	Linear	Meunier et al. (2007)

model which represents the phase transition. When there is the outer field h, the model becomes the asymmetric model, and there is no phase transition.

In this chapter we discussed the equilibrium voting model. The model which we discussed in Hisakado and Mori (2012) is the non-equilibrium model. The large r limit corresponds to the large N limit of the infinite-range model. We can observe the phase transition in both models.

In the case of non-equilibrium, we can obtain the same self-consistent equation. In this case the solution which is against the outer field may be selected. It is the characteristic point of the non-equilibrium model which we discuss in other chapters.

Appendix B Response Function of Animals

In this chapter we use three kinds of herders, digital, analog, and tanh type. In Sect. 3, we studied the Kirman's ant colony model. In this model the response function is analog. Here we have the question whether the real ant is analog herder. In this appendix we review the several experiments of animals about decision of the two choices. In Table 1 we show the list of the response function of the animals. These are the decision of two selections of animals. For the nonlinear response function, there is the spontaneous symmetry braking which corresponds to the phase transition. On the other hand, for linear response function, there is no phase transition. From the experiments some ants are analog herders, and some are tanh-type herders. About humans we presented in Chaps. 8 and 9. In the case of animals, the nonlinear response function is sometimes called quorum response.

References

Alfarano S, Lux T, Wagner F (2005) Estimation of agent-based models: the case of asymmetric herding model. Comput Econ 26(1):19–49

Anderson LR, Holt CA (1997) Information cascdes in the laboratory. Am Econ Rev 87(5):847–862

Bekers R, Deneubourg JL, Gross S (1992) Modulation of trial laying in the ant *Lasius niger* (Hymenoptera: Formicidae) and its role in the collective selection of a food source. J Insect Behav 6:751–759
Bikhchandani S, Hirshleifer D, Welch I (1992) A theory of Ffads, fashion, custom, and cultural changes as information cascades. J Polit Econ 100:992–1026
Böhm W (2000) The correlated random walk with boundaries: a combinatorial solution. J Appl Prob 37:470–479
Camazine S, Sbeyd J (1991) A model of collective nectar source selection by honey bees: Self-organization through simple rules. J Theor Biol 149:547–471
Castellano C, Fortunato S, Loreto V (2009) Statistical physics of social dynamics. Rev Mod Phys 81:591
Cont R, Bouchaud J (2000) Herd behavior and aggregate fluctuations in financial markets. Macroecon Dyn 4:170–196
Couzin ID, Krause J, James R, Ruxton GR, Franks NR (2002) Collective memory and spatial sorting in animal groups. J Theor Biol 218:1–11
Curty P, Marsili M (2006) Phase coexistence in a forecasting game. JSTAT P03013
Eguíluz V, Zimmermann M (2000) Transmission of information and herd behavior: an application to financial markets. Phys Rev Lett 85:5659–5662
Galam S (2008) A review of Galam models. Int J Mod Phys C 19:409–440
Geman S, Geman D (1984) Stochastic relaxation, Gibbs, distribution, and the Bayesian restoration of images. IEEE Trans PAMI 6:721–741
Hisakado M, Mori S (2010) Phase transition and information cascade in a voting model. J Phys A 43:315207
Hisakado M, Mori S (2011) Digital herders and phase transition in a voting model. J Phys A 44:275204
Hisakado M, Mori S (2012) Two kinds of phase transitions in a voting model. J Phys A 45:345002–345016
Hisakado M, Mori S (2015) Information cascade, Kirman's ant colony model and Ising model. Physica A 417:63–75
Hisakado M, Kitsukawa K, Mori S (2006) Correlated binomial models and correlation structures. J Phys A 39:1 5365–15378
Kadanoff L (1966) Scling laws for Ising model near T_c. Physics 2:263
Kirman A (1993) Ants, rationality, and recruitment. Q J Econ 108:137–156
Konno N (2002) Limit theorems and absorption problems for quantum random walks in one dimension. Quantum Inf Comput 2:578–595
Meunier H, Leca JB, Deneubourg JL, Petit O (2007) Group movement decisions in capuchin monkeys: the utility of an experimental study and a mathematical model to explore the relationship between individual and collective behaviours. Behavior 143:1511
Milgram S, Bickman L, Berkowitz L (1969) Note on the drawing power of clowds of different size. J Pers Soc Psychol 13:79–82
Mori S, Hisakado M, Takahashi T (2012) Phase transition to two-peaks phase in an information cascade experiment. Phys Rev E 86:26109
Partridge BL (1982) The structure and function of fish schools. Sci Am 245:90–99
Pratt SC, Mallon E, Sumpter DJL, Franks RN (2002) Quorum sensing, recruitment, and collective decision-making during colony emigration by the ant Leptothorax albipennis. Behav Ecol Sociobiol 52:117–127
Stauffer D (2002) Sociophysics: the Sznajd model and its applications. Comput Phys Commun 146(1):93–98
Tarde G (1890) Les lois de l'imitation. Felix Alcan. Paris
Ward AJ, Sumpter DJ, Cauzin ID, Hart PJ, Krause J (2008) Quorum decision-making facilitates information transfer in fish shoals. Proc Natl Acad Sci 205(19):6948–6953
Watts DJ, Dodds PS (2007) Influentials, networks, and public opinion formation. J Consum Res 34:441–458
Young (2011) The dynamics of social innovation. Proc Natl Acad Sci 108(4):21285–21291

Information Cascade and Networks

Masato Hisakado and Shintaro Mori

1 Introduction

Collective herding behavior can be seen in several fields. Such behaviors are observed in not only animals but also in human beings. They are interesting problems in several cross fields such as sociology (Tarde 1890), social psychology (Milgram et al. 1969), ethnology (Partridge 1982; Couzin et al. 2002), and economics. Statistical physics is an effective tool to analyze these macro phenomena of human beings, and the associated field is developed as sociophysics (Galam 1990). For example, in statistical physics, anomalous fluctuations in financial markets (Cont and Bouchaud 2000; Egíluz and Zimmermann 2000) and opinion dynamics (Stauffer 2002; Curty and Marsili 2006; Araújo et al. 2010) have been discussed.

In the case of human beings, to estimate public perception, people observe the actions of other individuals; then, they make a choice similar to that of others. It is also called social learning. Because it is usually sensible to do what other people are doing, collective herding behavior is assumed to be the result of a rational choice according to public perception. In ordinary cases, this is the collect strategy. But this approach can sometimes lead to arbitrary or even erroneous decisions as a macro phenomenon. This phenomenon is known as an information cascade (Bikhchandani et al. 1992).

How do people obtain public perception? In the previous paper, we discuss the case people obtain information from previous r voters using mean field approximations (Hisakado and Mori 2011). We call the lattice 1D extended lattice.

M. Hisakado (✉)
Nomura Holdings, Inc., Chiyoda-ku, Tokyo, Japan

S. Mori
Faculty of Science and Technology, Department of Mathematics and Physics, Hirosaki University, Hirosaki, Aomori, Japan

A.-H. Sato (ed.), *Applications of Data-Centric Science to Social Design*, Agent-Based Social Systems 14, https://doi.org/10.1007/978-981-10-7194-2_7

In the real world, people obtain information from their friends and influencers. The influencers become the hubs and affect many voters. Hence, we consider the voting model on several graphs, Barabási-Albert(BA) model, random graph, 1D extended lattice case, and the intermediate graph of these and compare results which are affected by networks (Barabáshi and Albert 1999).

In our previous paper, we introduced a sequential voting model that is similar to a Keynesian beauty contest (Keynes 1936; Hisakado and Mori 2010). At each time step t, one voter votes for one of two candidates. As public perception, the t-th voter can see all previous votes, i.e., $(t-1)$ votes. There are two types of voters – herders and independents – and two candidates.

Herders' behavior is known as the influence response function. Threshold rules have been derived for a variety of relevant theoretical scenarios as the influence response function. Some empirical and experimental evidence has shown the assumptions that individuals follow threshold rules when making decisions in the presence of social influence (Watt and Dodds 2007). This rule posits that individuals will switch from one to the other only when sufficiently many others have adopted the other. We call them digital herders. From our experiments, we observed that human beings exhibit a behavior between that of digital and analog herders (Mori et al. 2012). Analog herders vote for each candidate with the probabilities that are proportional to the candidates' votes. In this chapter, we discuss the digital and the analog herders cases to simplify the problems.

Here, we discuss a voting model with two candidates. We set two types of voters – independents and herders. The voting of independents is based on their fundamental values and rational. They collect information independently. On the other hand, the voting of herders is based on the number of previous votes, which is visible to them as public perceptions. In this study, we consider the case wherein a voter can see r previous votes which depends on several graphs.

In the upper limit of t, the independents cause the distribution of votes to converge to a Dirac measure against herders. This model contains two kinds of phase transitions. One is a transition of super and normal diffusions. This phase transition contains three phases – two super diffusion phases and a normal diffusion phase. In super diffusion phases, the voting rate converges to a Dirac measure slower than in a binomial distribution. In normal phase, the voting rate converges as in a binomial distribution. The other phase transition is referred to as information cascade transition. As the fraction of herders increases, the model features a phase transition beyond which a state where most voters make the correct choice coexists with one where most of them are wrong. The distribution of votes changes from one peak to two peaks. It is our purpose to clarify how the network of references affects these two phase transitions.

The remainder of this chapter is organized as follows. In Sect. 2, we introduce our voting model and mathematically define the two types of voters – independents and herders. In Sect. 3, we derive a stochastic differential equation and discuss the voting model on the random graph. In Sect. 4, we discuss the voting model on BA model. In Sect. 5, we discuss the model on fitness model. In the model, hubs which are stronger than BA model appear. In Sect. 6, we verify these transitions through numerical simulations and compare the influence from the networks. Finally, the conclusions are presented in Sect. 7.

2 Model

We model the voting of two candidates, C_0 and C_1; at time t, C_0 and C_1 have $c_0(t)$ and $c_1(t)$ votes, respectively. In each time step, one voter votes for one candidate; the voting is sequential. Hence, at time t, the t-th voter votes, after which the total number of votes is t. Voters are allowed to see r previous votes for each candidate; thus, they are aware of public perception. r is a constant number. The selections of r previous votes depends on networks.

There are two types of voters – independents and herders; we assume an infinite number of voters. The independents vote for C_0 and C_1 with probabilities $1 - q$ and q, respectively. Their votes are independent of others' votes, i.e., their votes are based on their fundamental values.

Here, we set C_0 as the wrong candidate and C_1 as the correct candidate in order to validate the performance of the herders. We can set $q \geq 0.5$, because we believe that independents vote for the correct candidate C_1 rather than for the wrong candidate C_0. In other words, we assume that the intelligence of the independents is virtually accurate.

On the other hand, the herders' votes are based on the number of previous r votes. Here, r does not always mean just previous r votes. We consider previous r votes which are selected by voters' network. At time t, the information of r previous votes are the number of votes for C_0 and C_1:$c_0^r(t)$ and $c_1^r(t)$, respectively. Hence, $c_0^r(t) + c_1^r(t) = r$ holds. If $r > t$, voters can see t previous votes for each candidate. In the limit $r \to \infty$, voters can see all previous votes. We define the number of all previous votes for C_0 and C_1 as $c_0^\infty(t) \equiv c_0(t)$ and $c_1^\infty(t) \equiv c_1(t)$. In real world, the number of references r depends of voters, but here we set r as constant.

In this chapter, herder is digital herder and analog herder (Hisakado and Mori 2010, 2011). Here, we define $c(t)_1^r/r = 1 - c(t)_0^r/r = z(t)$. Herder behavior is defined by the function of $f(z)$. We consider two types – (i) digital $f(z) = \theta(z - 1/2)$ and (ii) analog $f(z) = z$. Here, $\theta(z)$ is Heaviside function.

The independents and herders appear randomly and vote. We set the ratio of independents to herders as $(1 - p)/p$. In this study, we mainly focus on the upper limit of t. This refers to the voting of infinite voters.

We consider the voter can see r previous votes. It is the problem how the voter select r previous votes. The influence of the reference voters is understood as a voting model on networks. How the network affects the voting model is our problem. In this chapter, we analyze cases, random graph, Barabási-Albert(BA) model case, and their intermediate cases. In Fig. 1, we show three cases when $r = 2$. We discussed 1D extended lattice cases and showed that analog herder case is Kirman's ant colony model and the digital herder case is the kinetic Ising model (Hisakado and Mori 2015). A white (black) dot is a voter who voted to the candidate $C_0(C_1)$. Two arrows come into a dot. It means a voter refers to two voters

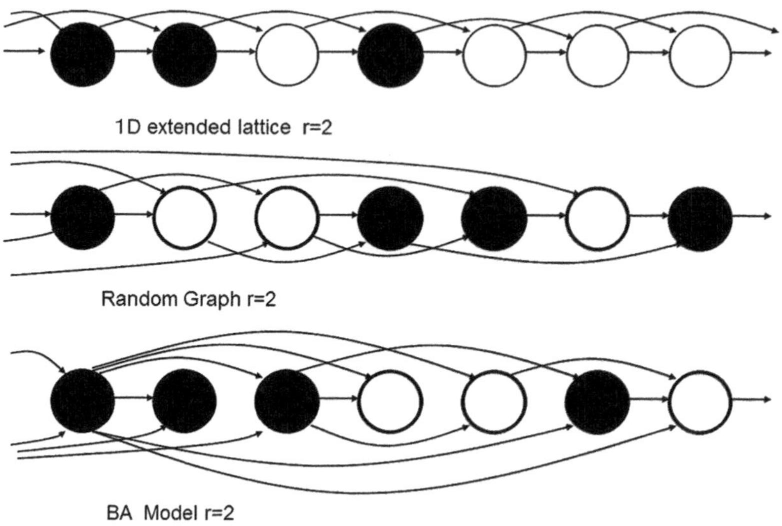

Fig. 1 Representation of graphs. It is directed acyclic graph (DAG). The graph consists of the arrows, and there is no cycle in the graph. We show extended 1D lattice, random graph, and BA model when $r = 2$. White (black) dot is a voter who voted to the candidate $C_0(C_1)$. Two arrows come into a dot. It means a voter refers to two voters when $r = 2$. In the case of extended 1D lattice, a voter refers to latest two voters. In the case of random graph, a voter refers to two previous voters who are selected randomly. In the case of BA model, a voter refers to two previous voters who are selected by the connectivity. Hence, there are voters who play the role of a hub in BA model. In the above figure, the first voter in network graph is a hub who influences many other voters

when $r = 2$. In the case of 1D extended lattice, a voter refers to latest two voters. In the case of random graph, a voter refers to two previous voters who are selected randomly. In the case of BA model, a voter refers to two previous voters who is selected by the connectivity which is introduced in BA model (Barabáshi and Albert 1999). BA model has the characteristics of scale-free network that contains hubs. The power index of BA model is three. Hence, there are voters who play a role of a hub. In the above figure, the first voter in BA model corresponds to a hub.

3 Random Graph

3.1 Digital Herder Case

We are interested in the limit $t \rightarrow \infty$. At time t t-th, voter selects r voters who have already voted. t-th voter can see the total votes of selected different r voters to obtain the information. In this section, we consider the case that the voter selects r votes randomly. Hence, this model is the voting model on the random graph.

Herders vote for a majority candidate; if $c_0^r(t) > c_1^r(t)$, herders vote for the candidate C_0. If $c_0^r(t) < c_1^r(t)$, herders vote for the candidate C_1. If $c_0^r(t) = c_1^r(t)$, herders vote for C_0 and C_1 with the same probability, i.e., 1/2. Here, at time t, the selected information of r previous votes are the number of votes for C_0 and C_1:$c_0^r(t)$ and $c_1^r(t)$, respectively. The herders in this section are known as digital herders.

We define $P_1^r(t)$ as the probability of that the t-th voter votes for C_1.

$$P_1^r(t) = \begin{Bmatrix} p + (1-p)q & ; c_1^r(t) > r/2; \\ p/2 + (1-p)q & ; c_1^r(t) = r/2; \\ (1-p)q & : c_1^r(t) < r/2. \end{Bmatrix} \tag{1}$$

In the scaling limit $t = c_0(t) + c_1(t) = c_0^\infty + c_1^\infty \to \infty$, we define

$$\frac{c_1(t)}{t} \Longrightarrow Z_\infty. \tag{2}$$

$Z(t)$ is the ratio of voters who vote for C_1 at t.

Here, we define π as the majority probability of binomial distributions of Z. In other words, the probability of $c_1^r(t) > 1/2$. When r is odd,

$$\pi(Z) = \sum_{g=\frac{r+1}{2}}^{r} \binom{r}{g} Z^g (1-Z)^{r-g} \equiv \Omega_r(Z). \tag{3}$$

When r is even, from the definition of the behavior of the herder,

$$\pi(Z) = \sum_{g=\frac{r}{2}+1}^{r} \binom{r}{g} Z^g (1-Z)^{r-g} + \frac{1}{2}\binom{r}{r/2} Z^g (1-Z)^{r/2}$$
$$= \sum_{g=\frac{r}{2}}^{r-1} \binom{r-1}{g} Z^g (1-Z)^{r-1-g} = \Omega_{r-1}(Z). \tag{4}$$

A majority probability π in even case becomes the odd case $r-1$. Here, after we consider only the odd case $r = 2n+1$, where $n = 0, 1, 2, \ldots$.

π can be calculated as follows:

$$\pi(Z) = \frac{(2n+1)!}{(n!)^2} \int_0^Z x^n (1-x)^n dx = \frac{1}{B(n+1, n+1)} \int_0^Z x^n (1-x)^n dx. \tag{5}$$

Equation (5) can be applied when the referred voters are selected overlap. In fact, the referred voters are not selected overlap. But in this paper, we use this approximation to study in large t limit.

We can rewrite (1) for random graph as

$$c_1(t) = k \rightarrow k+1 : P_{k,t} = p\pi(k/t) + (1-p)q,$$
$$c_1(t) = k \rightarrow k : Q_{k,t} = 1 - P_{k,t}, \tag{6}$$

We define a new variable Δ_t such that

$$\Delta_t = 2c_1(t) - t = c_1(t) - c_0(t). \tag{7}$$

We change the notation from k to Δ_t for convenience. Then, we have $|\Delta_t| = |2k - t| < t$. Thus, Δ_t holds within $\{-t, t\}$. Given $\Delta_t = u$, we obtain a random walk model:

$$\Delta_t = u \rightarrow u+1 : P_{u,t} = \pi(\frac{1}{2} + \frac{u}{2t})p + (1-p)q,$$
$$\Delta_t = u \rightarrow u-1 : Q_{u,t} = 1 - P_{u,t}.$$

We now consider the continuous limit $\epsilon \rightarrow 0$,

$$X_\tau = \epsilon \Delta_{[t/\epsilon]},$$
$$P(x,\tau) = \epsilon P(\Delta_t/\epsilon, t/\epsilon), \tag{8}$$

where $\tau = t/\epsilon$ and $x = \Delta_t/\epsilon$. Approaching the continuous limit, we can obtain the stochastic differential equation (see Appendix A):

$$\mathrm{d}X_\tau = [(1-p)(2q-1) - p + 2p\frac{(2n+1)!}{(n!)^2}\int_0^{\frac{1}{2}+\frac{X_\tau}{2\tau}} x^n(1-x)^n dx]\mathrm{d}\tau + \sqrt{\epsilon}. \tag{9}$$

In the case $r = 1$, the equation becomes

$$\mathrm{d}X_\tau = [(1-p)(2q-1) + p\frac{X_\tau}{\tau}]\mathrm{d}\tau + \sqrt{\epsilon}. \tag{10}$$

The voters vote for each candidate with the probabilities that are proportional to the candidates' votes. We refer to these herders as a kind of analog herders (Hisakado and Mori 2010).

We are interested in the behavior in the limit $\tau \rightarrow \infty$. The relation between X_∞ and the voting ratio to C_1 is $2Z - 1 = X_\infty/\tau$. We consider the solution $X_\infty \sim \tau^\alpha$, where $\alpha \leq 1$, since the maximum speed is τ when $q = 1$. The slow solution is $X_\infty \sim \tau^\alpha$, where $\alpha < 1$ is hidden by the fast solution $\alpha = 1$ in the upper limit of τ. Hence, we can assume a stationary solution as

$$X_\infty = \bar{v}\tau + (1-p)(2q-1)\tau, \tag{11}$$

where $\bar{v}$ is constant. Substituting (11) into (9), we can obtain

$$\bar{v} = -p + \frac{2p \cdot (2n+1)!}{(n!)^2} \int_0^{\frac{1}{2}+\frac{(1-p)(2q-1)}{2}+\frac{\bar{v}}{2}} x^n(1-x)^n dx \tag{12}$$

This is the self-consistent equation.

Equation (12) admits one solution below the critical point $p \leq p_c$ and three solutions for $p > p_c$. When $p \leq p_c$, we refer to the phase as the one-peak phase. When $p > p_c$, the upper and lower solutions are stable; on the other hand, the intermediate solution is unstable. Then, the two stable solutions attain a good and bad equilibria, and the distribution becomes the sum of the two Dirac measures. We refer to this phase as two-peak phase.

If $r = 2n + 1 \geq 3$, a phase transition occurs in the range $0 \leq p \leq 1$. If the voter obtains information from the above three voters, as the number of herders increases, the model features a phase transition beyond which a state where most voters make the correct choice coexists with one where most of them are wrong. If the voter obtains information from one or two voters, there is no phase transition. We refer to this transition as information cascade transition (Hisakado and Mori 2011).

Next, we consider the phase transition of convergence. This type of transition has been studied when herders are analog (Hisakado and Mori 2010). We expand X_τ around the solution $\bar{v}\tau + (1-p)(2q-1)\tau$.

$$X_\tau = \bar{v}\tau + (1-p)(2q-1)\tau + W_\tau. \tag{13}$$

Here, we set $X_\tau \gg W_\tau$, that is, $\tau \gg 1$. We rewrite (9) using (13) and obtain as follows:

$$\mathrm{d}W_\tau = p\frac{(2n+1)!}{(n!)^2 \cdot 2^{2n}}\frac{W_\tau}{\tau}[1 - \{\bar{v} + (1-p)(2q-1)\}^2]^n \mathrm{d}\tau + \sqrt{\epsilon} \tag{14}$$

We use relation (11) and consider the first term of the expansion. If we set $l = p(2n+1)!/\{(n!)^2 \cdot 2^{2n}\}[1 - \{\bar{v} + (1-p)(2q-1)\}^2]^n$, (54) is identical to (14).

From Appendix B, we can obtain the phase transition of convergence. The critical point p_{vc} is the solution of

$$p\frac{(2n+1)!}{(n!)^2 \cdot 2^{2n}}[1 - \{\bar{v} + (1-p)(2q-1)\}^2]^n = \frac{1}{2} \tag{15}$$

and the self-consistent equation (12).

Here, we consider the symmetric case, $q = 1/2$. The self-consistent equation (12) becomes

$$\bar{v} = -p + \frac{2p \cdot (2n+1)!}{(n!)^2} \int_0^{\frac{1}{2}+\frac{\bar{v}}{2}} x^n(1-x)^n dx. \tag{16}$$

RHS of (16) rises at $\bar{v}=0$. If $r=1,2$, there is only one solution $\bar{v}=0$ in all regions of p. In this case, Z has only one peak, at 0.5, which indicates the one-peak phase. In the case $r=1,2$, we do not observe information cascade transition (Hisakado and Mori 2011). On the other hand, in the case $r\geq 3$, there are two stable solutions and an unstable solution $\bar{v}=0$ above p_c. The votes ratio for C_1 attains a good or bad equilibrium. This is the so-called spontaneous symmetry breaking. In one sequence, Z is taken as $\bar{v}/2+1/2$ in the case of a good equilibrium, or as $-\bar{v}/2+1/2$ in the case of a bad equilibrium, where $\bar{v}$ is the solution of (16). This indicates the two-peak phase, and the critical point is $p_c=\frac{(n!)^2}{(2n+1)!}2^{2n}$ where gradient of the RHS of (16) at $\bar{v}=0$ is 1. In the case of $r=3$, $p_c=2/3$. As r increases, p_c moves toward 0. In the large limit $r\rightarrow\infty$, p_c becomes 0. It is consistent with case that herders obtain information from all previous voters (Hisakado and Mori 2016).

Next, we consider the normal-super transition of symmetric case $q=1/2$. We consider the case $r=2n+1\geq 3$. In this case, we observe information cascade transition. If $r\leq 2$, we do not observe information cascade transition, and we can only observe a part of the phases, as described below.

In the one-peak phase $p\leq p_c$, the only solution is $\bar{v}=0$. p_c is the critical point of the information cascade transition. The first critical point of convergence is $p_{vc1}=\frac{(n!)^2}{(2n+1)!}2^{2n-1}=\frac{p_c}{2}$. When $p\leq p_c$, p_{vc1} is the solution of (16) and (15). If $0<p<p_{vc1}$, the voting rate for C_1 becomes $1/2$, and the distribution converges as in a binomial distribution. If $p_c>p\geq p_{vc1}$, candidate C_1 gathers $1/2$ of all the votes in the scaled distributions, too. However, the voting rate converges slower than in a binomial distribution. We refer to these phases as super diffusion phases. There are two phases, $p=p_{vc1}$ and $p_c>p>p_{vc1}$; these phases differ in terms of their convergence speed. If $r\leq 2$, we can observe these three phases.

Above p_c, in the two-peak phase, we can obtain two stable solutions that are not $\bar{v}=0$. At p_c, $\bar{v}$ moves from 0 to one of these two stable solutions. In one voting sequence, the votes converge to one of these stable solutions. If $p_c<p\leq p_{vc2}$, the voting rate for C_1 becomes $\bar{v}/2+1/2$ or $-\bar{v}/2+1/2$, and the convergence occurs at a rate slower than that in a binomial distribution. Here, $\bar{v}$ is the solution of (16). We refer to this phase as a super diffusion phase. p_{vc2} is the second critical point of convergence from the super to the normal diffusion phase, and it is the solution of the simultaneous equations (16) and (15) when $p>p_c$. In the case $r=3$, we can obtain $p_{vc2}=5/6$ at $\bar{v}=\pm\sqrt{(3p-2)/p}$. If $p>p_{vc2}$, the voting rate for C_1 becomes $\bar{v}/2+1/2$ or $-\bar{v}/2+1/2$. But the distribution converges as in a binomial distribution. This is a normal diffusion phase. A total of six phases can be observed.

3.2 Analog Herder Case

The voters vote for each candidate with the probabilities that are proportional to the candidates' votes. We refer to these herders as a kind of analog herders (Hisakado and Mori 2010). We can write for analog herder case on random graph as

$$c_1(t) = k \to k+1 : P_{k,t} = p(k/t) + (1-p)q,$$
$$c_1(t) = k \to k : Q_{k,t} = 1 - P_{k,t}, \tag{17}$$

We define a new variable Δ_t such that

$$\Delta_t = 2c_1(t) - t = c_1(t) - c_0(t). \tag{18}$$

We change the notation from k to Δ_t for convenience. Then, we have $|\Delta_t| = |2k - t| < t$. Thus, Δ_t holds within $\{-t, t\}$. Given $\Delta_t = u$, we obtain a random walk model:

$$\Delta_t = u \to u+1 : P_{u,t} = (\frac{1}{2} + \frac{u}{2t})p + (1-p)q,$$
$$\Delta_t = u \to u-1 : Q_{u,t} = 1 - P_{u,t}.$$

We now consider the continuous limit $\epsilon \to 0$,

$$X_\tau = \epsilon \Delta_{[t/\epsilon]},$$
$$P(x,\tau) = \epsilon P(\Delta_t/\epsilon, t/\epsilon), \tag{19}$$

where $\tau = t/\epsilon$ and $x = \Delta_t/\epsilon$. Approaching the continuous limit, we can obtain the stochastic differential equation (see Appendix A):

$$\mathrm{d}X_\tau = [(1-p)(2q-1) + p\frac{X_\tau}{\tau}]\mathrm{d}\tau + \sqrt{\epsilon}. \tag{20}$$

We can assume a stationary solution as

$$X_\infty = \bar{\upsilon}\tau + (1-p)(2q-1)\tau, \tag{21}$$

where $\bar{\upsilon}$ is constant. Substituting (21) into (20), we can obtain $X_\infty = 2q - 1$. It means that there is no phase transition for this case. The average correct ratio is constant.

We expand X_τ around the solution $\bar{\upsilon}\tau + (1-p)(2q-1)\tau$.

$$X_\tau = \bar{\upsilon}\tau + (1-p)(2q-1)\tau + W_\tau. \tag{22}$$

Here, we set $X_\tau \gg W_\tau$, that is, $\tau \gg 1$. We rewrite (20) using (22) and obtain as follows:

$$\mathrm{d}W_\tau = p\frac{W_\tau}{\tau}\mathrm{d}\tau + \sqrt{\epsilon}. \tag{23}$$

From Appendix B, we can obtain the phase transition of convergence. The critical point $p_{\upsilon c} = 1/2$, and it is not dependent on q.

4 Barabási-Albert Model

4.1 Digital Herder Case

In this section, we consider the case that the voter selects r different voters who will be selected by the popularity. The popularity is proportional to the connectivity of voter i, such that $l_i/\sum_j l_j$. l_i is the sum of the number of voters whom voter i gave the information and whom voter i obtained information, referenced r voters. The total number of l_i after t-th voter voted is $\sum_j l_j = 2r(t-r+1)$ where $t \geq r$. l_i corresponds to the connectivity in BA model (Barabáshi and Albert 1999). Hence, the model is the voting model on BA model. The difference is the connectivity has color. The color depends on whether the voter voted to candidate C_1 or C_0. In Fig. 2, the voter who voted $C_1(C_0)$ is black(white). We define the total number of connectivity of the voters who voted the candidate $C_1(C_0)$ at t as $g_1(t)(g_0(t))$. Hence, $g_0(t)+g_1(t) = 2r(t-r+1)$ where $t \geq r$. In the scaling limit $g_0(t)+g_1(t) = 2r(t-r+1) \rightarrow \infty$, we define

$$\frac{g_1(t)}{2r(t-r+1)} \Longrightarrow \hat{Z}_\infty, \tag{24}$$

$\hat{Z}(t)$ is $g_1(t)/2r(t-r+1)$ at t.

We can write the evolution of connectivity

$$g_1(t) = \hat{k} \rightarrow \hat{k} + i :$$
$$(3r+1)/2 \leq i \leq 2r \quad P_{\hat{k},t}(i) = {}_rC_{2r-i}\hat{Z}^{i-r}(1-\hat{Z})^{2r-i}[(1-p)q+p],$$

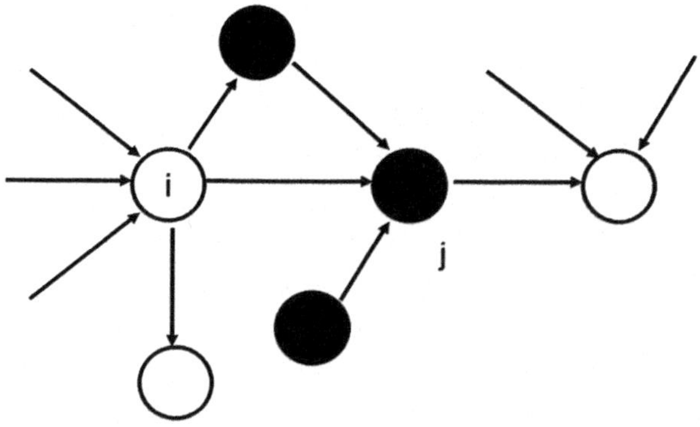

Fig. 2 The sample graph BA model with $r = 3$. The voter i refers to three voters and is referred by three voters. He/she is white, since he/she voted the candidate C_0. The connectivity of i is 6, and the color is white. The voter j refers to three voters and is referred by a voter. He/she is black, since he/she voted the candidate C_1. The connectivity of j is 4, and the color is black

$$
\begin{aligned}
r+1 \le i \le (3r-1)/2 \quad & P_{\hat{k},t}(i) = {}_rC_{2r-i}\hat{Z}^{i-r}(1-\hat{Z})^{2r-i}(1-p)q, \\
i = r \quad & P_{\hat{k},t}(i) = (1-\hat{Z})^r(1-p)q + \hat{Z}^r(1-p)(1-q), \\
(r+1)/2 \le i \le r-1 \quad & P_{\hat{k},t}(i) = {}_rC_{r-i}\hat{Z}^i(1-\hat{Z})^{r-i}(1-p)(1-q), \\
0 \le i \le (r-1)/2 \quad & P_{\hat{k},t}(i) = {}_rC_{r-i}\hat{Z}^i(1-\hat{Z})^{r-i}[(1-p)(1-q)+p].
\end{aligned} \tag{25}
$$

Here, we consider the self-consistent equations for connectivity in the large t limit.

$$
2r\hat{Z}_\infty = \sum_{i=1}^{2r} P_{\hat{k},t}(i) \cdot i = r(1-p)q + rp\pi(\hat{Z}_\infty) + r\hat{Z}_\infty. \tag{26}
$$

Hence, we can obtain

$$
\hat{Z}_\infty = (1-p)q + p\pi(\hat{Z}_\infty). \tag{27}
$$

On the other hand, the evolution equation for the voting ratio Z_∞ is

$$
Z_\infty = (1-p)q + p\pi(\hat{Z}_\infty). \tag{28}
$$

Comparing (27) and (28), we can obtain $Z_\infty \sim \hat{Z}_\infty$. It means the behavior of the voting ratio, Z_∞, is the same as the connectivity ratio, $\hat{Z}_\infty$. Hereafter, we analyze only behavior of the connectivity.

We define a new variable $\hat{\Delta}_t$ such that

$$
\hat{\Delta}_t = g_1(t) - r(t-r+1) = \frac{1}{2}\{g_1(t) - g_0(t)\}. \tag{29}
$$

We change the notation from $\hat{k}$ to $\hat{\Delta}_t$ for convenience. Thus, $\hat{\Delta}_t$ holds within $\{-r(t-r+1), r(t-r+1)\}$. Given $\hat{\Delta}_t = \hat{u}$, we obtain a random walk model:

$$
\begin{aligned}
& \hat{\Delta} = \hat{u} \to \hat{u} + i : \\
(r+1)/2 \le i \le r \quad & P_{\hat{u},t}(i) = {}_rC_{r-i}\hat{Z}^i(1-\hat{Z})^{r-i}[(1-p)q+p], \\
1 \le i \le (r-1)/2 \quad & P_{\hat{u},t}(i) = {}_rC_{r-i}\hat{Z}^i(1-\hat{Z})^{r-i}(1-p)q, \\
i = 0 \quad & P_{\hat{u},t}(i) = (1-\hat{Z})^r(1-p)q + \hat{Z}^r(1-p)(1-q), \\
(-r+1)/2 \le i \le -1 \quad & P_{\hat{u},t}(i) = {}_rC_{-i}\hat{Z}^{r+i}(1-\hat{Z})^{-i}(1-p)(1-q), \\
-r \le i \le -(r+1)/2 \quad & P_{\hat{u},t}(i) = {}_rC_{-i}\hat{Z}^{r+i}(1-\hat{Z})^{-i}[(1-p)(1-q)+p],
\end{aligned} \tag{30}
$$

where $\hat{Z} = \hat{k}/(2r(t-r+1)) = \hat{u}/(2r(t-r+1)) + 1/2$

We now consider the continuous limit $\epsilon \to 0$,

$$\hat{X}_\tau = \epsilon \hat{\Delta}_{[t/\epsilon]},$$
$$P(\hat{x}, \tau) = \epsilon P(\hat{\Delta}_t/\epsilon, t/\epsilon), \tag{31}$$

where $\tau = t/\epsilon$ and $\hat{x} = \hat{\Delta}_t/\epsilon$. Approaching the continuous limit, we can obtain the stochastic partial differential equation (see Appendix A):

$$\mathrm{d}\hat{X}_\tau = [r(1-p)q - \frac{r}{2} + \frac{\hat{X}_\tau}{2(\tau - r + 1)}$$
$$+ rp\frac{(2n+1)!}{(n!)^2}\int_0^{\frac{1}{2}+\frac{\hat{X}_\tau}{2r(\tau-r+1)}} x^n(1-x)^n dx]\mathrm{d}\tau + \sqrt{\epsilon}. \tag{32}$$

In the case $r = 1$, the equation becomes

$$\mathrm{d}\hat{X}_\tau = [(1-p)(q-\frac{1}{2}) + \frac{p+1}{2}\frac{\hat{X}_\tau}{\tau}]\mathrm{d}\tau + \sqrt{\epsilon}. \tag{33}$$

The voters also vote for each candidate with probabilities that are proportional to $\hat{X}_\tau$. But Eq. (33) is different from (10). The relation between the voting ratio for C_1 and $\hat{X}_\infty$ is

$$\frac{\hat{X}_\infty}{2r(\tau - r + 1)} = \hat{Z}_\infty - \frac{1}{2}. \tag{34}$$

We can assume a stationary solution as

$$\hat{X}_\infty = r\bar{\hat{v}}\tau + r(1-p)(2q-1)\tau, \tag{35}$$

where $\bar{\hat{v}}$ is constant. Since (34) and $0 \le \hat{Z} \le 1$, we can obtain

$$-1 \le \bar{\hat{v}} + (1-p)(2q-1) \le 1 \tag{36}$$

Substituting (35) into (32), we can obtain

$$\bar{\hat{v}} = -p + \frac{2p\cdot(2n+1)!}{(n!)^2}\int_0^{\frac{1}{2}+\frac{(1-p)(2q-1)}{2}+\frac{\bar{\hat{v}}}{2}} x^n(1-x)^n dx \tag{37}$$

This is the self-consistent equation, and it agrees to (12) which is obtained for random graph. Then, about the information cascade transition, the behavior in BA model case is the same as in the random graph case. When $p \le p_c$, we refer to the phase as the one-peak phase. Equation (37) admits three solutions for $p > p_c$. When $p > p_c$, the upper and lower solutions are stable; on the other hand, the intermediate solution is unstable. Then, the two stable solutions attain a good and bad equilibrium, and the distribution becomes the sum of the two Dirac measures. This phase is a two-peak phase. The phase transition point p_c in BA model case is the same as the random graph case. If $r = 2n + 1 \geq 3$, a phase transition occurs in the range $0 \leq p \leq 1$. If the voter obtains information from one or two voters, there is no phase transition.

Next, we consider the phase transition of convergence. We expand $\hat{X}_\tau$ around the solution $r\bar{\hat{v}}\tau + r(1-p)(2q-1)\tau$.

$$X_\tau = r\bar{\hat{v}}\tau + r(1-p)(2q-1)\tau + r\hat{W}_\tau. \tag{38}$$

Here, we set $X_\tau \gg W_\tau$. This indicates $\tau \gg 1$. We rewrite (32) using (38) and obtain as follows:

$$\mathrm{d}\hat{W}_\tau = [1 + p\frac{(2n+1)!}{(n!)^2 \cdot 2^{2n}}(1 - \{\bar{\hat{v}} + (1-p)(2q-1)\}^2)^n]\frac{W_\tau}{2\tau}\mathrm{d}\tau + \sqrt{\epsilon} \tag{39}$$

We use relation (37) and consider the first term of the expansion. If we set $l = 1/2[1 + p(2n+1)!/\{(n!)^2 \cdot 2^{2n}\}(1 - \{\bar{\hat{v}} + (1-p)(2q-1)\}^2)^n]$, (54) is identical to (39).

From Appendix B, we can obtain the phase transition of convergence. The critical point p_{vc} is the solution of

$$[1 + p\frac{(2n+1)!}{(n!)^2 \cdot 2^{2n}}(1 - \{\bar{\hat{v}} + (1-p)(2q-1)\}^2)^n]\frac{1}{2} = \frac{1}{2}, \tag{40}$$

and (37). RHS of (40) is not less than $1/2$. Using (36), we can obtain $p_{vc} = 0$. There is no normal phase in BA model case. The speed of convergence is always slower than the normal in all regions of p.

4.2 Analog Herder Case

In this subsection, we consider the analog herder case on BA model. We can write the evolution of connectivity for analog herder as

$$\begin{aligned}
&\hat{\Delta} = \hat{u} \rightarrow \hat{u} + i : \\
1 \leq i \leq r \quad &P_{\hat{u},t}(i) = {}_rC_{r-i}\hat{Z}^i(1-\hat{Z})^{r-i}[(1-p)q + \frac{i}{r}p], \\
i = 0 \quad &P_{\hat{u},t}(i) = (1-\hat{Z})^r(1-p)q + \hat{Z}^r(1-p)(1-q), \\
-r \leq i \leq -1 \; &P_{\hat{u},t}(i) = {}_rC_{-i}\hat{Z}^{r+i}(1-\hat{Z})^{-i}[(1-p)(1-q) - \frac{i}{r}p],
\end{aligned} \tag{41}$$

where $\hat{Z} = \hat{k}/(2r(t-r+1)) = \hat{u}/(2r(t-r+1)) + 1/2$.

We now consider the continuous limit $\epsilon \to 0$,

$$\hat{X}_\tau = \epsilon \hat{\Delta}_{[t/\epsilon]},$$
$$P(\hat{x}, \tau) = \epsilon P(\hat{\Delta}_t/\epsilon, t/\epsilon), \quad (42)$$

where $\tau = t/\epsilon$ and $\hat{x} = \hat{\Delta}_t/\epsilon$. Approaching the continuous limit, we can obtain the stochastic partial differential equation (see Appendix A):

$$\mathrm{d}\hat{X}_\tau = [r(1-p)(q-\frac{1}{2}) + \frac{p+1}{2}\frac{\hat{X}_\tau}{\tau}]\mathrm{d}\tau + \sqrt{\epsilon}. \quad (43)$$

We can assume a stationary solution as

$$\hat{X}_\infty = r\bar{\hat{v}}\tau + r(1-p)(2q-1)\tau, \quad (44)$$

where $\bar{\hat{v}}$ is constant. Substituting (44) into (43), we can obtain $\hat{X}_\infty = r(2q-1)\tau$. It means that there is no phase transition for this case and the average correct ratio is constant. Herders cannot amplify the average correct ratio.

Next, we consider the phase transition of convergence. We expand $\hat{X}_\tau$ around the solution $r\bar{\hat{v}}\tau + r(1-p)(2q-1)\tau$.

$$X_\tau = r\bar{\hat{v}}\tau + r(1-p)(2q-1)\tau + r\hat{W}_\tau. \quad (45)$$

Here, we set $X_\tau \gg W_\tau$. This indicates $\tau \gg 1$. We rewrite (43) using (45) and obtain as follows:

$$\mathrm{d}\hat{W}_\tau = (1+p)\frac{W_\tau}{2\tau}\mathrm{d}\tau + \sqrt{\epsilon} \quad (46)$$

We use relation (44) and consider the first term of the expansion.

From Appendix B, we can obtain the phase transition of convergence. The critical point is $p_{vc} = 0$. There is no normal phase in BA model case. The speed of convergence is always slower than the normal in all regions of p.

5 Fitness Model

In this section, we consider fitness model on BA model (Bianconi and Barabáshi 2001a,b). We set the fitness of each voter, and the fitness is the weight of the connectivity. In this model, stronger hubs appear. The power index of BA model is three. The power index of fitness model is below three. The problem is whether the stronger hubs affect the phase transition point.

The popularity is proportional to the weighted connectivity of voter i, such that $\eta_i l_i / \sum_j \eta_j l_j$. l_i is the sum of the number of voters whom voter i gave the information and whom voter i obtained information, referenced r voters. η_i is the fitness of voter i. The total number of weighted l_i after t-th voter voted is $\sum_j l_j = 2r(t - r + 1)\eta$ where $t \geq r$ and $\bar{\eta}$ is the average of the fitness. l_i corresponds to the connectivity in BA model. The color depends on whether the voter voted to candidate C_1 or C_0 as Sect. 4. We define the total number of weighted connectivity of the voters who voted the candidate $C_1(C_0)$ at t as $g_1(t)(g_0(t))$. Hence, $g_0(t) + g_1(t) = 2r\bar{\eta}(t - r + 1)$ where $t \geq r$. In the scaling limit $g_0(t) + g_1(t) = 2\bar{\eta}r(t - r + 1) \to \infty$, we define

$$\frac{g_1(t)}{2r\eta(t - r + 1)} \Longrightarrow \hat{Z}_\infty, \tag{47}$$

$\hat{Z}(t)$ is $g_1(t)\bar{\eta}/2r(t - r + 1)$ at t.

Here, we consider the self-consistent equations for connectivity in the large t limit. We set the condition that i-th voter votes and $\hat{Z}_\infty$ does not change as equilibrium. We can obtain using (25)

$$r\bar{\eta}\hat{Z}_\infty + r\eta_i\hat{Z}_\infty = r(1 - p)q\eta_i + rp\pi(\hat{Z}_\infty)\eta_i + r\bar{\eta}\hat{Z}_\infty. \tag{48}$$

Hence, we can obtain

$$\hat{Z}_\infty = (1 - p)q + p\pi(\hat{Z}_\infty). \tag{49}$$

It is the same as (27). It means that the phase diagram for fitness model is the same as the standard BA model. In this mean, we can conclude that the role of hub is limited.

6 Numerical Simulations

To confirm the analytic results, we performed 10^6 Monte Carlo (MC) simulations up to $t = 10^3$. In Fig. 3, we perform simulations for the symmetric independent voter case, i.e., $q = 1/2$. Convergence of distribution in random graph case and BA model case for the symmetric (a) $q = 0.5$ and (b) $q = 0.8$ cases. The reference number is $r = 3$. The horizontal axis represents the ratio of herders p, and the vertical axis represents the speed of convergence γ. We define γ as $Var(Z) = \tau^{-\gamma}$, where $Var(z)$ is the variance of $Z.\gamma = 1$ is the normal phase and $0 < \gamma < 1$ is the super diffusion phase.

Figure 3a is for the symmetric case $q = 0.5$. As discussed in the previous sections, the critical point of the convergence of the transition from normal diffusion to super diffusion is $p_{vc} \sim 0$ for BA model case. It means there is no normal phase

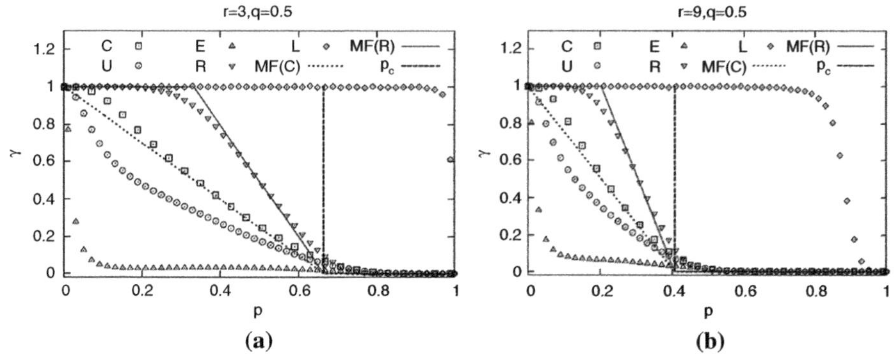

Fig. 3 Convergence of distribution in random graph case and BA model case for the symmetric (**a**) $q = 0.5$ and (**b**) $q = 0.8$ cases. The reference number is $r = 3$. The horizontal axis represents the ratio of herders p, and the vertical axis represents the speed of convergence γ. (**a**) is for the symmetric case $q = 0.5$. (**b**) is for the asymmetric case $q = 0.8$

for BA model case. The critical point is theoretically $p_{vc1} \sim 1/2p_{vc} = 1/3$ for random case, where p_c is the critical point of information cascade transition.

At the critical point of information cascade transition, the distribution splits in two delta functions, and exponent γ becomes 0. For the BA model and random graph case, the critical point $p_c = 2/3$ is the same.

In Fig. 3b, we consider the asymmetric cases wherein $q = 0.8$ for random graph and BA model cases. The critical point of the convergence of transition from normal diffusion to super diffusion is $p_{vc1} \sim 0$ for BA model case. It means there is no normal phase for this case. The critical point of information cascade transition is the same in the case of random graph and BA model cases.

7 Concluding Remarks

We investigated a voting model which contains collective herding behaviors on the random graph, BA model, and fitness case. In this chapter, we investigated the difference of phases which depend on network. The voter referred to the information from the network. In the BA model and fitness model cases, the network is similar to real networks which have hubs. In the continuous limit, we could obtain stochastic differential equations. Using the stochastic differential equations, we analyze the difference of the models.

The model has two kinds of phase transitions. One is information cascade transition, which is similar to the phase transition of Ising model. As the herders increased, the model featured a phase transition beyond which a state where most voters make the correct choice coexists with one where most of them are wrong. In this transition, the distribution of votes changed from the one-peak phase to the two-peak phase.

The other transition was the transition of the convergence between super and normal diffusions. In the one-peak phase, if herders increased, the variance converged slower than in a binomial distribution. This is the transition from normal diffusion to super diffusion. In the two-peak phase, we can also find this phase transition, and the sequential voting converged to one of the two peaks.

In the case of random graph, we can observe all phases. In the case of BA model and fitness model, we can observe only super phase in one-peak and two-peak phase. On the other hand, in the case of 1D extended lattice phase, there is only one-peak phase.

The critical point p_c in the random graph case, the BA model case, and the fitness model is the same. The difference can be observed in normal and super phases. In the case of BA model case, there is no super phase. The convergence speed is slower than the normal case. In the case of random case, the super phase and normal phase coexist. In the case of fitness model whose hubs are stronger than the BA model, the phase is the same as BA model. In conclusion, the influence of hubs can be seen in the convergence speed only and cannot be seen in the phase transition between one-peak phase and two-peak phase.

In Watt and Dodds (2007), the "influential hypothesis" was discussed. The hypothesis means that influential or hubs are important to the formation of public opinion. In our model, the network does not affect the critical point of information cascade transition. It is a phase transition beyond which a state where most voters make the correct choice coexists with one where most of them are wrong. It means the hubs or influencers do not affect the conclusions of votes in large t limit. In this chapter, hubs affected only to the critical point of super-normal transitions in large t limit. The phase transition is transition of the speed of the convergence. Hence, hubs affect the standard deviation of the votes but cannot change the conclusions. In this mean, we can conclude that the influence of hub is limited. On the other hand, the response function is important. The response function decides the information cascade transition. In this chapter, we have studied the analog and digital response function cases. We discuss the response functions from the viewpoints of experiments and empirical data in other chapters.

Appendix A Derivation of Stochastic Differential Equation

We use $\delta X_\tau = X_{\tau+\epsilon} - X_\tau$ and ζ_τ, a standard i.i.d. Gaussian sequence; our objective is to identify the drift f_τ and the variance g_τ^2 such that

$$\delta X_\tau = f_\tau(X_\tau)\epsilon + \sqrt{\epsilon} g_\tau(X_\tau)\zeta_{\tau+\epsilon}. \tag{50}$$

Given $X_\tau = x$, using the transition probabilities of Δ_n, we get

$$\mathrm{E}(\delta X_\tau) = \epsilon \mathrm{E}(\Delta_{[\tau/\epsilon]+1} - \Delta_{[\tau/\epsilon]}) = \epsilon(2p_{[\frac{l/\epsilon+\tau/\epsilon}{2}],\tau/\epsilon} - 1)$$

$$= \epsilon[(1-p)(2q-1)-p+2p\frac{(2n+1)!}{(n!)^2}\int_0^{\frac{1}{2}+\frac{X_\tau}{2\tau}} x^n(1-x)^n dx]. \quad (51)$$

Then, the drift term is $f_\tau(x) = (1-p)(2q-1) + p\tanh(\lambda x/2\tau)$. Moreover,

$$\sigma^2(\delta X_\tau) = \epsilon^2[1^2 p_{[\frac{l/\epsilon+\tau/\epsilon}{2}],\tau/\epsilon} + (-1)^2(1 - p_{[\frac{l/\epsilon+\tau/\epsilon}{2}],\tau/\epsilon})] = \epsilon^2, \quad (52)$$

such that $g_{\epsilon,\tau}(x) = \sqrt{\epsilon}$. We can obtain X_τ such that it obeys a diffusion equation with small additive noise:

$$\mathrm{d}X_\tau = [(1-p)(2q-1)-p+2p\frac{(2n+1)!}{(n!)^2}\int_0^{\frac{1}{2}+\frac{X_\tau}{2\tau}} x^n(1-x)^n dx]\mathrm{d}\tau + \sqrt{\epsilon}. \quad (53)$$

Appendix B Behavior of Solutions of Stochastic Differential Equation

We consider the stochastic differential equation

$$\mathrm{d}x_\tau = (\frac{l x_\tau}{\tau})\mathrm{d}\tau + \sqrt{\epsilon}, \quad (54)$$

where $\tau \geq 1$.

Let σ_1^2 be the variance of x_1. If x_1 is Gaussian ($x_1 \sim \mathrm{N}(x_1, \sigma_1^2)$) or deterministic ($x_1 \sim \delta_{x1}$), the law of x_τ ensures that the Gaussian is in accordance with density

$$p_\tau(x) \sim \frac{1}{\sqrt{2\pi}\sigma_\tau}\mathrm{e}^{-(x-\mu_\tau)^2/2\sigma_\tau^2}, \quad (55)$$

where $\mu_\tau = \mathrm{E}(x_\tau)$ is the expected value of x_τ and $\sigma_\tau^2 \equiv \nu_\tau$ is its variance. If $\Phi_\tau(\xi) = \log(\mathrm{e}^{\mathrm{i}\xi x_\tau})$ is the logarithm of the characteristic function of the law of x_τ, we have

$$\partial_\tau \Phi_\tau(\xi) = \frac{l}{\tau}\xi\partial_\xi \Phi_\tau(\xi) - \frac{\epsilon}{2}\xi^2, \quad (56)$$

and

$$\Phi_\tau(\xi) = \mathrm{i}\xi\mu_\tau - \frac{\xi^2}{2}\nu_\tau. \quad (57)$$

Identifying the real and imaginary parts of $\Phi_\tau(\xi)$, we obtain the dynamics of μ_τ as

$$\dot{\mu}_\tau = \frac{l}{\tau}\mu_\tau. \quad (58)$$

The solution for μ_τ is

$$\mu_\tau = x_1 \tau^l. \tag{59}$$

The dynamics of ν_τ are given by the Riccati equation

$$\dot{\nu}_\tau = \frac{2l}{\tau}\nu_\tau + \epsilon. \tag{60}$$

If $\nu \neq 1/2$, we get

$$\nu_\tau = \nu_1 \tau^{2l} + \frac{\epsilon}{1-2l}(\tau - \tau^{2l}). \tag{61}$$

If $l = 1/2$, we get

$$\nu_\tau = \nu_1 \tau + \epsilon \tau \log \tau. \tag{62}$$

We can summarize the temporal behavior of the variance as

$$\nu_\tau \sim \frac{\epsilon}{1-2l}\tau \qquad \text{if} \quad l < \frac{1}{2}, \tag{63}$$

$$\nu_\tau \sim (\nu_1 + \frac{\epsilon}{2l-1})\tau^{2l} \qquad \text{if} \quad l > \frac{1}{2}, \tag{64}$$

$$\nu_\tau \sim \epsilon \tau \log(\tau) \qquad \text{if} \quad l = \frac{1}{2}. \tag{65}$$

This model has three phases. If $l > 1/2$ or $l = 1/2$, x_τ/τ converges slower than in a binomial distribution. These phases are the super diffusion phases. If $0 < p < 1/2$, x_τ/τ converges as in a binomial distribution. This is the normal phase (Hisakado and Mori 2010).

References

Araújo NAM, Andrade JS Jr, Herrmann HJ (2010) Tactical voting inplurality elections. PLoS One 5:e12446

Barabáshi A-L, Albert R (1999) Emergence of scaling in random networks. Science 286:509–512

Bianconi G, Barabáshi A-L (2001a) Competition and multiscaling in evolving networks. Eur Phys Lett 54:436–442

Bianconi G, Barabáshi A-L (2001b) Bose-Eisnstein condensation in complex networks. Phys Rev Lett 86:5632–5635

Bikhchandani S, Hirshleifer D, Welch I (1992) A theory of fads, fashion, custom, and cultural changes as information cascades. J Polit Econ 100:992–1026

Cont R, Bouchaud J (2000) Herd behavior and aggregate fluctuations in financial markets. Macroecon Dyn 4:170–196

Couzin ID, Krause J, James R, Ruxton GR, Franks NR (2002) Collective memory and spatial sorting in animal groups. J Theor Biol 218:1–11

Curty P, Marsili M (2006) Phase coexistence in a forecasting game. JSTAT P03013

Eguíluz V, Zimmermann M (2000) Transmission of information and herd behavior: an application to financila markets. Phys Rev Lett 85:5659–5662

Galam G (1990) Social paradoxes of majority rule voting and renormalization group. Stat Phys 61:943–951

Hisakado M, Mori S (2010) Phase transition and information cascade in a voting model. J Phys A 43:315207

Hisakado M, Mori S (2011) Digital herders and phase transition in a voting model. J Phys A 22:275204

Hisakado M, Mori S (2015) Information cascade, Kirman's ant colony model and Ising model. Physica A 417:63–75

Hisakado M, Mori S (2016) Phase transiton of information cascade on network. Physica A 450:570–584

Keynes JM (1936) General theory of employment interest and money. Palgrave Macmillan, London

Milgram S, Bickman L, Berkowitz L (1969) note on the drawing power of crowds of different size. J Per Soc Psyco 13:79–82

Mori S, Hisakado M, Takahashi T (2012) Phase transition to two-peaks phase in an information cascade voting experiment. Phys Rev E 86:026109–026118

Partridge BL (1982) The structure and function of fish schools. Sci Am 245:90–99

Stauffer D (2002) Sociophysics: the Sznajd model and its applications. Comput Phys Commun 146(1):93–98

Tarde G (1890) Les lois de l'imitation. Felix Alcan, Paris

Watts DJ, Dodds PS (2007) Influentials, networks, and public opinion formation. J Consum Res 34:441–458

The Pitman-Yor Process and Choice Behavior

Masato Hisakado and Shintaro Mori

1 Introduction

In physics, equilibrium states are comparatively well understood, whereas non-equilibrium states continue to attract much attention (Privman 1997; Hinrichsen 2000; Mantegna and Stanley 2008). The latter states pose several interesting problems, and clarifying and classifying the nature of non-equilibrium stationary states continues to be a central research theme (Sasamoto and Spohn 2010). In other disciplines, the non-equilibrium stationary state is referred to. as the equilibrium state. The ecology literature highlights that the equilibrium state in a zero-sum model, in which the total number of individuals is constant, is an important process (Hubbell 2001; Hisakado et al. 2018). The economics literature discusses the equilibrium state in which companies survive competitive conditions (Aoki 2002; Fujiwara et al. 2004).

The Ewens sampling formula is a one-parameter probability distribution on the set of all partitions of an integer (Ewens 1990). The Pitman sampling formula is a two-parameter extension of the Ewens sampling formula (Pitman 2006). The Pitman-Yor process (Pitman and Yor 1997) and a generalized Pólya urn (Yamato and Shibuya 2001) are the non-equilibrium stochastic processes that derive the Pitman sampling formula (Pitman 2006). These processes permit new entries of individuals and an increasing number of them. In a similar non-equilibrium process in which the number of species or vertices increases, a power-law distribution can be obtained (Yule 1925; Simon 1955; Barabashi and Albert 1999).

M. Hisakado (✉)
Nomura Holdings, Inc., Chiyoda-ku, Tokyo, Japan

S. Mori
Faculty of Science and Technology, Department of Mathematics and Physics,
Hirosaki University, Hirosaki, Aomori, Japan

A.-H. Sato (ed.), *Applications of Data-Centric Science to Social Design*,
Agent-Based Social Systems 14, https://doi.org/10.1007/978-981-10-7194-2_8

We introduced a sequential voting model in previous studies (Hisakado and Mori 2015). At each time step t, one voter chooses one of two candidates. In addition, the t-th voter can see the previous r votes and, thus, is given access to public perception. When the voters vote for a candidate with a probability that is proportional to the previous referable votes and there are two candidates, the model can be considered as Kirman's ant colony model (Kirman 1993). In these previous studies, a beta-binomial distribution was derived as the equilibrium distribution of the r referable votes in the stationary state of the process (Hisakado et al. 2006). If we assume that voters can refer to all votes, the process becomes a non-equilibrium process. The equilibrium distribution and the probability distribution in the non-equilibrium process are the same (Hisakado and Mori 2015).

The response function is important for opinion dynamics, and decision-making depends on social influence. In this study, we consider the case of analog herders who vote for a candidate with a probability that is proportional to the referable votes. We refer to the response function in this case as an analog type.

On the other hand, threshold rules have been used to influence response functions in a variety of relevant theoretical scenarios (Hisakado and Mori 2011). This rule posits that individuals will choose one of two choices only when a sufficient number of other individuals have adopted that choice. We refer to such individuals as digital herders. From our experiments, we observe that people's individual behavior falls between that of digital herders and that of analog herders. In this study, we show that people behave as analog herders when posting to a bulletin board system.

We extend Kirman's ant colony model when the number of candidates is greater than two and not fixed (Kirman 1993; Hisakado and Mori 2015). The model is a finite reference version of the Pitman-Yor process and the generalized Pólya urn model (Pitman and Yor 1997; Yamato and Shibuya 2001). We derive the Pitman sampling formula as an equilibrium distribution. As a comprehensive example of the model, we analyze time series data for posts on 2ch.net. In the former case, votes and candidates in the voting model correspond to posts on bulletin boards and the bulletin boards' threads. When r is small, the posting process is described by the voting model.

The remainder of this paper is organized as follows. In Sect. 2 we introduce a voting model, and we derive the Pitman sampling formula as an equilibrium distribution of votes. In Sect. 3 we present the characteristics of several parameters. In Sect. 4 we study time series data for posts on 2ch.net using the voting model, and we conclude in the last section.

2 Model

We examine choice behavior using a voting model with candidates $C_1, C_2, \cdots$. At time t, candidate C_j has $c_j(t)$ votes. At each time step, a voter votes for one candidate, and the voting is sequential. Thus, at time t, the tth voter votes, after which the total number of votes is t. Voters are allowed to see the r previous votes for each candidate, where r is a constant, and, thus, voters are aware of public perception. The candidates are allowed to both enter and exit. The voter votes for the new candidate C_i with probability $(\theta + K_r\alpha)/(\theta + r)$, where r is the number of referred votes and K_r is the number of candidates who have more than one vote in the last r votes. α and θ are parameters. If a candidate does not have more than one vote in the last r votes, he/she exits. i in C_i is the number of candidates who have appeared in the past plus one. The number of candidates at $t = 1$ is one, and there is only one candidate C_1.

In terms of the Chinese restaurant process or Hoppe's urn process, we describe the voting process as follows (Pitman 2006; Yamato and Shibuya 2001). At first, there is an urn with θ black balls in it. In each step, one ball is drawn from the urn, and two balls are placed back into the urn. In the first turn, a black ball is drawn, and a ball of color 1 and the black ball are placed back into the urn. In subsequent turns, if the drawn ball is black, a ball of another color that has not appeared in the past and the black ball are returned to the urn, and if the drawn ball is not black, the ball is duplicated, and the two balls are placed back into the urn. The difference between this voting model and the Chinese restaurant process is that the voter refers to only the recently added r balls and the black balls.

We illustrate the parameter space in Fig. 1. θ is the parameter that controls the overall probability of adding a new candidate, and α is the parameter that adjusts the entry probability of new candidates according to the number of candidates $K_r = K$. When $\alpha < 0$, the constraint $\theta = -\alpha K$ exists.

We consider the case in which voters are analog herders. If $c_j(t) \geq 1$, the transition is

$$c_j(t) = k \rightarrow k + 1 : P_{j,k,t:l,t-r} = \frac{-\alpha + (k - l)}{\theta + r}, \tag{1}$$

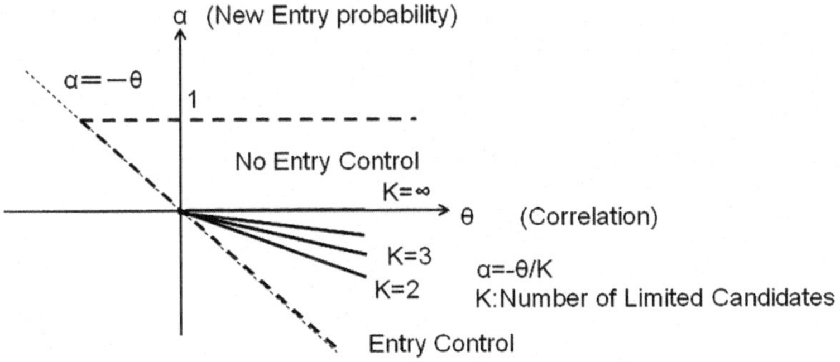

Fig. 1 Parameter space α and θ. α is the parameter that adjusts the entry probability of new candidates as per the number of candidates, K_r. θ is the parameter that controls the overall probability of adding a new candidate. The dotted line is not included in the parameter space. When $\alpha < 0$, the constraint $\theta = \alpha K$ exists, and K_r cannot exceed K

where $P_{j,k,t:l,t-r}$s are the probabilities of the process. The number of votes for C_j at $(t-r)$ is $c_j(t-r)=l$. Hence, if $(k-l)=0$, candidate C_j exits the system.

The process of a new candidate C_j entering is

$$c_j(t)=0\rightarrow 1: P_{j,k,t:l,t-r}=\frac{K_r\alpha+\theta}{\theta+r}, \tag{2}$$

where the number of candidates who have more than 0 votes is K_r.

When $\alpha\geq 0$, from (1) and (2), the constraints $\theta+\alpha>0$ and $1>\alpha\geq 0$ exist. (See Fig. 1.) There is no upper limit on the number of candidates. When $\alpha>0$, the probability of a new entry increases with an increase in K_r. When $\alpha=0$, the probability of a new entry is constant. When $\alpha<0$, from (2), the constraint $\alpha K+\theta=0$ exists. The probability of a new entry decreases with an increase in K_r, and K is the upper limit of K_r. The number of candidates with more than one vote does not exceed K. This model is similar to the model that does not allow candidate entry, which is further discussed in Appendix A.

The distribution of $c_j(t)$ as the partition of integer t follows the Pitman sampling formula in the generalized Pólya urn model (Pitman 2006; Yamato and Shibuya 2001). This is a non-equilibrium process, and the number of votes increases. We focus not on the snapshot $c_j(t)$ but on the time series of state $c_j(t)-c_j(t-r)$. This is an equilibrium process, and the number of total votes is constant.

We consider a hopping rate among $(r+1)$ states $\hat{k}_j=k-l$, $\hat{k}_j=0,1,\cdots,r$, and, here, we focus on the state. At each t, the vote at time $(t-r)$ is deleted, and a new one is obtained. $\hat{k}_j$ is the number of votes that candidate C_j obtained in the previous r votes.

First, we consider the case $\hat{k}_j>1$. The transition is

$$\begin{aligned}
\hat{k}_j\rightarrow\hat{k}_j+1: P_{\hat{k}_j,\hat{k}_j+1,t}&=\frac{r-\hat{k}_j}{r}\frac{-\alpha+\hat{k}_j}{\theta+r-1},\\
\hat{k}_j\rightarrow\hat{k}_j-1: P_{\hat{k}_j,\hat{k}_j-1,t}&=\frac{\hat{k}_j}{r}\frac{(\theta+\alpha)+(r-\hat{k}_j-1)}{\theta+r-1},\\
\hat{k}_j\rightarrow\hat{k}_j: P_{\hat{k}_j,\hat{k}_j,t}&=1-P_{\hat{k},\hat{k}-1,t}-P_{\hat{k},\hat{k}+1,t}.
\end{aligned}\tag{3}$$

$P_{\hat{k}_j,\hat{k}_j\pm 1,t}$ and $P_{\hat{k}_j,\hat{k}_j,t}$ are the probabilities of the process. $P_{\hat{k}_j,\hat{k}_j\pm 1,t}$ is the product of the probabilities of exit and entry.

We consider hopping from candidate C_i to C_j.

$$\begin{aligned}
\hat{k}_i\rightarrow\hat{k}_i-1,\hat{k}_j\rightarrow\hat{k}_j+1: P_{\hat{k}_i\rightarrow\hat{k}_i-1,\hat{k}_j\rightarrow\hat{k}_j+1,t}&=\frac{\hat{k}_i}{r}\frac{-\alpha+\hat{k}_j}{\theta+r-1},\\
\hat{k}_i-1\rightarrow\hat{k}_i,\hat{k}_j+1\rightarrow\hat{k}_j: P_{\hat{k}_i-1\rightarrow\hat{k}_i,\hat{k}_j+1\rightarrow\hat{k}_j,t}&=\frac{\hat{k}_j+1}{r}\frac{-\alpha+\hat{k}_i-1}{\theta+r-1}.
\end{aligned}\tag{4}$$

Here, we define $\mu_r(\hat{k}, t)$ as the distribution function of state $\hat{k}$ at time t. The number of all states is $(r+1)$. Using the fact that the process is reversible, in the equilibrium, we have

$$\frac{\mu_r(\hat{k}_i, \hat{k}_j, t)}{\mu_r(\hat{k}_i - 1, \hat{k}_j + 1, t)} = \frac{\hat{k}_j + 1}{\hat{k}_i} \frac{-\alpha + \hat{k}_i - 1}{-\alpha + \hat{k}_j}. \tag{5}$$

We separate indexes i and j and obtain

$$\begin{aligned}
\frac{\mu_r^i(\hat{k}_i, t)}{\mu_r^i(\hat{k}_i - 1, t)} &= \frac{-\alpha + \hat{k}_i - 1}{\hat{k}_i} c, \\
\frac{\mu_r^j(\hat{k}_j + 1, t)}{\mu_r^j(\hat{k}_j, t)} &= \frac{-\alpha + \hat{k}_j}{\hat{k}_j + 1} c,
\end{aligned} \tag{6}$$

where c is a constant.

In the equilibrium, the number of candidates with $\hat{k}_j > 0$ is K_r. We ignore candidates with $\hat{k}_j = 0$ and change the number of candidates and votes from $C_j, \hat{k}_j$ to $\tilde{C}_m, \tilde{k}_m$, where $m = 1, \cdots, K_r$, wherein $\tilde{k}_m > 0$.

We can write the distribution as

$$\mu_r(\tilde{\boldsymbol{a}}, \infty) = \binom{\theta + r - 1}{r}^{-1} \prod_{m=1}^{K_r} \frac{(1-\alpha)^{[\tilde{k}_m]}}{\tilde{k}_m!} \mu_r(\tilde{\boldsymbol{a}} = \underbrace{(1, 1, \cdots, 1)}_{K_r}, \infty), \tag{7}$$

where $\tilde{\boldsymbol{a}} = (\tilde{k}_1, \cdots, \tilde{k}_{K_r})$ and $x^{[n]} = x(x+1)\cdots(x+n-1)$, which is the Pochhammer symbol.

Given (6), we can obtain the equilibrium condition between $\tilde{\boldsymbol{a}} = \underbrace{(1, 1, \cdots, 1)}_{K_r} = \mathbf{1}$ and $\tilde{\boldsymbol{a}} = \underbrace{(0, 0, \cdots, 0)}_{K_r} = \mathbf{0}$:

$$\frac{\mu_r^i(\tilde{\boldsymbol{a}} = (\overbrace{1, \cdots, 1}^{n}, 0, \cdots, 0), \infty)}{\mu_r^i(\tilde{\boldsymbol{a}} = (\underbrace{1, \cdots, 1}_{n-1}, 0, \cdots, 0), \infty)} = \frac{\theta + (n-1)\alpha}{n} c, \tag{8}$$

where $n = 1, \cdots, K_r$. Hence, we can obtain

$$\mu_r(\tilde{\boldsymbol{a}} = \mathbf{1}, \infty) = \prod_{m=1}^{K_r} \frac{\theta + (m-1)\alpha}{m} \mu_r(\tilde{\boldsymbol{a}} = \mathbf{0}, \infty) = \frac{(\theta)^{[K_r:\alpha]}}{K_r!}, \tag{9}$$

where $x^{[n:\alpha]} = x(x+\alpha)\cdots(x+(n-1)\alpha)$. Therefore, we can write (7) as

$$\mu_r(\tilde{\boldsymbol{a}}, \infty) = \binom{\theta + r - 1}{r}^{-1} \frac{\theta^{[K_r:\alpha]}}{K_r!} \prod_{j=1}^{r} \left(\frac{(1-\alpha)^{[j-1]}}{j!} \right)^{a_j}, \tag{10}$$

where a_j is the number of candidates who have j votes. Therefore, the number of candidates $\sum_{j=1}^{r} a_j = K_r$ and that of votes $\sum_{j=1}^{r} j a_j = r$ are related. Hereafter, we use a partition vector $\hat{\boldsymbol{a}} = (a_1, \cdots, a_r)$.

We consider the partitions of the integer K_r. To normalize, we add the following combination term, $K_r!/a_1!\cdots a_r!$:

$$\mu_r(\hat{\boldsymbol{a}}, \infty) = \frac{r!\theta^{[K_r:\alpha]}}{\theta^{[r]}} \prod_{j=1}^{r} \left(\frac{(1-\alpha)^{[j-1]}}{j!} \right)^{a_j} \frac{1}{a_j!}. \tag{11}$$

Equation (11) is simply a Pitman sampling formula (Pitman 2006). In the limit $\alpha = 0$, we can obtain the Ewens sampling formula (Ewens 1990):

$$\mu_r(\hat{\boldsymbol{a}}, \infty) = \frac{r!}{\theta^{[r]}} \prod_{j=1}^{r} \left(\frac{\theta}{j} \right)^{a_j} \frac{1}{a_j!}. \tag{12}$$

The Ewens sampling formula for the equilibrium process is presented in Aoki (2002).

3 Four Regions in the Parameter Space

In this section, we characterize four regions in the parameter space. The parameter θ denotes the intensity of correlations, and α refers to the intensity of competition. In the upper plane, newcomers increase with a rise in α. In the lower plane, the probability of the minimum growing increases with a decrease in α.

We consider both an increase and a decrease in $\hat{k}_j$. The probability of a decrease in votes for candidate j is $\hat{k}_j/r$, and that of an increase in votes for candidate j is $(-\alpha + \hat{k}_j)/(\theta + r - 1)$, as shown in (1). We consider the condition in which the probability of an increase is larger than that of a decrease:

$$\alpha \leq -\frac{\hat{k}_j}{r}(\theta - 1). \tag{13}$$

In this region, the number of votes increases on average.

We divide the parameter space into four regions, $I \sim IV$, as shown in Fig. 2a. To clarify the regions, we define zone $U_{\hat{k}_j}$ as the region where the probability of an increase in votes is larger than that of a decrease. $U_{\hat{k}_j}$ is defined in (13). U_r is $\alpha \leq -(\theta - 1)$, and U_0 is $\alpha \leq 0$.

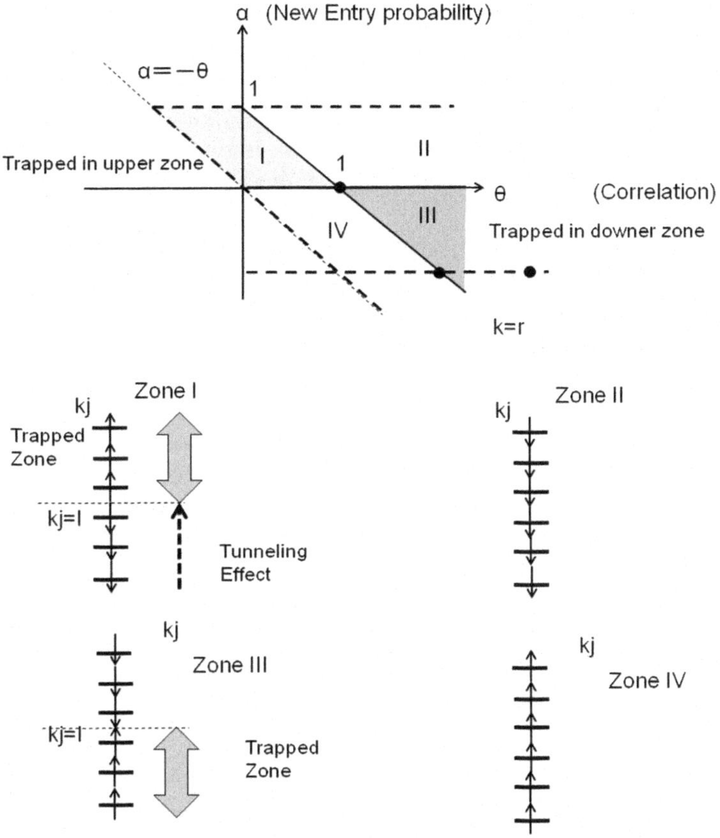

Fig. 2 Four regions in the parameter space and the average increase and decrease in each region

We define zone I, where $\alpha > 0$ and $\alpha < (1-\theta)$, as U_r. In zone I, $U_r \supset U_{r-1} \supset \cdots \supset U_0$. We consider the case in which the parameter set $\boldsymbol{x} = (\theta, \alpha)$ is $\boldsymbol{x} \in U_l$ and $\boldsymbol{x} \notin U_{l-1}$. In this case, $k_j \geq l$ is the increasing zone and $k_j < l$ the decreasing zone. If a candidate has more than l votes, he/she can increase the number of votes and maintain his/her position. On the other hand, it is difficult to increase votes if the candidate has less than l votes. We show the average trend in Fig. 2b. In this region, the leader in the trapped zone has an advantage. On the other hand, the competition intensifies for newcomers as α increases.

We define zone II as $\alpha > 0$ and $\alpha \geq (1-\theta)$. In zone II, $\boldsymbol{x} \notin U_0, \cdots, U_r$. Furthermore, it is difficult to increase the number of votes for every candidate and to be a stable leader. The zone becomes more competitively intense with an increase in α. In other words, it is possible to adjust the competitive intensity and protect newcomers by adjusting α, which denotes the number of newcomers.

In the plane in the lower half, there is a capacity limit and no newcomers. We define zone III as $\alpha < 0$ and $\alpha > (1-\theta)$. When $\alpha < 0$, $U_0 \supset U_1 \supset \cdots \supset U_r$. We consider the case in which the parameter set $\boldsymbol{x} = (\theta, \alpha)$ is $\boldsymbol{x} \in U_l$ and $\boldsymbol{x} \notin U_{l+1}$. In this case, $\hat{k}_j > l$ is the decreasing zone and $\hat{k}_j \leq l$ the increasing zone. It is easy to increase the number of votes to $\hat{k}_j = l$, but it is difficult to increase the number of votes above $\hat{k}_j = l + 1$. In this region, it is also difficult to be a stable leader.

We define zone IV as $\alpha < 0$ and $\alpha \leq (1 - \theta)$. In this zone, $\boldsymbol{x} \in U_0, \cdots, U_r$. It is easy to increase the votes for each candidate. In addition, this zone is competitive when the number of members is fixed.

Next, we consider the Ewens sampling formula on the θ axis. When $\alpha = 0$ and $\theta = 1$, the probabilities of an increase and decrease both become $\hat{k}_j/r$. For any $\hat{k}_j$, the probabilities of an increase and decrease are equal. Thus, the correlation becomes $\rho = 1/2$ (see Appendix A), and there is a uniform random permutation. The probability of an increase or a decrease is proportional to the number of partitions.

When $\alpha = 0$ and $\theta < 1$ within the boundaries of zone IV, if candidates can enter, the number of votes easily increases. In this zone, the correlation is high. When $\alpha = 0$ and $\theta > 1$ within the boundaries of zone II, it is difficult to increase the number of votes. Here, there is a low correlation. In summary, there are numerous candidates who have few votes.

4 Data Analysis of a Bulletin Board System

In this section, we examine the data of posts to a bulletin board system (BBS), 2ch.net. 2ch.net is the largest BBS in Japan and covers a wide range of topics. Each bulletin board is separated by a field unit or a category, such as, for example, news, food and culture, and net relations. Each category is further divided into genres, or boards, and each board contains numerous threads, which are segregated by topics that belong to the board. Writing and viewing boards is done on a thread. There are about 900 boards on 2ch.net. It is possible to make anonymous posts on all threads.

We study the time series of posts on the following ten boards: business news, East Asia news, live news, music news, breaking news, digital camera, game, entertainment, international affairs, and press. We label these boards as the No. 1, No. 2, $\cdots$ No. 10 boards, respectively. The first five boards fall under the news category. Each board has several hundred threads, and managers maintain the number of threads by removing old ones and replacing them with new threads. The duration of a post on a thread is set to 5 days for the No. 4 and No. 5 boards. As a result, the lifetime of a thread is generally about a few days. One cannot post more than 1,000 posts to a thread. Threads that no longer allow posts are deleted from the thread lists of the boards, and managers prepare a new sequential thread using the same thread title. The lifetime of a thread can therefore be longer than the abovementioned duration, as the postable duration rule applies to descendant threads with a new start date. We identify sequential descendant threads from a common ancestor thread as one thread. Table 1 summarizes the statistics of the threads of the ten boards.

Table 1 Statistics of 2ch.net post data. The observation period, the total number of threads N, the average number of posts per thread T/N, and its standard deviation (S.D.) are presented in the third, fourth, fifth, and sixth columns, respectively. w_{Max} in the seventh column lists the maximum number of posts on a thread. The numerical value in the eighth column indicates the average lifetime of a thread [in days]. The lifetime is defined as the difference between the last and first post date. s_H is in the last column and indicates the time horizon, which is defined in Eq. (14)

No.	Board Name	Obs. Period	N	T/N	S.D.	w_{Max}	Lifetime	s_H[sec]
1	Business News	Aug. 10, 2009–Dec. 31, 2009	8,248	140	290	7,707	7.5	260.0
2	East Asia News	Mar. 8, 2009–Aug.5, 2009	8,225	388	1,022	27,966	7.5	205.4
3	Live News	Mar. 8, 2009–Aug. 5, 2009	15,307	53	333	30,443	2.0	95.1
4	Music News	Mar. 8, 2009–Aug. 5, 2009	23,000	332	1,123	78,388	2.8	140.2
5	Breaking News	Mar. 8, 2009–Dec.31, 2009	33,677	658	1,497	113,220	2.9	94.5
6	Digital Camera	Aug. 10, 2009–Dec. 31, 2009	835	527	1,530	33,494	251.4	
7	Game	Mar. 8, 2009–Aug. 10, 2009	1,371	241	286	2,043	132.6	
8	Entertainment	Aug. 10, 2009–Dec. 31, 2009	1,464	134	1188	30999	35.4	
9	Int. Affairs	Aug. 10, 2009–Dec. 31, 2009	688	233	241	1,000	1,123.8	
10	Press	Aug. 10, 2009–Dec. 31, 2009	1,011	235	296	2,182	358.5	

The total number of posts on each board is about 0.2–22 million. Each thread has an average of several hundred posts, and the standard deviation is large. The maximum number of posts w_{Max} is 50–100 times larger than the average. The average lifetime of a thread on news boards is several days, which is derived from the strict rule defining the period within which a post can be made on threads. There is no strict rule for the remaining five boards, and the average lifetimes are considerably longer than those of the news boards.

We label threads by $n \in \{1, \cdots, N\}$, and N is the total number of threads that appear on the board. We describe the tth post to the board at $s[sec]$ by the thread number n and s as $(n(t), s(t)), t = 1, \cdots, T$. We measure the post time s by setting the time of the first post time on the board as zero seconds. We present the scatter plot of the time series post data $(n(t), s(t)), t = 1, \cdots, T$ in the (n, s) plane of the No. 5 and No. 8 boards in Fig. 3. Because the threads have a strict finite lifetime of 5 days on the No. 5 board, the plot shows a narrow strip pattern. Some threads have a longer lifetime because they have a long family tree from ancestors to descendants. The No. 8 board shows a wide strip pattern, which indicates that the number of threads is considerably large.

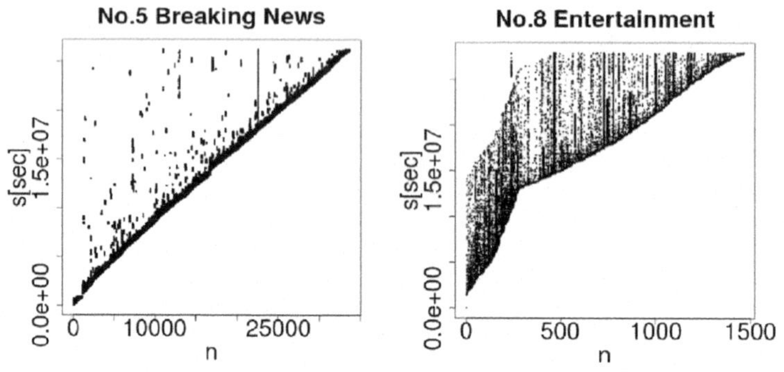

Fig. 3 Scatterplot of post data $(n(t), s(t))$ for the No. 5 and No. 8 boards. Each dot corresponds to a post

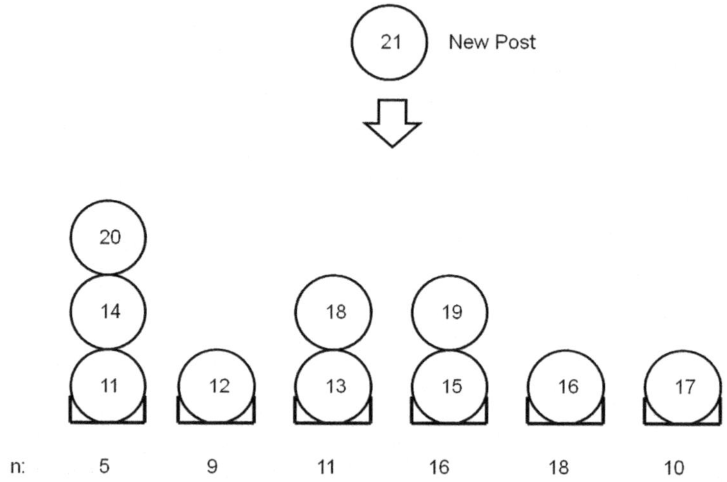

Fig. 4 Posts, threads, and a board. Posts are shown as balls and are labeled as t. There are $K_r = 6$ threads with a non-zero number of posts. n shows the threads numbers. The "21st post? refers to the previous 10 posts, $t = 11, 12, \cdots .10$. The thread numbers are $n = 5, 9, 11, 16, 18, 10$ and appear $(k_5, k_9, k_{11}, k_{16}, k_{18}, k_{10}) = (3, 1, 2, 2, 1, 1)$ times, respectively. The multiplicities $\boldsymbol{a}$ are $a_1 = 3, a_2 = 2$ and $a_3 = 1$. $a_1 + a_2 + a_3 = K_r = 6$ and $1 \cdot a_1 + 2 \cdots a_2 + 3 \cdot a_3 = 10$ hold. The probability of a post on a thread with two posts is $2 \cdot \frac{2-\alpha}{\theta+10}$. The probability of a thread not appearing in ten posts is $\frac{\theta+6\cdot\alpha}{\theta+10}$

4.1 Correlation Function $C(\tau)$ for Equilibrium r

We identify threads and posts as candidates and votes in the voting model in Fig. 4. As previously shown, when voting occurs with reference to the previous r votes, the stationary distribution of the r consecutive previous votes obeys the Pitman sampling formula in (11). In this case, the correlation between $n(t)$ and $n(t-\tau)$ for the voting lag τ does not decay for $\tau < r$, as the distribution is stationary in

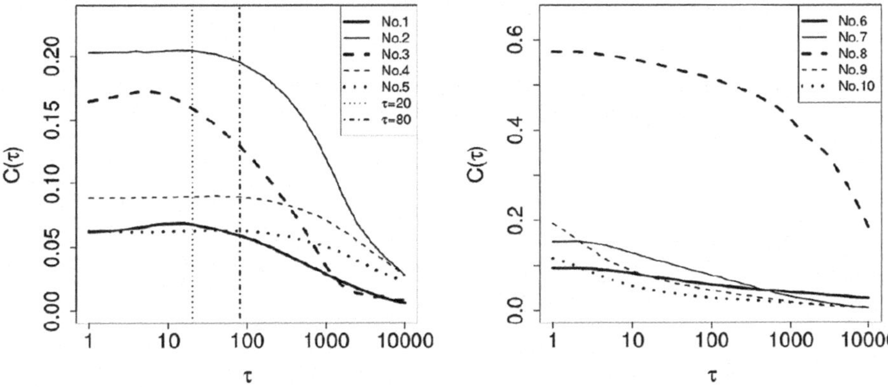

Fig. 5 Plot of $C(\tau)$ vs. τ. The left (right) panel shows the results for the first (remaining) five boards. τ on the x-axis indicates the voting lag, and $C(\tau)$ denotes the autocorrelation function between $n(t)$ and $n(t+\tau)$

r consecutive votes. In the range $\tau > r$, $C(\tau)$ dumps, so if a post on the board is described by this voting model with reference r, $C(\tau)$ should demonstrate this feature. We adopt the expectation value of the coincidence of $n(t)$ and $n(t-\tau)$ as the correlation between $n(t)$ and $n(t-\tau)$,

$$C(\tau) \equiv \mathrm{E}(\delta_{n(t),n(t-\tau)}).$$

We assume that $C(\tau)$ does not depend on t, and we estimate it using time series data $\{n(t)\}, t = 1, \cdots, T$ as

$$C(\tau) = \frac{1}{T-\tau}\sum_{t=1}^{T-\tau} \delta_{n(t),n(t+\tau)}.$$

Figure 5 illustrates the semilogarithmic plot $C(\tau)$ vs. τ. The properties of $C(\tau)$ of a voting model with finite r are summarized in Mori and Hisakado (2015). We see a plateau structure in which $C(\tau)$ does not decrease with τ for the first five news boards in the left panel. For the No. 4 and No. 5 boards, $C(\tau)$ is almost constant for $\tau \leq \tau_c \simeq 80$. As for the other boards, $C(\tau)$ decreases for $\tau \geq \tau_c = 20$ for the No. 1 and No. 2 boards and for smaller values of $\tau_c \sim 5$ for the No. 3 board. In these news boards, $C(\tau)$ is almost constant among posts within τ_c. On the other hand, $C(\tau)$ decreases with τ for the latter boards in the right panel, with the exception of the No. 8 board. As for the No. 8 board, $C(\tau)$ is large for large τ and gradually decreases, indicating that the board has special features.

We apply the voting model to the posts on 2ch.net in the boards for the news category with $r < \tau_c$. We adopt $r = 80$ for the No. 4 and No. 5 boards, $r = 20$ for the No. 1 and No. 2 boards, and $r = 5$ for the No. 3 board. To interpret τ_c, we highlight the response times of board users. We believe that a user needs several minutes to respond to posts. Next, the post should be random for a short time interval, and the probability of a post on a thread is roughly estimated as the post ratio in the previous posts. In the last column of Table 1, we indicate the time horizon $s_H[sec]$ for τ_c, which is defined as the mean duration between posts multiplied by τ_c, as

$$s_H \equiv \frac{s(T) - s(1)}{T - 1} \cdot \tau_c. \tag{14}$$

s_H is about 1.5–4 min, which is possibly the requisite time duration to respond to posts.

4.2 Estimation of the Parameters θ and α

We use time series data $\{n(t)\}, t = 1, \cdots, T, n \in \{1, \ldots, N\}$ and estimate the model parameters θ and α using the maximum likelihood principle. In the model, the probability of a post on a thread that appears in the past r posts $\hat{k}$ times is defined as in (1). The probability of a post to $a_{\hat{k}}$ threads with $\hat{k}$ is

$$P_{Existing}(\hat{k}) = a_{\hat{k}} \cdot \frac{\hat{k} - \alpha}{\theta + r}. \tag{15}$$

The probability of a post on a new thread that does not appear in the past r posts depends on the number of threads K_r in the past r posts and is defined in (2):

$$P_{New}(K_r) = \frac{K_r\alpha + \theta}{\theta + r}. \tag{16}$$

For $t \in [1\times 10^4, T-r]$, we choose $t_n, n = 1, \cdots S$ randomly and study the following $r+1$ sequence, $n(t_n), n(t_n+1), \cdots, n(t_n+r-1)$. We estimate the number of threads K_r and the number of threads with $\hat{k}$ posts $a_{\hat{k}}$ in the past r posts. $\sum_{\hat{k}} a_{\hat{k}} = K_r$ holds. If thread $n(t_n + r)$ does not exist in the K_r threads, the likelihood is $P_{New}(K_r)$. If the thread exists and thread $n(t_n + r)$ appears $\hat{k}$ times, the likelihood is $P_{Existing}(\hat{k})$. The likelihood of S sample is then estimated by the products of these likelihoods for all $s = 1, \cdots, S$. We adopt $S = 2 \times 10^5$. In addition, we fit the parameters using the maximum likelihood principle for the distribution of the partitions of $a_{\hat{k}}$ with the Pitman sampling formula in (11).

The estimated values for the parameters are summarized in Table 2. We adopted $r = 20$ for the No. 1 and No. 2 boards, $r = 5$ for the No. 3 board, and $r = 80$ for the No. 4 and No. 5 boards by the correlation analysis. The standard errors are estimated using the square root of the negative eigenvalue for the Hessian of the log likelihood. For $r = 80$, we only show the results by fitting with probabilistic rules.

Table 2 Fitting results of θ and α for probabilistic rules in (15) and (16). We use the maximum likelihood principle, and the sample number S is 2×10^5. We show the estimates for the No. 1 and No. 2 boards with $r = 20$, for the No. 3 board with $r = 5$, and for the No. 4 and No. 5 boards with $r = 80$ in the third and fourth columns. For $r = 5$ and $r = 20$, we show the goodness-of-fit results using the Pitman sampling formula in (11) in the fifth and sixth columns. We adopt the same samples for the two fittings. The standard error (S.E.) in the last digit of the estimate is provided in parentheses

		Fit with probabilistic rules		Fit with Pitman's distribution	
No.	r	θ(S.E.)	α(S.E.)	θ(S.E.)	α(S.E.)
1	20	8.8(2)	0.37(1)	8.7(0)	0.390(2)
2	20	2.2(1)	0.42(1)	2.0(0)	0.418(2)
3	5	1.7(0)	0.560(4)	1.3(0)	0.623(3)
4	80	10.0(3)	0.35(1)	NA	NA
5	80	11.9(4)	0.28(1)	NA	NA

To verify the probabilistic rules, we directly estimate $P_{Exissting}(\hat{k})$ and $P_{New}(K_r)$. We calculate the number of threads with post times $\hat{k}$ for the past r posts and denote it as $N(\hat{k})$. In addition, we count the number of times a post is made on an exiting thread with posts $\hat{k}$ and denote it as $N_{post}(\hat{k})$. The estimator for $P_{Existing}(\hat{k})$ is

$$\hat{P}_{Existing}(\hat{k}) = \frac{N_{Post}(\hat{k})}{N(\hat{k})}. \tag{17}$$

Likewise, we count the number of threads K_r and the number of times a post is made on a new thread when the number of threads is K_r. We denote them as $N_{K_r}(K_r)$ and $N_{New}(K_r)$, respectively. The estimator for $P_{New}(K_r)$ is then denoted as

$$\hat{P}_{New}(K_r) = \frac{N_{New}(K_r)}{N_{K_r}(K_r)}. \tag{18}$$

Figure 6 presents the estimates for $\hat{P}_{Existing}(\hat{k})$ and $\hat{P}_{new}(K_r)$. We also plot $P_{Existing}(\hat{k})$ and $P_{New}(K_r)$ in (15) and (16) with fitted values for θ and α in Table 2. The estimated results for the maximum likelihood fit well with the results from the estimators $\hat{P}_{Existing}(\hat{k})$ and $\hat{P}_{New}(K_r)$. The parameters fall in zone II, which was introduced in the previous section. In this zone, it is difficult for a leader to appear.

4.3 Distribution of K_r and $\hat{k}$

We compare (11) and the probability mass function for K_r using the fitted parameters in Table 2 and those of an empirical distribution. The probability mass function $P_r(K_r)$ for the number of candidates K_r with reference r is given as

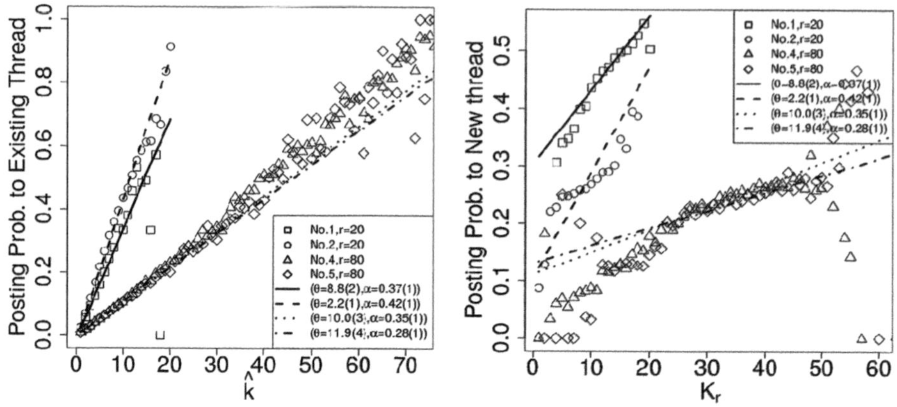

Fig. 6 Plots of $P_{Existing}(\hat{k})$ vs. $\hat{k}$ and $P_{New}(K_r)$ vs. K_r. The symbols denote the estimated results using $\hat{P}_{Existing}(\hat{k})$ in (17) and $\hat{P}_{K_r}(K_r)$ in (18). The lines denote the plots of (15) and (16) with fitted parameters for θ and α in Table 2

$$P_r(K_r) = \frac{\theta^{[r:\alpha]}}{\theta^{[r]}} c(r, K_r, \alpha)\alpha^{-K_r}, \tag{19}$$

where $c(r, K_r, \alpha)$ is the generalized Stirling number or the C-numbers (Pitman 2006). As for the probability mass function for the post times $\hat{k}$, we calculate the number of posts $\hat{k}_{1st} \geq \hat{k}_{2nd} \geq \hat{k}_{3rd}$ for the most popular three threads in the past r posts in addition to all post times $\hat{k}$ for all threads. We plot the results in Fig. 7.

As we can see, the fitting results are good. The distributions of $\hat{k}_{1st}$, $\hat{k}_{2nd}$, $\hat{k}_{3rd}$, $\hat{k}$, and K_r are well described by the Pitman sampling formula (11) and $P_r(K_r)$ with fitted parameters for θ, α in Table 2.

4.4 Distribution of Total Votes $c_n(T)$

In this subsection, we discuss the distribution of total votes $c_n(T)$ for thread n, where $c_n(t) = \sum_{t'=1}^{t} \delta_{n(t'),n}$. Figure 8 shows the semilogarithmic plot of the cumulative distribution $P(k) \equiv \mathrm{P}(c_n(T) \geq k)$ vs. k. The left (right) panel depicts the results for the first (remaining) five boards. The dotted line denotes the cumulative distribution of the log-normal distribution with the same mean and variance. As is clearly shown, the results show good fits. In the previous subsection, we confirm that the distribution of posts on the four boards obey the equilibrium Pitman sampling formula. The difference between the equilibrium and non-equilibrium suggests that the posting process for a large r is not described by the voting process.

In the remaining boards, we confirm that the distribution of the number of posts on the No. 8 board follows the power-law distribution and the power-law index is 1.65, $P(k) \sim k^{-(1.65-1)}$. We have seen that the board has a long memory in Fig. 5.

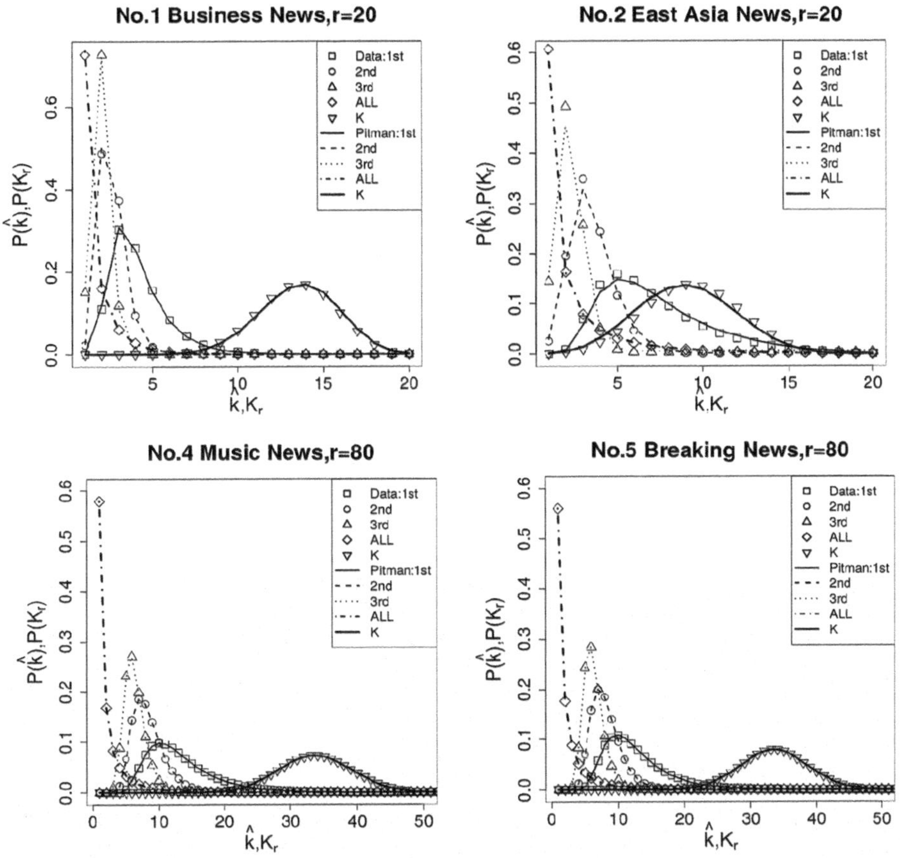

Fig. 7 Plots of the distribution of post times $\hat{k}$ and number of threads K_r In the past r posts for data with symbols and for (11) and (19). We sort $\hat{k}$ in descending order and select the largest three values $\hat{k}_1 \geq \hat{k}_2 \geq \hat{k}_3$. The symbols ∘, and △ denote the empirical distributions of $\hat{k}_1$, $\hat{k}_2$, and $\hat{k}_3$, respectively. ⋄ and ∇ show the distribution of all $\hat{k}$ and K_r. The lines indicate the plots of $P_r(K_r)$ in (19) and the distributions of the ordered $\hat{k}_j$ and $\hat{k}$, which are calculated using the Pitman sampling formula in eq. (11). We adopt $r = 20(80)$ for the No. 1 and No. 2 (No. 4 and No. 5) boards. The parameters for (11) and (19) are presented in Table 2. We adopt the second set for $r = 20$ and the first set for $r = 80$

Furthermore, we can confirm that the probability of a post is proportional to the number of posts and that of a new thread is proportional to the number of threads for a large r in Fig. 9. As r increases, the latter dependence disappears, and the process is described by the Yule process (Yule 1925). As the power-law exponent is less than two, the fitness model for evolving networks might be a better candidate to describe the posting process (Bianconi and Barabási 2001; Hisakado and Mori 2016).

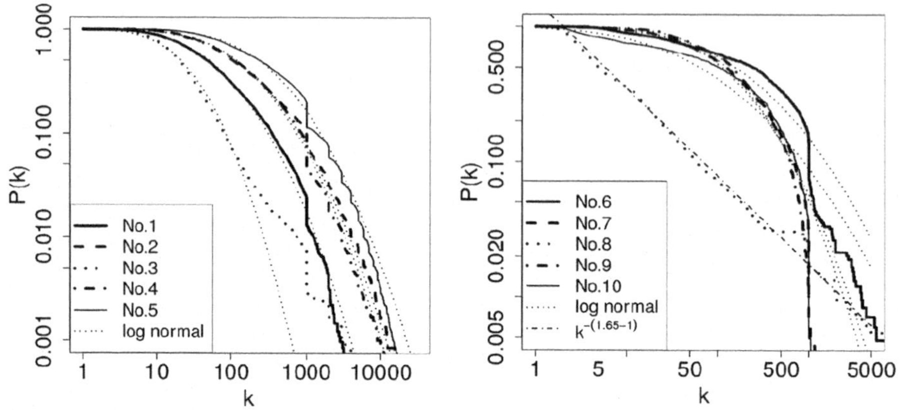

Fig. 8 Plot of $P(k) \equiv \mathrm{P}(c_n(T) \geq k)$ vs. k. The left (right) panel presents the results for the first (remaining) five boards. The means and standard deviations of c_j for the first five boards are (3.93, 1.41), (4.79, 1.48), (3.03, 1.10), (4.76, 1.45), and (5.47, 1.47)

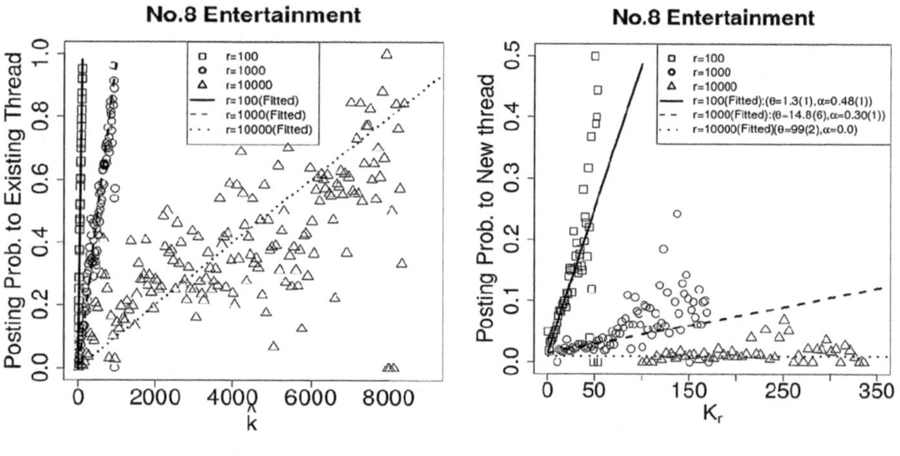

Fig. 9 Plot of $P_{Existing}(\hat{k})$ vs. $P_{New}(K_r)$ for the No. 8 board. The symbols show the estimated results using $\hat{P}_{Existing}(\hat{k})$ in (17) and $\hat{P}_{K_r}(K_r)$ in (18). The lines denote the plots of (15) and (16)

5 Concluding Remarks

In this chapter, we discuss choice behavior using a voting model comprising voters and candidates. Voters vote for a candidate with a probability that is proportional to the ratio of previous votes, which is visible to the voters. In addition, voters can obtain information from a finite number r of the most recent previous voters.

In the large t limit, the system is equilibrated, and the partition of r votes follows the Pitman sampling formula. Kirman's ant colony model is a special case that corresponds to the number of states $K = 2$. The equilibrium probability distribution and the non-equilibrium probability distribution for $t = r$ are the same. We propose using this voting model for the posting process of a BBS, 2ch.net, where users can select one of many threads to make a post. We explore how this choice depends on the last r posts and the distribution of the last r posts across boards. We conclude that the posting data in the news category is described by the voting model. The equilibrium time or time horizon s_H is about 1.5–4 min. Up to this time horizon, the probability of posting on a thread is proportional to the ratio of posts on the thread, in other words analog herders.

Recently the monopoly of information is a hot topic. A few largest companies like GAFA gather most of data. In Fig. 2 we divided four regions, I, II, III, and IV, in the parameter space. Here we consider that the candidates are enterprises and the voters are users. The lower half is the membership region, high barriers to enter. The barriers to enter become lower and lower for the enterprises in the Internet era. On the other hand, for users the enterprises which gather larger data are useful platforms because of the convenience. It is called network externality. In Fig. 3 we can observe the winner got all in the zone I. The monopoly will be advanced here after.

Appendix A Fixed Number of Candidates Case

We model the voting of K candidates, $C_1 \cdots C_K$. At time t, candidate C_j has $c_j(t)$ votes. In this appendix, we consider the case in which the number of candidates K is fixed, that is, no new entry is allowed. In each time step, one voter votes for one candidate; the voting is sequential. Hence, at time t, the tth voter votes, after which the total number of votes is t. Voters are allowed to see r previous votes for each candidate and, thus, are aware of public perception. r is a constant number. We consider the case in which all voters vote for the candidate with a probability proportional to the previous votes ratio, which is visible to the voters.

The transition is

$$c_j(t) = k \rightarrow k+1 : P_{j,k,t:l,t-r} = \frac{\frac{q_j(1-\rho)}{\rho} + (k-l)}{\frac{1-\rho}{\rho} + r} = \frac{\beta_j + (k-l)}{\theta + r}, \tag{20}$$

where $c_j(t-r) = l$, ρ is the correlation coefficient and q_j is the initial constant of the jth candidate (Hisakado et al. 2006). ρ is the correlation of the beta-binomial model. The constraint $\sum_{j=1}^{K} q_j = 1$ exists. We define $\theta = (1-\rho)/\rho$ and $\beta_j = q_j(1-\rho)/\rho$. $P_{j,k,t:l,t-r}$ denotes the probabilities of the process. The voting ratio for C_j at $t-r$ is $c_j(t-r) = l$. We consider the case $\beta_j \geq 0$ from $P_{j,k,t:l,t-r} > 0$ and the constraint $\sum_j \beta_j = \theta$. When $\beta_j = \beta$, the constraint becomes $\beta K = \theta$.

We consider the hopping rate among $(r+1)$ states $\hat{k}_j = k - l$, $\hat{k}_j = 0, 1, \cdots, r$. In each step of t, the vote at time $(t-r)$ is deleted, and a new vote is obtained. $\hat{k}$ is the number of votes candidate C_j obtained in the latest r votes. In case $K = 2$, the model becomes Kirman's ant colony model (Kirman 1993). The dynamic evolution of the process is given by

$$\hat{k}_j \to \hat{k}_j + 1 : P_{\hat{k}_j, \hat{k}_j+1,t} = \frac{r - \hat{k}_j}{r} \frac{\beta_j + \hat{k}_j}{\theta + r - 1},$$

$$\hat{k}_j \to \hat{k}_j - 1 : P_{\hat{k}_j, \hat{k}_j-1,t} = \frac{\hat{k}_j}{r} \frac{(\theta - \beta_j) + (r - 1 - \hat{k}_j)}{\theta + r - 1},$$

$$\hat{k}_j \to \hat{k}_j : P_{\hat{k}_j, \hat{k}_j,t} = 1 - P_{\hat{k},\hat{k}-1,t} - P_{\hat{k},\hat{k}+1,t}.$$

$P_{\hat{k}_j,\hat{k}_j \pm 1,t}$ are the probabilities of the process and the products of exit votes and new entry votes.

We consider hopping from candidate C_i to C_j.

$$\hat{k}_i \to \hat{k}_i - 1, \hat{k}_j \to \hat{k}_j + 1 : P_{\hat{k}_i \to \hat{k}_i - 1, \hat{k}_j \to \hat{k}_j + 1,t} = \frac{\hat{k}_i}{r} \frac{\beta_j + \hat{k}_j}{\theta + r - 1},$$

$$\hat{k}_i - 1 \to \hat{k}_i, \hat{k}_j + 1 \to \hat{k}_j : P_{\hat{k}_i - 1 \to \hat{k}_i, \hat{k}_j + 1 \to \hat{k}_j,t} = \frac{\hat{k}_j + 1}{r} \frac{\beta_i + \hat{k}_i - 1}{\theta + r - 1}.$$

Here, we define $\mu_r(\hat{k}, t)$ as a distribution function of the state $\hat{k}$ at time t. The number of all states is $(r + 1)$. Given that the process is reversible, we have

$$\frac{\mu_r(\hat{k}_i, \hat{k}_j, t)}{\mu_r(\hat{k}_i - 1, \hat{k}_j + 1, t)} = \frac{\hat{k}_j + 1}{\hat{k}_i} \frac{\beta_i + \hat{k}_i - 1}{\beta_j + \hat{k}_j}. \tag{21}$$

We can separate indexes i and j and obtain

$$\frac{\mu_r^i(\hat{k}_i, t)}{\mu_r^i(\hat{k}_i - 1, t)} = \frac{\beta_i + \hat{k}_i - 1}{\hat{k}_i} c$$
$$\frac{\mu_r^j(\hat{k}_j + 1, t)}{\mu_r^j(\hat{k}_j, t)} = \frac{\beta_j + \hat{k}_j}{\hat{k}_j + 1} c, \tag{22}$$

where c is a constant. Using (22) sequentially, in the limit $t \to \infty$, we can obtain the equilibrium distribution, which can be written as

$$\mu_r(\boldsymbol{a}, \infty) = \binom{\theta + r - 1}{r}^{-1} \prod_{j=1}^{K} \binom{\beta_j + \hat{k}_j - 1}{\hat{k}_j}, \tag{23}$$

where $\boldsymbol{a} = (\hat{k}_1, \hat{k}_2, \cdots, \hat{k}_K)$. This distribution is written as

$$\mu_r(\boldsymbol{a}, \infty) = \frac{r!}{\theta^{[r]}} \prod_{i=1}^{K} \frac{\beta_i^{[\hat{k}_i]}}{\hat{k}_i!}, \tag{24}$$

where $x^{[n]} = x(x+1)\cdots(x+n-1)$. This is the Dirichlet-multinomial distribution.

Here, we set $\beta_j = \beta$. The relation $\beta = -\alpha$ exists, where α is the parameter used in the main text. We write (23) as

$$\mu_r(\hat{\boldsymbol{a}}, \infty) = \binom{\theta + r - 1}{r}^{-1} \prod_{j=1}^{r} \binom{\beta + j - 1}{j}^{a_j}, \tag{25}$$

where a_j is the number of candidates for whom j voters voted and $\hat{\boldsymbol{a}} = (a_1, \cdots, a_r)$. Hence, the relations $\sum_{i=1}^{r} a_i = K_r < K$ and $\sum_{i=1}^{r} i a_i = r$ exist. Here, we define K_r as the number of candidates who have more than one vote. $(K - K_r)$ candidates have no vote.

We consider the partitions of integer K_r. To normalize, we add the term of combination: $K!/a_1! \cdots a_r!(K - K_r)!$. We obtain

$$\begin{aligned}\mu_r(\hat{\boldsymbol{a}}, \infty) &= \frac{K!}{a_1! \cdots a_r!(K - K_r)!} \binom{\theta + r - 1}{r}^{-1} \prod_{j=1}^{r} \binom{\beta + j - 1}{j}^{a_j} \\ &= \frac{r! \theta^{[K_r:-\beta]}}{\theta^{[r]}} \prod_{j=1}^{r} (\frac{(1+\beta)^{[j-1]}}{j!})^{a_j} \frac{1}{a_j!},\end{aligned} \tag{26}$$

where $x^{[n:-\beta]} = x(x - \beta)\cdots(x - (n-1)\beta)$. We use the relation $\theta = K\beta$. Equation (26) is simply the Pitman sampling formula (Pitman 2006).

In the limit $\beta \to 0$ and $K \to \infty$, subject to a fixed $\theta = \beta K$, we can obtain the Ewens sampling formula. In this case, the sum of the probabilities that a candidate who has zero votes can obtain one vote is $\theta/(\theta + r)$. This case is the same as $\alpha = 0$ in Sect. 2.

Appendix B Partition Number and Pitman Distribution

A partition of positive integer is called integer partition problem. It is the number of the representations that positive integer n as the sum of the positive integer. Here we write the number $p(n)$. For example, 3 is presented as 3, 2+1, 1+1+1, and $p(3) = 3$.

The generating function of $p(n)$ is known as

$$\begin{aligned}\sum_{n=0}^{\infty} p(n)x^n &= 1 + x + 2x^2 + 3x^3 + 5x^4 + 7x^5 + 11x^6 + \ldots \\ &= \prod_{k=1}^{\infty} \left(\frac{1}{1 - x^k}\right) = (1 + x + x^2 \cdots)(1 + x^2 + x^4 + \cdots) \\ &\quad (1 + x^3 + \cdots).\end{aligned} \tag{27}$$

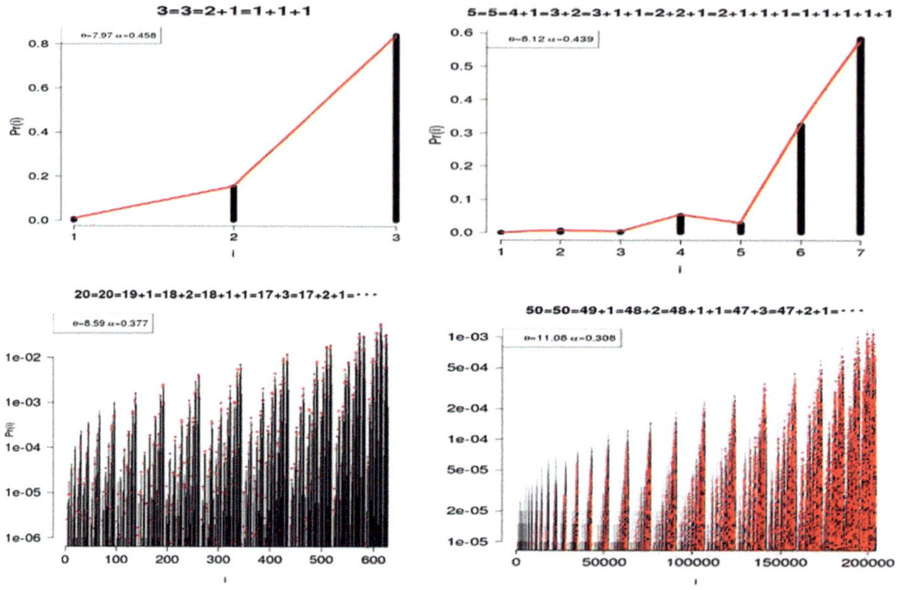

Fig. 10 The black bar is the Pitman distribution. The red line is the empirical data of 2ch. Here we show $n = 3, 5, 20, 50$ cases

We can confirm the combinatorics of n as the coefficients of x^n. The Pitman distribution represents the distribution for each representation. We show the Pitman distribution and the votes for 2ch in Fig. 10.

The red line is the business data which fits well to the Pitman distribution. The parameter is calibrated by the historical data, when n is less than the range of the auto correlation which we discussed in the Sect. 4.1. We can confirm 2ch data is represented by the Pitman distribution.

References

Aoki M (2002) Modeling aggregate behavior and fluctuations in economics. Cambridge University Press, London

Barabási A, Albert R (1999) Emergence of scaling in random networks. Science 286:509–512

Bianconi G, Barabási A-L (2001) Competition and multiscaling in evolving networks. Europhys Lett 54:436–442

Ewens WJ (1990) In: Lessard S (ed) Population genetics theory – the past and future. Kluwer Academic, London

Fujiwara Y, Guilmi CD, Aoyama H, Gallegati M, Souma W (2004) Do Pareto-Zipf and Bibrat laws hold true? An analysis with European firms. Physica A 335:197

Hinrichsen H (2000) Non-equilibrium critical phenomena and phase transitions into absorbing sates. Adv Phys 49:815–958

Hisakado M, Sano F, Mori S (2018) Pitman-Yor process and an empirical study of choice behavior. J Phys Soc Jpn 87(2):024002-2419
Hisakado M, Mori S (2011) Digital herders and phase transition in a voting model. J Phys A 44:275204
Hisakado M, Mori S (2015) Information cascade, Kirman's ant colony model and Ising model. Physica A 417:63–75
Hisakado M, Mori S (2016) Phase transition of information cascade on network. Physica A 450:570–584
Hisakado M, Kitsukawa K, Mori S (2006) Correlated binomial models and correlation structures. J Phys A 39:15365–15378
Hubbell SP (2001) A unified natural theory of biodiversity and biogeography. Princeton University Press, Princeton
Kirman A (1993) Ants, rationality, and recruitment. Quart J Eco 108:137–156
Mantegna RN, Stanley HE (2007) Introduction to econophysics: correlations and complexity in finance. Cambridge University Press, Cambridge
Mori S, Hisakado M (2015) Finite-size scaling analysis of binary stochastic process and universality classes of information cascade phase transition. J Phys Soc Jpn 84:54001–54013.
Privman V (ed) (1997) Nonequilibrium statistical mechanics in one dimension. Cambridge University Press, Cambridge
Pitman J (2006) Combinatorial stochastic processes. Springer, Berlin
Pitman J, Yor M (1997) The two-parameter Poisson-Dirichlet distribution derived from a stable subordinator. Ann Prob 25(2):855–900
Sasamoto T, Spohn H (2010) The one-dimensional KPZ equation: an exact solution and its universality. Phys Rev Lett 104:230602
Simon HA (1955) On a class of skew distribution functions. Biometrica 42:425–440
Yamato H, Shibuya M (2001) Pitman's model in random partition. RIMS 1240:64–73
Yule G (1925) A mathematical theory of evolution, based on the calculations of Dr. J. C. Willis, F.R.S. Philos Trans R Soc Lond B 213:21

Domino Effect in Information Cascade

Shintaro Mori and Masato Hisakado

1 Information Cascade

Social learning is a learning process by observing others' behaviors and adopting them (Rendell et al. 2010). As the cost of social learning is smaller than individual learning, many kinds of animals and human adopt the process. However, it is also known to be error-prone. The obtained information might be outdated and wrong. In information cascade, where individuals make sequential decisions after observing the actions of those ahead of them and follow the majority's behavior of the preceding individuals without regard to their own private information, we observe such sub-optimal case (Bikhchandani et al. 1992). When information cascade starts, the majority behavior continues in the sequence. Even if it is sub-optimal, it never switches to the optimal one as the private information is lost in later decisions. Some individuals might recognize that the majority's behavior is based only on a few individuals' private information. They might choose the minority's choice based on their private information. The uniformity is broken and there might occur a switch to an optimal behavior. The perception of the "shallowness" of the depth of the private information that piled up in the majority's choice is one mechanism of the fragility of information cascade. It explains the localized and short-lived nature of fluctuations in culture, like fads and fashion (Devenow and Welch 1996). Even if the uniformity is broken, it is another problem that the minority's choice becomes the majority.

S. Mori (✉)
Faculty of Science and Technology, Department of Mathematics and Physics, Hirosaki University, Hirosaki, Aomori, Japan

M. Hisakado
Nomura Holdings, Inc., Chiyoda-ku, Tokyo, Japan

A.-H. Sato (ed.), *Applications of Data-Centric Science to Social Design*, Agent-Based Social Systems 14, https://doi.org/10.1007/978-981-10-7194-2_9

As the first individual's choice greatly affects the majority's choice of the early individuals that trigger the information cascade, domino effect is the key concept in information cascade (Mori and Hisakado 2015a,b). The central question about information cascade is when and how the domino effect propagates along the individuals' choice sequence. If many individuals recognize that the true "depth" of an information cascade is shallow, the information cascade is fragile. The uniform behavior is broken and domino effect disappears. However, if the network externality of the majority's choice reinforces as more majority adopter appears, the conformity in the information cascade is robust, and the domino effect continues forever.

In this chapter, we address the central question of information cascade by studying the domino effect of several simple models. We explain the necessary condition for the occurrence of an infinitely long cascade. Based on the results, one can understand the micro-macro features of the information cascade experiments that are presented in later chapters.

2 General Setting and Measure of Domino Effect

We start from the general setting of information cascade and introduce a measure of the domino effect. Assume that there are two options, A and B, one of which is chosen to be correct with equal probability. Without loss of generality, we label the correct (incorrect) option as 1 (0). Each individual has private information about the true option. We denote the order of the individual in the sequential choice as $t = 1, 2, \cdots$ and individual t's private information and decision as $S(t) \in \{0, 1\}$ and $X(t) \in \{0, 1\}$, respectively. In information cascade experiment, $S(t)$ is random which takes 1 with probability q.

$$P(S(t) = 1) = q \text{ and } P(S(t) = 0) = 1 - q.$$

The summary statistics of how many individuals among first t have chosen correct and incorrect option are denoted as $N_1(t)$ and $N_0(t)$, respectively.

$$N_1(t) = \sum_s X(s) \text{ and } N_0(t) = t - N_0(t).$$

We denote the difference in the number of correct and incorrect choices up to the t-th individual as $N(t) = N_1(t) - N_0(t)$. Individual $t + 1$ observes these summary statistics and make decision $X(t+1)$ based on $(S(t+1), N_1(t), N_0(t))$. Information cascade models are determined by the probabilistic rule for $X(t + 1)$. It is given by the conditional probability $P(X(t + 1) = x|S(t + 1), N_1(t), N_0(t)), x \in \{0, 1\}$.

The domino effect of information cascade is measured by the correlation function $C(t)$, which is defined as the difference in the conditional probabilities (Mori and Hisakado 2015b).

$$C(t) = \Pr(X(t+1) = 1|X(1) = 1) - \Pr(X(t+1) = 1|X(1) = 0).$$

$C(t)$ can also be defined as the covariance of $X(1)$ and $X(t+1)$ divided by the variance of $X(1)$.

$$C(t) = \frac{\text{Cov}(X(1), X(t+1))}{\text{V}(X(1))}.$$

The first definition makes it easy to understand $C(t)$ as domino effect. Or more intuitively, we can rewrite it as

$$C(t) = \Pr(X(t+1) = 0|X(1) = 0) - \Pr(X(t+1) = 0|X(1) = 1).$$

This definition indicates that $C(t)$ measures how the first individual's choice $X(1)$ affects the later individual $t+1$'s choice $X(t+1)$. In general, by the change from $X(1) = 1$ to $X(1) = 0$, the conditional probability for $P(X(t+1) = 0|X(1))$ increases, and the difference between the two conditional probabilities is positive. If domino effect disappears as the later individuals choose independently from the first individual's choice, $C(t)$ decays to zero. If it remains and the domino effect propagates to infinitely later individuals, $C(t)$ in the limit $t \to \infty$ is positive. We denote it as c.

$$c \equiv \lim_{t\to\infty} C(t).$$

The central problem in information cascade is the clarification of the condition for $c > 0$.

3 Simple Models

We study several simple models of information cascade and classify the asymptotic behavior of $C(t)$ and the limit value c. We clarify the condition when the domino effect disappears ($c = 0$) or remains forever ($c > 0$). We also study models that show the phase transition between $c = 0$ and $c > 0$ by the change of the model's parameters. We explain that it is a non-equilibrium phase transition and is akin to percolation transition.

3.1 Pólya Urn

The Pólya urn is a simple stochastic process for contagion where it is taken into account by a reinforcement mechanism (Pólya 1931). As the social learning process is a kind of contagion process, we study Pólya urn as a model of information cascade. There are initially R_0 red balls and B_0 blue balls in an urn. At each step, one draws a ball randomly from the urn and duplicates it. Then, one returns the balls, and the probability of selecting a ball of the same color is reinforced. As the process is repeated infinitely, the ratio of red balls in the urn z becomes random and obeys the beta distribution Beta (R_0, B_0).

In the context of information cascade, the red balls correspond to correct choice, and the blue balls correspond to incorrect choice. Individuals draw a ball from the urn, and the color of the ball is his private information. He chooses the same option with the color of the ball.

$$X(t) = S(t)$$

The probability that individual $t+1$ chooses the correct option is

$$P(X(t+1) = S(t+1) = 1) = \frac{R_0 + N_1(t)}{t + R_0 + B_0}$$

Here, $N_1(t)$ is the number of individuals who have drawn red balls until individual t. The correlation function $C(t)$ does not decay with t, and the limit value c is $1/(R_0 + B_0 + 1)$.

$$C(t) = \frac{1}{R_0 + B_0 + 1}$$

The domino effect continues forever.

3.2 BHW's Basic Cascade Model

We study the "basic model" of information cascade (Bikhchandani et al. 1992). We assume that the first individual chooses 1 (0) if his private signal is 1 (0).

$$X(1) = S(1).$$

The second individual can infer the first individual's signal $S(1)$ from $X(1)$. If the first individual chooses 1 (0), the second individual chooses 1 (0) if his signal is 1 (0). If his signal contradicts the first individual's choice, we assume he chooses the same option as his signal $S(2)$, which is different from the tiebreaking convention in the "basic model," where the individual chooses 1 or 0 with equal probability. We summarize the decision rules for the first two individuals as

$$X(t) = S(t), t = 1, 2.$$

There are three situations for the third individual: (1) If both predecessors have chosen 1, irrespective of his signal $S(3)$, he chooses 1. The following individuals also choose 1, and a correct cascade, which is called up cascade, starts. (2) If both have chosen 0, an incorrect cascade, or down cascade, starts. (3) One has chosen 1, and the other has chosen 0. The third individual is in the same situation as the first individual, and he chooses the option matching his private information, $X(3) = S(3)$.

The probability that both of the first two individuals receive correct (incorrect) signals is $q^2((1-q)^2)$, so an up (down) cascade starts with probability $q^2((1-q)^2)$. From the above discussion, if $N(t) \geq 2(\leq -2)$, individual $t+1$ chooses $X(t+1) = 1(0)$ irrespective of his private information $S(t+1)$. And up (down) cascade starts from $t+1$.

$$X(t+1) = \theta(N(t)) \text{ if } |N(t)| \geq 2$$

If $|N(t)| \leq 1$, individual $t+1$ trusts his private information and chooses $X(t+1) = S(t+1)$.

$$X(t+1) = S(t+1) \text{ if } |N(t)| \leq 1$$

As information cascade starts for $|N(t)| \geq 2$, we identify all states with $N(t) \geq 2(\leq -2)$ as $N(t) = 2(-2)$, and there remain five states, $N(t) \in \{-2, -1, 0, 1, 2\}$. If t is even, there are three states, $N(t) \in \{-2, 0, 2\}$, and there are four states, $N(t) \in \{-2, -1, 1, 2\}$, if t is odd. Figure 1 illustrates the model. In the figure, we also show the probabilistic rule for the transition between states. At $t = 0$, $N(0) = 0$, and it jumps to $N(t) = 1(-1)$ with probability q $(1-q)$. From $t = 1$ to

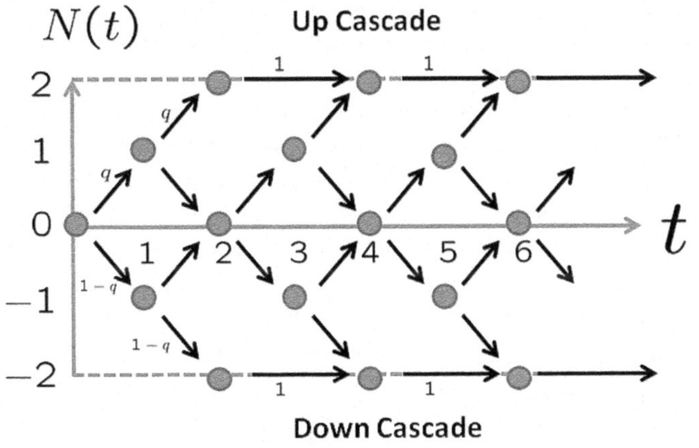

Fig. 1 Simple model of information cascade. As information cascade starts for $|N(t)| \geq 2$, We identify all states with $N(t) \geq 2(\leq -2)$ as $N(t) = 2(-2)$. States $N(t) \in \{2, 1, 0, -1, -2\}$ and probabilities for $X(t) \in \{0, 1\}$

$t = 2$, the same rule applies, and $N(t)$ increases (decreases) by 1 with probability q $(1-q)$. If $N(t) = 2(-2)$ at $t = 2$, an up (down) cascade starts. Later individuals choose 1 (0) for $t \geq 3$, and $N(t)$ remains $2(-2)$. If $N(t) = 0$ at $t = 2$, the third individual chooses 1 with probability q. In general, if $|N(t)| \leq 1$, $N(t)$ increases (decreases) by 1 with probability q $(1-q)$. The problem is a random walk model with absorbing walls at $N(t) = \pm 2$. As t increases, the probability that the random walk is absorbed in the walls increases. In the limit $t \to \infty$, all random walks are absorbed in the walls. The state $N(t) = 0$ for even t is absorbed into the state $N(t+2) = 2$ with probability $q^2/(q^2 + (1-q)^2)$ and is absorbed into the state $N(t+2) = -2$ with probability $(1-q)^2/(q^2 + (1-q)^2)$. The probability for an up cascade in the limit $t \to \infty$, which we denote by $P_2(\infty)$, is then given as

$$P_2(\infty) \equiv \Pr(N(\infty) = 2) = \frac{q^2}{q^2 + (1 - q^2)}. \tag{1}$$

In the up (down) cascade, individuals always choose 1 (0), and $P_2(\infty)$ is the limit value for the probability of the correct choice. It is greater than q for $q > 1/2$, and the deviation shows an increase in the accuracy from that of the signal $S(t)$. $P_2(\infty) - q$ is a measure of the collective intelligence (left figure in Fig. 2).

$C(t)$ is estimated as

$$C(2n) = \frac{q(1-q)}{q^2 + (1-q)^2} + \frac{(1-2q)^2}{2(q^2 + (1-q)^2)}(\sqrt{(2q(1-q))})^{2n},$$
$$C(2n+1) = C(2n).$$

$C(t)$ decays exponentially with t and the limit value is the first term.

$$c = \lim_{t\to\infty} C(t) = \frac{q(1-q)}{q^2 + (1-q)^2}. \tag{2}$$

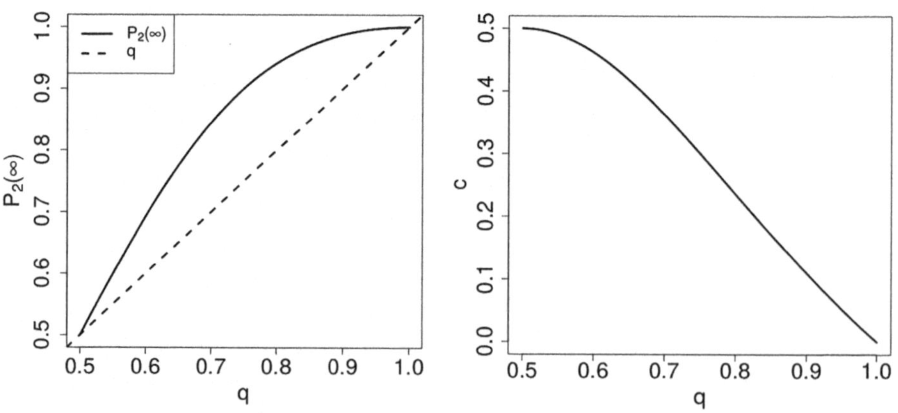

Fig. 2 Plot of $P_2(\infty)$ (Eq. (1)) and c (Eq. (2)) vs. q

c changes continuously with q, and it takes zero at $q = 1$ (right figure in Fig. 2). The results indicate that the domino effect continues forever irrespective of the value of q. The first individual's choice affects the probability of the occurrence of the up and down cascades.

3.3 Correlated Random Walk

If the individuals refer to the previous individual only, $X(t+1)$ depends on $X(t)$ and $S(t+1)$. In the case, the system is described as a correlated random walk (CRW) (Mori and Hisakado 2015a). If they refer to their private information only, $X(t+1) = S(t+1)$ holds, and the probability that $X(t+1)$ takes 1 is q. If they choose the same choice with the previous individuals, $X(t+1) = X(t)$ holds. We interpolated the two limits by a parameter p and introduce the next model for $X(t+1)$.

$$P(X(t+1) = 1|X(t) = x, S(t+1) = s) = (1-p) \cdot s + p \cdot x.$$

If one takes the average over $S(t+1)$ with probabilities q and $1-q$, we have

$$P(X(t+1) = 1|X(t) = x) = (1-p) \cdot q + p \cdot x.$$

$C(t)$ is then estimated as

$$C(t) = p^t.$$

$C(t)$ decays exponentially to zero with the increase of t, and the domino effect disappears. We introduce correlation length ξ to characterize the length scale of the correlation function as $C(t) \propto e^{-t/\xi}$. As $C(t) = p^t = e^{t \log p}$, we have

$$\xi = -1/\log(p).$$

If $p < 1$, ξ is finite and $C(t)$ decays exponentially with t. If $p = 1$, ξ diverges and $c = \lim_t C(t) = 1$. The model has two limits, $c = 0$ and $c = 1$, and the critical value of p is $p_c = 1$.

3.4 Majority Rule Model

Individuals choose the majority choice of the previous r individuals with probability p or refers to his private signal with probability $1-p$. For simplicity, we assume r is an odd number greater than three, $r \in \{3, 5, 7 \cdots\}$. The probabilistic rule for $X(t+1), t \geq r$ is written as

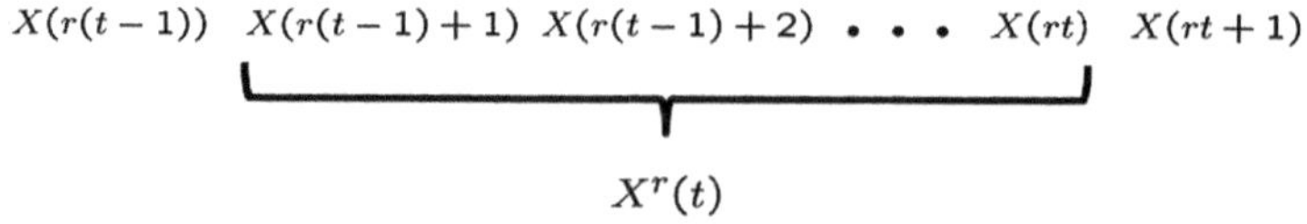

Fig. 3 Real-space renormalization by coarse graining from $X(r(t-1)+1), \cdots, X(rt)$ to $X^r(t)$

$$P\left(X(t+1)=1|\sum_{s=t-r+1}^{t} X(s)=x\right)=(1-p)\cdot q+p\cdot\theta(x-r/2).$$

When r is finite, the correlation length ξ is finite, and the domino effect does vanish for $p<1$. We can map the model to CRW by applying real-space renormalization transformation. Here, real-space renormalization transformation is a kind of coarse-graining process, and we group r variables $X(r(t-1)+1), \cdots, X(rt)$ into $X^r(t)$ by the majority rule as depicted in Fig. 3.

$$X^r(t)=\theta(X(r(t-1)+1)+\cdots+X(rt)-r/2).$$

rT variables $X(1), \cdots, X(rT)$ are then transformed into T variables $X^r(1), \cdots, X^r(T)$. We modify the above probabilistic rule to the next one for $X(rt+s), s=1, \cdots, r$ as

$$P(X(rt+s)=1|X^r(t)=x)=(1-p)\cdot q+p\cdot x.$$

$X(rt+s), s=1, \cdots, r$, which are transformed into $X^r(t+1)$, depends not on the nearest r individuals but the r individuals $X(r(t-1)+s'), s'=1, \cdots, r$, which are transformed into $X^r(t)$. $X^r(t+1)$ is composed of $X(rt+1), \cdots, X(r(t+1))$, and they take 1 with probability $(1-p)q+px$. The probability that $X^r(t)$ takes 1 becomes

$$P(X^r(t+1)=1|X^r(t)=x)=(1-p_r)\cdot q_r+p_r\cdot x.$$

The parameters q_r and p_r are

$$q_r=\frac{\pi_r((1-p)q)}{\pi_r((1-p)q)+\pi_r((1-p)(1-q))},$$
$$p_r=1-(\pi_r((1-p)q)+\pi_r((1-p)(1-q))).$$

Here, $\pi_r(x)$ is defined as $\pi_r(x)=\sum_{n=(r+1)/2}^{r} {}_rC_n\cdot x^n(1-x)^{r-n}$, and ${}_rC_n$ is the binomial coefficient. $\pi_r(x)$ gives the majority probability $P(n>r/2)$ if n obeys binomial distribution B(r,x). In the limit $r\to\infty$, $\pi_r(x)$ behaves as $\theta(x-1/2)$.

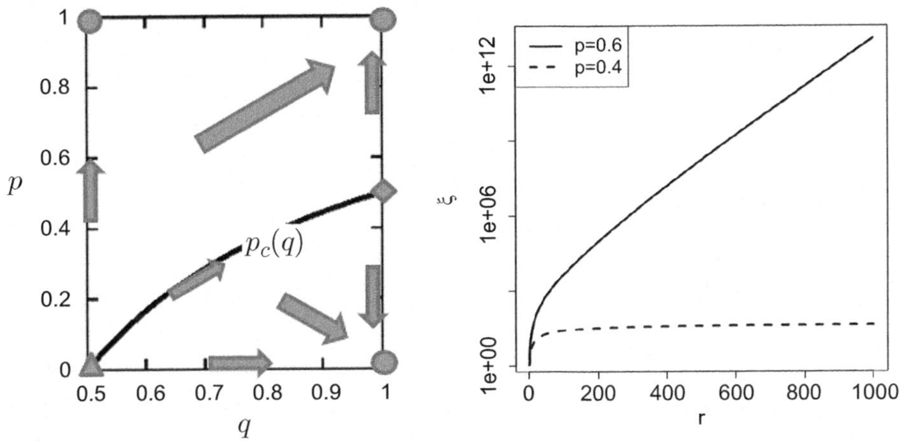

Fig. 4 (Left) Plot of transformation $(q, p) \to (q_r, p_r)$ under the renormalization transformation $\{X(r(t-1)+1), \cdots, X(rt)\} \to X^r(t)$. Arrows indicate the direction of movement from (q, p) to (q_r, p_r). There are three stable fixed points at (1, 1), (1, 0), (1/2, 1) (filled circles) and two unstable fixed points at (1/2, 0), (1, 1/2) (filled triangle and diamond, respectively). (Right) r dependence of ξ for $p = 0.6 > p_c = 0.5$ and $p = 0.4 < p_c = 0.5$

We study the transformation $(q, p) \to (q_r, p_r)$ under the renormalization transformation. The transformation has three stable fixed points (∘), one unstable fixed point (△), and one marginally stable fixed point (⋄), which are summarized in the left figure of Fig. 4. When $q > 1/2$, (q, p) on the line $p = 1 - 1/2q$ moves along it under the transformation because (q_r, p_r) also satisfies $p_r = 1 - 1/2q_r$. We call this line the critical line and denote $p_c(q) = 1 - 1/2q$. In the limit $r \to \infty$, we have $\lim_{r\to\infty}(q_r, p_r) = (1, 1/2)$. If $p < p_c(q)$, we have $\lim_{r\to\infty}(q_r, p_r) = (1, 0)$. If $p > p_c(q)$, we have $\lim_{r\to\infty}(q_r, p_r) = (1, 1)$. The marginal fixed point at (1, 1/2) has one stable and one unstable direction. If $q = 1/2$, $\lim_{r\to\infty}(q_r, p_r) = (1/2, 1)$ for $p > 0$. (1/2, 0) is an unstable fixed point.

There are two important points. The first one is that if r is finite, $p_r < 1$ for $p < 1$. The correlation length for $X^r(t)$ is $1/\log(p_r)$. As $X(r(t-1)+1), \cdots, X(rt)$ are grouped into $X^r(t)$; the correlation length for the original $X(t)$ is

$$\xi = -r/\log p_r$$

ξ is finite for $p < 1$ and $r < \infty$. $C(t)$ decays exponentially with t and $c = 0$. The domino effect $C(t)$ disappears for large t. The second point is the difference in the dependence of ξ on r for $p < p_c(q)$ and $p > p_c(q)$. When $p < p_c(q)$, p_r goes to 0 in the limit $r \to \infty$. When $p > p_c(q)$, p_r goes to 1 in the limit $r \to \infty$. As $\xi \propto -\log p_r$, ξ diverges in the limit. If ξ diverges, $C(t)$ does not decay exponentially and it might occur that $c > 0$. The right figure in Fig. 4 shows the dependence of ξ on r. We set $q = 1$ and $p_c(q) = 1/2$. When $p = 0.4 < p_c(q)$, ξ is finite in the limit $r \to \infty$. $C(t)$ decays exponentially with t even for the limit

$r \to \infty$. When $p = 0.6 > p_c(q)$, ξ diverges exponentially with r. This indicates that $C(t)$ does not decay exponentially. In the next section, we calculate $C(t)$ in the limit $r \to \infty$ and study its asymptotic behavior for $p > p_c(q)$, $p < p_c(q)$ and $p = p_c(q)$.

4 Nonlinear Polya Urn: Majority Rule Case

We consider the $r \to \infty$ limit of the majority rule model in the previous section (Hisakado and Mori 2011; Mori and Hisakado 2015a). More precisely, individual $t+1$ refers to all his previous t individuals.

$$P(X(t+1) = 1|N_1(t) = n) = (1-p) \cdot q + p \cdot \theta(n - t/2).$$

We write the ratio of correct choice as $z(t) \equiv N_1(t)/t$. The probabilistic rule for $X(t+1)$ is rewritten as

$$P(X(t+1) = 1|z(t) = z) = f(z) = (1-p) \cdot q + p \cdot \theta(z - 1/2).$$

We can interpret the individuals as the mixture of independent voters who decide based on their private information and majority voters who choose the majority choice with ratios $1-p$ and p. $f(z)$ describes the probability $P(X(t+1) = 1|z(t) = z)$ as function of z.

Depending on the value of p, the number of the solution for $f(z) = z$ or fixed point of the map $f : z \to z$ changes. Figure 5 shows the plots of $f(z)$ for $q = 0.8$ with $p = 0.1$ (left) and $p = 0.6$ (right). $p_c = 1 - 1/2q = 3/8$. When $z > 1/2$, $f(z) = (1-p)q + p = q + p(1-q) \equiv z_+$. When $z < 1/2$, $f(z) = (1-p)q = q - pq \equiv z_-$. As z_+ is always greater than q, $f(z)$ crosses the diagonal at (z_+, z_+). If $p < p_c$ and $z < 1/2$, $z_- > (1-p_c)q = 1/2$, and $f(z)$ does not cross the

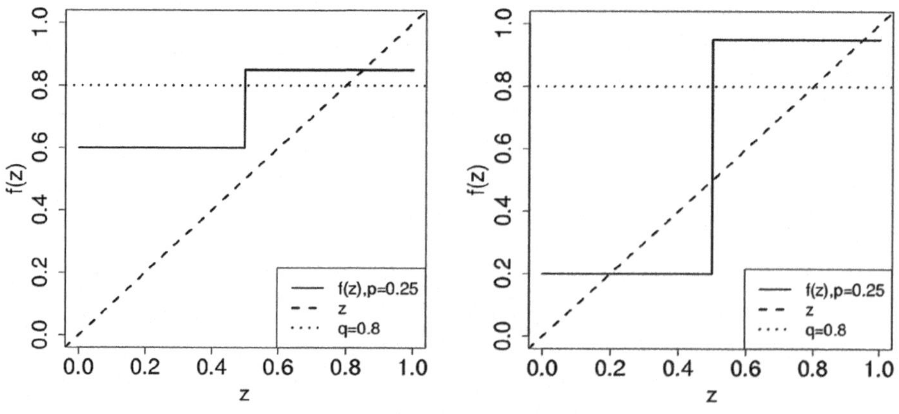

Fig. 5 Plots of $f(z)$ vs. z for $q = 0.8$ and $p_c = 3/8$. The left figure shows the plot for $p = 0.1 < p_c$. The right figure shows the plot for $p = 0.6 > p_c$

diagonal. On the other hand, if $p > p_c$, $z_- < 1/2$ and $f(z)$ crosses the diagonal at (z_-, z_-). If $p = p_c$, $f(z)$ touches the diagonal at $z = 1/2 \equiv z_t$ and crosses the diagonal at (z_+, z_+). It is known that the z_+ and z_- are stable fixed points and the probability for the convergence of $z(t)$ to these points are positive (Hill et al. 1980). On the other hand, for $p = p_c$, the fixed point at z_t is called touch point, and it is unstable (Pemantle 1991, 2007).

From the results, we can anticipate the asymptotic behavior of $C(t)$ (Mori and Hisakado 2015a). For $p < p_c(q)$, there is only one stable state at z_+, and $z(t)$ always converges to z_+ irrespective of the value of $X(1)$. The initial condition $X(1) = x$ does not affect the probability $\mathrm{P}(X(t+1)|X(1) = x)$ in the limit $t \to \infty$ and $c = 0$. The domino effect disappears after many individual choices. On the other hand, if $p > p_c(q)$, as there are two stable states at $z_\pm$, the probability to the convergence to $z_\pm$ depends on the initial value $X(1) = x$. If $z(t)$ converges to $z_\pm$, $\mathrm{P}(X(t+1) = 1|z(t) \to z_\pm) = f(z_\pm)$. As $f(z_+) = z_+ > f(z_-) = z_-$, c should remain positive in the limit $t \to \infty$.

As $f(z)$ changes discontinuously at $z = 1/2$, it is necessary to estimate the probability that $z(t)$ enters into $z < 1/2$ and $z > 1/2$ and $z = 1/2$ under the initial condition $X(1) = x$ for the estimation of $C(t)$. We call these domains as regions I, II, and III. If $z(t)$ stays in region $J = \{I, II, III\}$, the probability that $X(t+1)$ takes 0 is $1 - z_- = A$, $1 - z_+ = C$, $1 - ((1-p)q + p/2) = B$, respectively. We write them as q_J, $J \in \{I, II, III\}$. We denote the probabilities that $z(t)$ stays in region $J \in \{I, II, III\}$ with initial condition $X(1) = x$ as $P_J(t|X(1) = x)$. $C(t)$ is written as

$$C(t) = \sum_J q_J \left(P_J(t|X(1) = 1) - P_J(t|X(1) = 0)\right)$$

We introduce a tilted two-dimensional square lattice (m, n), $m, n \in \{0, 1, \cdots\}$ with the origin at the top of it. We describe the stochastic process as a directed path on the lattice (Fig. 6). We map $X(t)$, $t \in \{1, 2, \cdots\}$ to a path on the lattice as $(m, n) = (C_1, C_0)$ with $C_1 = \sum_{s=1}^{t} X(s)$ and $C_0 = t - C_1$. The path starts from the top $[(0, 0)]$, and if $X(t) = 1(0)$, it moves to the lower right (left). The "directed" means that the path always goes downward.

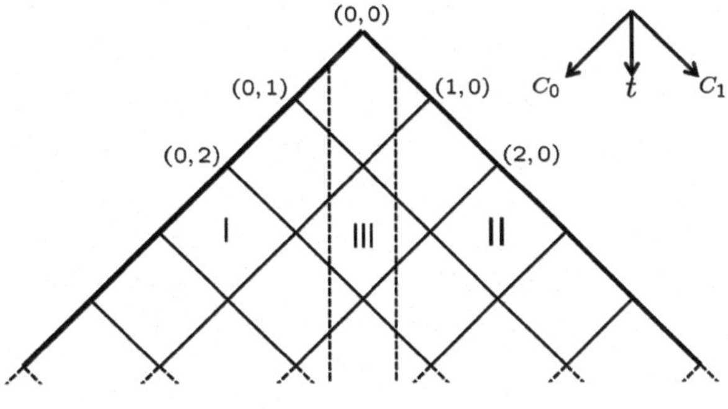

Fig. 6 Directed path representation of $\{X(t)\}$ as $(C_1, C_0) = (m, n)$

If $q = 1$ and $z_+ = 1$, when $z(t)$ enters into region II, it cannot return to region III. The wall $n = m - 1$ in II becomes an absorbing wall for the path. If the initial condition is $X(1) = 1$, it is in region II and $P_{II}(t|X(1) = 1) = 1$ holds. The remaining quantities are $P_J(t|X(1) = 0)$. We denote the number of paths from $(0, 0)$ to (s, s) in I and III that enter III k times as $A_s^k = k(2s - k - 1)!/s!(s - k)!$. The probability that a path starts from (0, 1) and reaches (s, s) is then given as

$$\Pr(z(t = 2s) = 1/2|X(1) = 0) = \sum_{k=1}^{s} A_s^k \cdot p^{s-1}(1 - p)^s/2^{k-1}.$$

For large t, the $k = 1$ term is dominant, we have

$$P_{\text{III}}(t = 2s|X(1) = 0) \simeq \frac{2}{p} \cdot \frac{e}{8\sqrt{\pi}} s^{-3/2}(4p(1 - p))^s \quad \text{for } t >> 1. \tag{3}$$

The probability that a path enters II from (s, s) to $(s + 1, s)$ is $(1 - p/2)$, and we have the following expression.

$$P_{\text{II}}(t = 2s + 1|X(1) = 0) = \sum_{u=1}^{s} P_{\text{III}}(2u|X(1) = 0) \cdot (1 - p/2).$$

$P_I(t|X(1) = 0) \simeq 1 - P_{II}(t|X(1) = 0)$ holds as $P_{III}(t|X(1) = 0)$ is negligibly small for large t. $\Pr(X(t + 1) = 0) = p$ for $z(t) < 1/2$ and we have

$$C(t) = p \cdot P_{\text{I}}(t|X(1) = 0) = p \cdot (1 - P_{\text{II}}(t|X(1) = 0)) \quad \text{for } t >> 1.$$

The limit value c is then estimated as

$$c = \begin{cases} 0 & p \le 1/2, \\ \frac{4p-2}{(1+p)} & p \ge 1/2. \end{cases}$$

For $p < p_c(1) = 1/2$, $C(t)$ decays exponentially for large t and $c = 0$. For $p > p_c$, $c > 0$ and the domino effect remains forever. c plays the role of the order parameter of the continuous phase transition. We estimate the critical exponents β for the order parameter c by expanding c around $p = p_c(1) = 1/2$. Then we have $\beta = 1$ for $c \propto |p - 1/2|^{\beta}$.

In order to understand the nature of the phase transition, we rewrite the expression for $C(t)$ using c as

$$C(t = 2s) = c + p \cdot \sum_{u=s+1}^{\infty} P_{\text{III}}(2u|X(1) = 0) \cdot (1 - p/2).$$

Putting the asymptotic form of Eq. (3) for $P_{III}(2u|X(1)=0)$ in the expression, we obtain

$$C(t) \sim c + p \cdot \int_t^\infty ds \frac{e}{8\sqrt{\pi}} s^{-3/2} e^{-s/\xi(p)} \cdot (1-p/2).$$

Here, we define $\xi = -/\log\sqrt{4p(1-p)}$. At $p = p_c(1) = 1/2$, $\xi = \infty$ and $C(t)$ behaves as

$$C(t) \sim \int_t^\infty s^{-3/2} ds \propto t^{-1/2}.$$

$C(t)$ behaves as $t^{-\alpha}$ with $\alpha = 1/2$. The domino effect vanishes; however, the disappearance is extremely slow. About $\nu_{||}$ of the critical exponent for the divergence of $\xi \propto |p - p_c|^{\nu_{II}}$ as $p \to p_c$, we expand ξ around $p_c = 1/2$ and obtain $\nu_{||} = 2$.

The above directed path picture suggests the relation between the information cascade phase transition and a phase transition to absorbing states. The phase transition to absorbing state is a non-equilibrium phase transition. As its typical example, directed percolation is well known. For $q = 1$, the wall $n = m - 1$ in II becomes an absorbing wall, and a directed path does not return to I if it touches the wall. Region II becomes an absorbing state for the path. $C(t)$ is the survival probability of the path in I, which is the order parameter of the phase transition into absorbing states. We can assume the following scaling form for $C(t)$ in the critical region for $q > 1/2$.

$$C(t) \propto t^{-\alpha} \cdot g(t/\xi), \text{ and } \alpha = 1/2.$$

Here, we assume that $C(t)$ is scaled by ξ with a universal scaling function g. In the limit $t \to \infty$ for $p \gtrsim p_c(q)$, $\lim_{t\to\infty} C(t) = c > 0$. Because $g(x)$ should behave as $g(x) \sim x^\alpha$ to cancel $t^{-\alpha}$, we have

$$c \propto \xi^{-\alpha} \propto |p - p_c(q)|^{\alpha\nu_{||}}.$$

We have the scaling relation $\beta = \nu_{||} \cdot \alpha$, which has been shown to hold for $q = 1$ as $1 = 2 \cdot 1/2$. The scaling relation suggests that only two exponents, β and $\nu_{||}$, might be sufficient to characterize the universality class of the information cascade phase transition.

The result in the previous section suggests that the phase transition for general $q > 1/2$ is governed by the fixed point $(1, 1/2)$ of the renormalization transformation $(q, p) \to (q_r, p_r)$. In particular, at $p = p_c(q)$, $q_I = 1/2$ and it is a simple symmetric random walk. In II and III, $q_J > 1/2$, $J \in \{II, III\}$, and it is not symmetric. The nonrecurrence probability that a simple random walk does not return to the diagonal up to t behaves as $t^{-1/2}$. $C(t)$ is proportional to the probability that the random walker remains in I because it is difficult for the random walker to return from II to I even for $q < 1$. Thus, $C(t) \propto P_I(t|X(1) =$

0) $\sim t^{-1/2}$ holds generally at the critical point. Furthermore, because the paths of the random walker in I are concentrated around $z \sim 1/2$, the same asymptotic behaviors $\Pr(X(t') = 0|X(t) = 0) \sim |t' - t|^{-1/2}$ should hold for $t, t' > 0$. This suggests that the translational invariance is recovered at $p = p_c$ and $C(t, t') \equiv \mathrm{Cov}(X(t), X(t'))/\mathrm{V}(X(t)) \propto |t - t'|^{-1/2}$.

We can obtain the scaling form for $C(t)$ as

$$C(t) = bt^{-1/2}g(t/\xi).$$

b and ξ are estimated as

$$b = \sqrt{\frac{8}{\pi}}\left(\frac{2q-1}{4q-1}\right)$$
$$\xi^{-1} = -\ln\sqrt{4(p+(1-p)(1-q))((1-p)q)}$$

$g(x)$ is given as

$$g(x) = \begin{cases} \sqrt{4\pi x} + \frac{x^{1/2}}{2}\int_x^\infty u^{-3/2}e^{-u}du & p > p_c(q_*), \\ \frac{x^{1/2}}{2}\int_x^\infty u^{-3/2}e^{-u}du & p < p_c(q_*). \end{cases} \tag{4}$$

We can estimate $C(t)$ for $t \leq 10^5$ and $q_* = 0.6$ by numerically integrating the master equation for the system. Figure 7 shows the results. The empty symbols

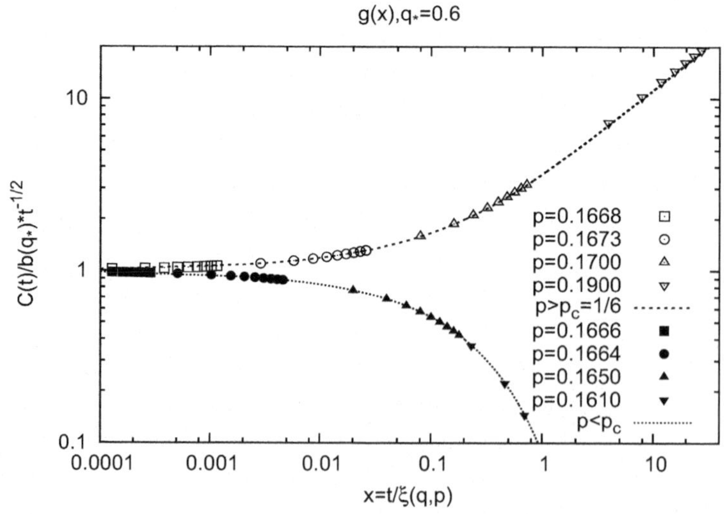

Fig. 7 Plot of $C(t)/bt^{-1/2}$ vs. t/ξ with empty symbols indicating $p > p_c(q) = 1/6$ and filled symbols indicating $p < p_c(q)$. We adopt $q = 0.6$ and $10^4 \leq t \leq 10^5$. The lines show the results of Eq. (4)

indicate the results for $p > p_c(q)$. We have adopted the value of p in the vicinity of $p_c(q) = 1/6$ and $t \in [10^4, 4 \times 10^5]$. As can be clearly seen, the data obtained for different p and t values lies on the curve of Eq. (4). The filled symbols indicate the results for $p < p_c(q)$. The curve of Eq. (4) describes the data.

4.1 Analog Herder Model

Individual $t + 1$ refers to all his previous t individuals, and the probabilistic rule for $X(t + 1)$ is written as

$$P(X(t + 1) = 1|z(t) = z) = f(z) = (1 - p) \cdot q + p \cdot z.$$

We can interpret the individuals as the mixture of independent voters who decide based on their private information and herding voters who choose the majority choice with the probability of the ratio of the individuals of the option (Hisakado and Mori 2010). In the case $p = 1$, $f(z)$ is the limit $t \to \infty$ of $P(X(t + 1) = 1)$ of the Pólya urn. We call such proportional herding voter as analog herder.

The model has only one equilibrium at q, which is the probability that the independent voter chooses the correct option (Fig. 8). $C(t)$ satisfies the recursive relation.

$$C(t) = \frac{t - 1 + p}{t} C(t - 1).$$

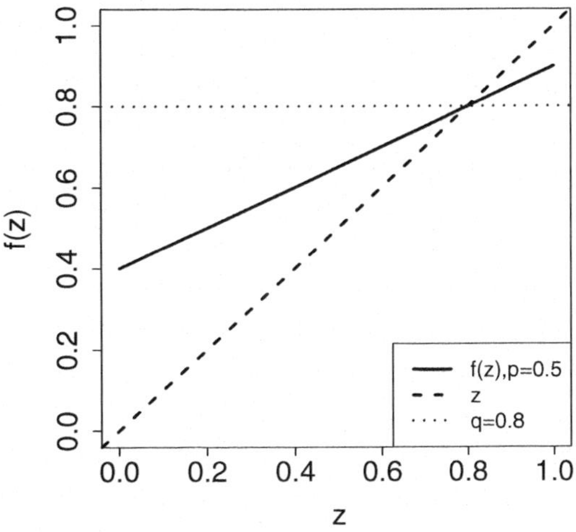

Fig. 8 Plots of $f(z)$ vs. z for $q = 0.8$ and $p = 1/2$

We solve it and obtain

$$C(t) = \prod_{s=1}^{t} \frac{s-1+p}{s} \propto t^{p-1}.$$

The limit value c is zero and the domino effect disappears with power-law decay.

5 Nonlinear Polya Urn: Logistic-Type Case

We study a model where the majority rule $\theta(z - 1/2)$ is replaced with sigmoidal continuous function. Individual $t + 1$ refers to all previous t individuals, and the probabilistic rule for $X(t + 1)$ is rewritten as

$$P(X(t+1) = 1 | N(t)/t = z) = f(z) = \frac{1}{2}\left(\tanh[J(2z-1) + h] + 1\right).$$

The choice of $f(z)$ is arbitrary, and we adopt the above form, which is familiar in the field of physics (Stanley 1971; Mori and Hisakado 2015b). The fixed point of $f(z)$ is a solution to $f(z) = z$. Using the mapping $m = 2z - 1$, we obtain the self-consistent equation $m = \tanh(J \cdot m + h)$ for the magnetization m in the mean-field Ising model. Here, we consider only the case for which $J \geq 0$. Because of the symmetry under $(X, h) \leftrightarrow (1 - X, -h)$, we also assume that $h \geq 0$.

5.1 Fixed Point and Touch Point

The number of fixed points for $f(z)$ depends on (J, h). There is a threshold value $J = J_c(h)$ as a function of h (Fig. 9). For $J < J_c(h)$, there is only one fixed point

Fig. 9 Plot of $J_c(h)$ in (J, h)

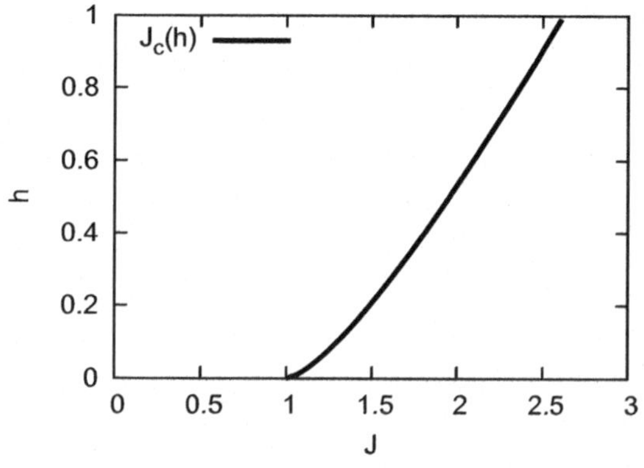

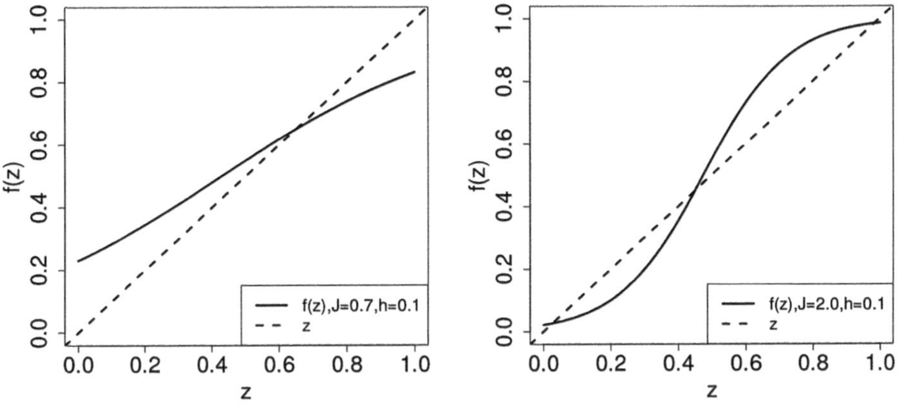

Fig. 10 Plot of $f(z)$ vs. z for $J < J_c(h)$ (left) and $J > J_c(h)$ (right). We adopt $J = 0.7, h = 0.1$ and $J = 2.0, h = 0.1$. The intersection between $y = f(z)$ and $y = z$ is the fixed point of $f(z)$. $J < J_c(h)$, with one fixed point at $z = z_+$ (left). $J > J_c(h)$, with three fixed points at $z \in \{z_-, z_u, z_+\}$ (right)

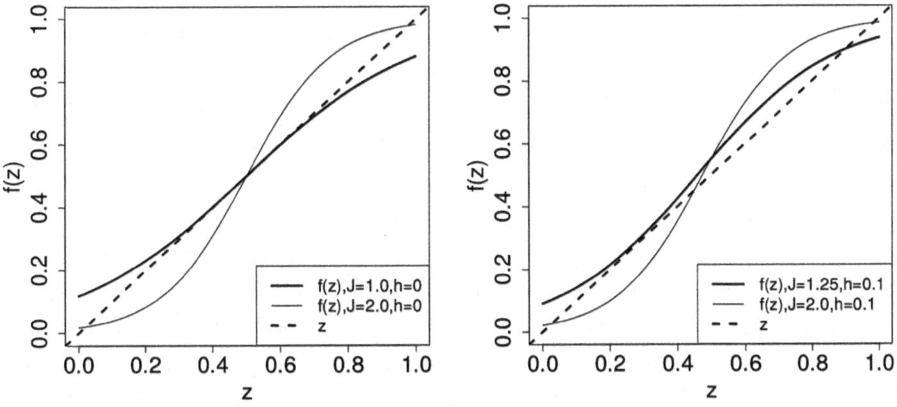

Fig. 11 Plot of $f(z)$ vs. z. $J = 1.0, h = 0.0$ and $J = 2.0, h = 0.0$ in the left figure. There is one fixed point at $z_t = 1/2$ for $J = J_c(0)$. $J = 1.25, h = 0.1$ and $J = 2.0, h = 0.1$ in the right figure. There is one fixed point at z_+ for $J = J_c(h)$ and one touch point at $z = z_t$

at $z = z_+$ (left figure in Fig. 10). By increasing J, $f(z)$ becomes tangential to the diagonal z at z_t for $J = J_c(h)$. For $h > 0$, $z_t \neq z_+$, and both z_t and z_+ are stable (right figure in Fig. 11). For $h = 0$, $z_t = z_+$, and it is also stable (left figure in Fig. 11). In both cases, the slope of the curve at z_t is equal to one. For $J > J_c(h)$, there are three fixed points, and we denote them as $z_- < z_u < z_+$; $z_\pm$ is stable, and z_u is unstable (right figure in Fig. 10). We denote the slope of $f(z)$ at $z_\pm, z_t$ as $l_\pm \equiv f'(z_\pm), l_t = f'(z_t) = 1$. As $z_\pm$ is stable and downcrossing, $l_\pm < 1$.

We note a crucial difference between $h = 0$ and $h > 0$. For $h = 0$, the touchpoint at $z_t = 1/2$ for $J = J_c(0) = 1$ coincides with the stable fixed point at $z_+ = 1/2$

for $J < J_c(0)$. It splits into the two stable fixed points at $z = z_\pm$ for $J > J_c(0)$ (right figure in Fig. 11). $z_\pm$ continuously moves away from $z_t = z_+$ and $z_+ - z_- \propto (J - J_c)^{1/2}$ for $|J - J_c| << 1$ as in the case of the mean-field Ising model. On the other hand, for $h > 0$, the touchpoint z_t appears at a different position from z_+ for $J = J_c(h)$ (left figure in Fig. 11). As J increases from $J_c(h)$, z_t splits into z_- and z_u. The change from $J < J_c(h)$ to $J > J_c(h)$ is discontinuous. This difference suggests that the phase transition is continuous for $h = 0$ and discontinuous for $h > 0$.

5.2 $C(t)$ *for* $J \neq J_c(h)$

The asymptotic behavior of $C(t)$ depends on (J, h). If $J < J_c(h)$, there is one stable fixed point at z_+, and $z(t)$ converges to z_+ through the power-law relation $\mathrm{E}(z(t) - z_+) \propto t^{l_+ - 1}$. Here, the expectation value $\mathrm{E}(A)$ of a certain quantity A is defined as the ensemble average over the paths of the stochastic process. If $J > J_c(h)$, there is another stable fixed point at z_-. Both $z_\pm$ are stable, and $z(t)$ converges to one of the fixed points. The convergence of $z(t)$ to $z_\pm$ also exhibits a power-law behavior $\mathrm{E}(z(t) - z_\pm) \propto t^{l_\pm - 1}$. Here, the ensemble average is the average over the samples that converges to $z_\pm$. We assume that the probability that $z(t)$ converges to one of the $z_\pm$ depends on $X(1)$, and we denote this as

$$p_\pm(x) \equiv \Pr(z(t) \to z_\pm | X(1) = x).$$

For $J < J_c(h)$, $z(t)$ always converges to z_+ irrespective of the value of $X(1) = x$, and $p_+(x) = 1$ holds. In this case, we set $p_-(x) = 0$. Regarding the asymptotic behavior of the convergence of $z(t) \to z_\pm$, which also depends on $X(1)$, we assume

$$\mathrm{E}(z(t) - z_\pm | X(1) = x) \simeq W_\pm(x) t^{l_\pm - 1}$$

We write the dependence of the coefficients $W_\pm(x)$ on the value of $X(1)$ explicitly. Using these behaviors and notations, we estimate the asymptotic behavior of $C(t)$ as

$$\begin{aligned}
C(t) &= \Pr(X(t+1) = 1 | X(1) = 1) - \Pr(X(t+1) = 1 | X(1) = 0) \\
&= \mathrm{E}(f(z(t)) | X(1) = 1) - \mathrm{E}(f(z(t)) | X(1) = 0) \\
&= \sum_{x=0}^{1} (-1)^{x-1} \{\mathrm{E}(f(z(t)) | x) \Pr(z(t) \to z_+ | x) \\
&\quad + \mathrm{E}(f(z(t)) | x) \Pr(z(t) \to z_- | x)\}
\end{aligned}$$

$$
\begin{aligned}
&\simeq \sum_{x=0}^{1}(-1)^{x-1}\{(q_{+}+l_{+}\mathrm{E}(z-z_{+}|x))p_{+}(x)+(q_{-}+l_{-}\mathrm{E}(z-z_{-}|x))p_{-}(x)\} \\
&= \sum_{x=0}^{1}(-1)^{x-1}\Big\{(q_{+}+l_{+}W_{+}(x)t^{l_{+}-1})p_{+}(x) \\
&\quad +(q_{-}+l_{-}W_{-}(x)t^{l_{-}-1})p_{-}(x)\Big\} \\
&= \sum_{x=0}^{1}\Big[q_{+}p_{+}(x)+q_{-}p_{-}(x)+(l_{+}W_{+}(x)p_{+}(x)t^{l_{+}-1} \\
&\quad +l_{-}W_{-}(x)p_{-}(x)t^{l_{-}-1})\Big](-1)^{x-1}.
\end{aligned} \tag{5}
$$

Here, we expand $f(z)$ as

$$
f(z) = f(z_{\pm}+l_{\pm}\cdot(z-z_{\pm})) \simeq q_{\pm}+l_{\pm}\cdot(z-z_{\pm}).
$$

Given that $p_{+}(x)+p_{-}(x)=1$ for $x=0,1$, the limit value $c \equiv \lim_{t\to\infty} C(t)$ is estimated to be

$$
c = (q_{+}-q_{-})(p_{+}(1)-p_{+}(0)).
$$

For $J < J_c(h)$, $p_{+}(x)=1$ and $c=0$. For $J > J_c(h)$, the probability for the convergence of $z(t)$ to z_{-} is positive. It is natural to assume that $p_{+}(1) > p_{+}(0)$ and $c > 0$ for $J > J_c(h)$.

The asymptotic behavior of $C(t)$ is governed by the term with the largest value among $\{l_{+}, l_{-}\}$ for $J > J_c(h)$. We define $l_{\max}$ as

$$
l_{\max} \equiv \begin{cases} l_{+} \ , & J < J_c(h), \\ \mathrm{Max}\{l_{+}, l_{-}\} \ , & J > J_c(h). \end{cases}
$$

We summarize the asymptotic behavior of $C(t)$ as

$$
C(t) \simeq c + c'\cdot t^{l-1} \ \text{and} \ l = l_{\max}.
$$

Here, we write the coefficient of the term proportional to t^{l-1} as c'. If $J > J_c(h)$, the constant term c is the leading term. If $J < J_c(h)$, the power-law term $c'\cdot t^{l-1}$ is the leading term.

5.3 $J = J_c(h)$

On the boundary $J = J_c(h)$, there are two stable points z_+ and z_t for $h > 0$. As z_t is stable, the probability for the convergence of $z(t)$ to z_t is positive. It is natural to assume that $p_+(1) > p_+(0)$ and $c > 0$. If $h = 0$, there is only one stable point at $z_+ = z_t = 1/2$ and $c = 0$. As $l_{\max} = l_t = 1$, we anticipate that $|C(t) - c|$ becomes a decreasing function of $\ln t$. One possibility is a power-law behavior of $\ln t$ such as

$$C(t) \simeq c + c' \cdot (\ln t)^{-\alpha}.$$

We derive α by a simple heuristic argument. At first, we consider the case of $h = 0$. There is only one stable touchpoint at z_t, and $z(t)$ converges to it. Equation (5) suggests that the asymptotic behavior of $C(t)$ is governed by $\mathrm{E}(z - z_t|x)$ as z_t is the only stable state. As $f(z_t) = z_t$ and $f'(z_t) = 1$ at z_t, $f(z)$ can be approximated in the vicinity of z_t as

$$f(z) = -\delta(z - z_t)^3 + z.$$

Here δ is a positive constant, as z_t is stable (left figure in Fig. 11). The time evolution of $\mathrm{E}(z - z_t|x)$ is given as

$$\mathrm{E}(z(t+1) - z_t|x) - \mathrm{E}(z(t) - z_t|x) = \frac{1}{t+1}\mathrm{E}(f(z(t)) - z(t)|x) \simeq -\frac{\delta}{t}\mathrm{E}((z - z_t)^3|x).$$

Here the denominator $t + 1$ in the middle of the equation comes from the fact that there occurs a $\frac{1}{t+1}$ change in $\mathrm{E}(z(t)|x)$ for $X(t + 1) \in \{0, 1\}$. We also assume $\mathrm{E}((z(t) - z_t)^3|x) \simeq \mathrm{E}(z(t) - z_t|x)^3$, and the equation can be written as

$$\frac{d}{dt}\mathrm{E}(z(t) - z_t|x) = -\frac{\delta}{t}\mathrm{E}(z(t) - z_t|x)^3.$$

The solution to this shows the next asymptotic behavior

$$\mathrm{E}(z - z_t|x) \propto (\ln t)^{-1/2},$$

and we obtain $\alpha = 1/2$.

Likewise, for $h > 0$, there are two stable states q_+ and q_t. The subleading term in $C(t)$ is governed by $\mathrm{E}(z - z_t|x)$. We can approximate $q(z)$ in the vicinity of z_t to be

$$f(z) = \delta(z - z_t)^2 + z.$$

Here δ is a positive constant (right figure in Fig. 11). If $z(t) > z_t$, $z(t)$ moves toward the right-hand direction, on average, and converges to z_+. We only need to consider the case $z(t) < z_t$ and $z(t)$ converges to z_t. In the case, $\mathrm{E}(z(t) - z_t|x)$ obeys the next differential equation.

$$\frac{d}{dt}\mathrm{E}(z(t) - z_t|x) = \frac{\delta}{t}\mathrm{E}(z(t) - z_t|x)^2.$$

The solution shows the next asymptotic behavior

$$\mathrm{E}(z(t) - z_t|x) \propto (\ln t)^{-1},$$

and we obtain $\alpha = 1$.

The system shows a continuous phase transition for $h = 0$. As we have seen, the analysis does not depend on the precise form of $f(z)$. As far as $f(z)$ has Z_2-symmetry, $f(1-z) = 1 - f(z)$, and $y = f(z)$ are tangential to $y = z$ at $z_t = 1/2$, a similar continuous phase transition occurs (Mori and Hisakado 2015b). In order to discuss the universality class of the phase transition, we compare the scaling properties of $C(t)$ for two models. One model is the logistic model. For $h = 0$, $J_c(0) = 1$. Another model adopts the next $f_r(z)$ with three parameters r, $p \in [0, 1], q \in [1/2, 1]$.

$$f_r(z) = (1-p)\cdot q + p \cdot \pi_r(z)$$

$$\pi_r(z) = \sum_{s=(r+1)/2}^{r} {}_rC_s \cdot z^s(1-z)^{r-s}$$

r is odd number greater than or equal to three, $r \in \{3, 5, 7\cdots\}$. In the limit $r \to \infty$, $q_r(z)$ reduces to that of the majority model, $(1-p)\cdot q + p\cdot\theta(z - 1/2)$. This model corresponds to the mean-field approximation of the majority rule model, where the individual chooses the majority of r randomly chosen previous variables with a probability of p. For $q = 1/2$, $f_r(z)$ has Z_2-symmetry, and the threshold value $p_c(r)$ is determined by the condition $1 = f_r'(z_t = 1/2) = p_c(r)\cdot\pi_r'(1/2)$. p_c is explicitly given as

$$p_c(r) = \frac{[(r-1)/2!]^2 2^{r-1}}{r!}.$$

$p_c(r)$ is a decreasing function of r and $\lim_{r\to\infty} p_c(r) = 0$, which is compatible with $p_c = 1 - 1/2q_* = 0$ of the digital model with $q_* = 1/2$.

At $J = J_c(0)$, $C(t)$ obeys a power law of $\ln t$ as $C(t) \simeq c' \ln t^{-\alpha'}$ with $\alpha' = 1/2$ for the former model. Below $J_c(0) = 1$, $C(t) \propto t^{l-1}$ with $l = l_+ = J$. We set $\Delta J = J_c - J = 1 - J$. The expression for the exponent of $C(t) \propto t^{l-1}$ is given by

$$1 - l = \Delta J \quad , \quad J < J_c(0) = 1.$$

For $J > 1$, $C(t) - c \propto t^{l-1}$ with $l = l_+ = l_- = f'(z_+)$. As in the case of the estimation of the critical exponents for the mean-field Ising model (Stanley 1971), we solve $z_+ = f(z_+)$ with the assumption $\Delta J \equiv J - 1 << 1$. We obtain $z_+ - 1 = \sqrt{3\Delta J}$ and estimate l as

$$1 - l = \frac{1}{2}\Delta J \quad , \quad J > J_c(0) = 1.$$

As J approaches $J_c(0)$ from below and above of J_c, $1 - l$ approaches 0. At $J = J_c$, $C(t)$ obeys a power law of $\ln t$. As $t^{-(1-l)} = e^{-(1-l)\ln t}$, we can regard $1/(1-l)$ as the correlation length in the scale of $\ln t$. We assume the phenomenological scaling ansatz for $C(t)$ to be

$$C(t) = \ln t^{-\alpha} g((1-l)\ln t). \tag{6}$$

In the ansatz, $\ln t$ dependence of $C(t)$ is scaled with $1/(1-l)$. $g(x)$ is the universal function and is finite at $x = 0$. For $J = J_c$ and $l = 1$, $C(t) = g(0)\ln t^{-\alpha}$ with $g(0) = c'$. In the limit $x \to \infty$, in order to compensate the $\ln t^{-\alpha}$ term, $g(x)$ should behave as $g(x) \propto x^{\alpha}$ for $J > J_c(0)$. Then $C(t)$ behaves as $\lim_{t\to\infty} C(t) \propto (1-l)^{\alpha} = \Delta J^{\alpha}$. The critical exponent β for $c \propto \Delta J^{\beta}$ coincides with α. The exponent $\nu_{||}$ is defined for the divergence of $1/(1-l)$ as

$$1/(1-l) \propto \Delta J^{-\nu_{||}}.$$

As $(1-l) \propto \Delta J$, we obtain $\nu_{||} = 1$. The scaling relation $\beta = \alpha \cdot \nu_{||}$ holds.

For the latter model with $f_r(z)$, $l = p \cdot f_r'(z_t = 1/2)$ for $p < p_c(r)$. As $1 = p_c \cdot f_r'(1/2)$ holds, the correlation length $1/(1-l)$ is estimated to be

$$1/(1-l) = \frac{1}{\pi_r'(1/2)(p_c(r) - p)}$$

and diverges as $1/(1-l) \propto \Delta p^{-\nu_{||}}$ with $\nu_{||} = 1$. Here, we define $\Delta p \equiv p_c(r) - p$. For $p > p_c$, we assume $\Delta p = p - p_c(r) << 1$ and estimate z_+ by solving $z_+ = q_r(z_+)$ up to $O(\Delta z^3) = O((z_+ - 1/2)^3)$. One may then show that $1 - l = f_r'(z_+) \propto \Delta p$, and we obtain $\nu_{||} = 1$. If we assume the scaling form for $C(t)$ to be $C(t) \simeq \ln t^{-\alpha} \cdot g(\ln t \cdot (1-l))$ and define β as $c \propto \Delta p^{\beta}$, we obtain $\beta = \alpha \cdot \nu_{||}$.

If the two models share the same value for α and β, this suggests that they are in the same universality class.

5.4 *Numerical Calculation of $g(x)$*

We estimate the universal function $g(x)$ assumed in Eq. (6) numerically. For the former logistic model with $h = 0$, we estimate that $\alpha' = 1/2$. $J_c(0) = 1$ and we

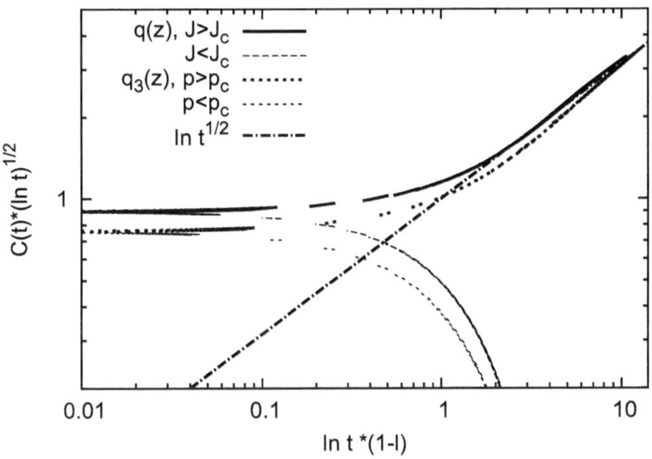

Fig. 12 Plot of $\ln t^{1/2} \cdot C(t)$ vs. $(l - 1)\ln t$

estimate $C(t)$ for $t \le 4 \times 10^5$ and $2/3 < J < 3/2$. For the latter model with $r = 3$ and $q = 1/2$, $p_c(3) = 2/3$. We estimate $C(t)$ for $t \le 4 \times 10^5$ and $4/9 < p < 1$. Using data for $C(t)$ between $10^4 \le t \le 4 \times 10^5$, we determine $g(x)$ to be

$$g(x) = \ln t^{1/2} \cdot C(t) \quad , \quad x = (1 - l)\ln t.$$

$g(x)$ should be smooth near $x = 0$ and $g(0) = c'$. For a sufficiently large J, $c \simeq 1$ and $l = 0$. $g(x)$ should behave as $x^{1/2}$ for sufficiently large x values. For $J < 1$, $g(x)$ should decrease exponentially.

Figure 12 shows the results of this analysis. The thick continuous and thick dashed lines indicate the results for $J > J_c(0)$ and $p > p_c(3)$, respectively. As can be clearly seen in this figure, the data obtained for different J and t values and for different p and t values lie on two curves, which represent $g(x)$ in the phase with $c > 0$ for both models. For large x, $g(x) \simeq x^{1/2}$. The thin continuous and thin dashed lines indicate the results for $J < J_c(0)$ and $p < p_c(3)$, respectively. The data lie on two curves, which represent $g(x)$ in the phase with $c = 0$ for both models. $g(x)$ can be seen to decay exponentially. The results indicate that the scaling ansatz in Eq. (6) holds with $\alpha = 1/2$.

6 Summary

We explain the domino effect in information cascade by studying several simple information cascade models. It was argued that the information cascade is fragile as the depth of cumulative private information is shallow and the later individual might choose the minority choice by trusting his private information. In order to test the conjecture, we propose to measure the domino effect in information cascade by estimating the correlation function $C(t)$ between the first individual's choice and the later individual's choice.

If the individual recognizes that the depth of the majority choice is shallow and might choose the minority choice based on his private information, the sequential choice process is modeled as correlated random walk (CRW) or CRW with finite memory length r, where the individual refers to previous r individual's choice. In the case, $C(t)$ decays exponentially, and the majority choice should become minority choice after many individuals' choices. If $q > 1/2$ and the information of the individual voter contain some information, the majority choice converges to the correct choice. Even so, the correlation length r might exponentially diverge with r if the ratio p of the individuals who choose the majority choice in the r choices is larger than some critical values p_c. If the precision of the private information is $q > 1/2$, p_c is given as $p_c(q) = 1 - 1/2q$.

If the network externality to enforce the majority choice increases with the number of votes, which occurs when individuals do not recognize the shallowness of the depth of the majority choice, the sequential choice process is described by nonlinear Pólya urn. In the case, the probabilistic rule of the sequential choice is described by a function $f(z)$, and the equilibrium of the system is given as the stable fixed point of f. If there is one equilibrium, $C(t)$ decays with t and the domino effect disappears. Even if the system is in the down cascade, it can be reverted to the up cascade. If there are two equilibria, the limit value $c = \lim_{t\to\infty} C(t)$ is positive, and the first individual's choice propagates to the infinitely later individuals. The change in the number of equilibrium is a non-equilibrium phase transition. If f is Z_2-symmetric, the phase transition is continuous. The asymptotic behavior $C(t)$ is described by the scaling form as

$$C(t) \propto (\ln t)^{-\alpha} g(\ln t/\xi).$$

$g(x)$ is the universal function and the scaling relation $\beta = \nu_{||} \cdot \alpha$, holds. If f is not Z_2-symmetric, c changes discontinuously from $c = 0$ in the one-equilibrium phase to $c > 0$ in two-equilibrium phase.

In later chapters, we estimate $f(z)$ using experimental data of information cascade and the time series data of horse race betting market.

References

Bikhchandani S, Hirshleifer D, Welch I (1992) A theory of fads, fashion, custom, and cultural changes as informational cascades. J Polit Econ 100:992–1026

Devenow A, Welch I (1996) Rational herding in nancial economics. Euro Econ Rev 40:603–615

Hill B, Lane D, Sudderth W (1980) A strong law for some generalized urn processes. Ann Probab 8:214–226

Hisakado M, Mori S (2010) Phase transition and information cascade in voting model. J Phys A Math Theor 43:315207–315219

Hisakado M, Mori S (2011) Digital Herders and phase transition in a voting model. J Phys A Math Theor 44:275204–275220
Mori S, Hisakado M (2015a) Finite-size scaling analysis of binary stochastic processes and universality classes of information cascade phase transition. J Phys Soc Jpn 84:054001–054013
Mori S, Hisakado M (2015b) Correlation function for generalized Pólya urns: finite-size scaling analysis. Phys Rev E92:052112–052121
Pemantle R (1991) When are touchpoints limits for generalized Pólya urns? Proc Am Math Soc 113:235–243
Pemantle R (2007) A survey of random processes with reinforcement. Pobab Surv 4:1–79
Pólya G (1931) Sur quelques points de la théorie des probabilités. Ann Inst Henri Poincar 1:117–161
Rendell L, Boyd R, Cownden D, Enquist M, Eriksson K, Feldman MW, Fogarty L, Ghirlanda S, Lillicrap T, Laland KN (2010) Why copy others? Insights from the social learning strategies tornament. Science 328:208–213
Stanley HE (1971) Introduction to phase transitions and critical phenomena. Oxford University Press, London

Information Cascade Experiment: General Knowledge Quiz

Shintaro Mori and Masato Hisakado

1 Information Cascade as Social Contagion Process

Because of the progress in information communication technology, we often rely on social information in decision-making, and the social contagion process has long been extensively studied. The Pólya urn is a simple stochastic process in which contagion is taken into account by a reinforcement mechanism (Pólya 1931). There are initially R_0 red balls and B_0 blue balls in an urn. At each step, one draws a ball randomly from the urn and duplicates it. Then, one returns the balls, and the probability of selecting a ball of the same color is strengthened. As the process is repeated infinitely, the ratio z of red balls in the urn becomes random and obeys the beta distribution $\mathrm{Beta}(R_0, B_0)$. In the process, information of the first draw propagates and affects the later draws. The correlation between the color of the first ball and that of a ball chosen later is $1/(R_0 + B_0 + 1)$.

As the Pólya urn process is very simple, and there are many reinforcement phenomena in nature and the social environment, many variants of the process have been proposed under the name of nonlinear Pólya urn (Pemantle 2007). In the process, the choice of the ball is described by a nonlinear function $f(z)$ of the ratio of red balls z (Hill et al. 1980). In contrast to the original Pólya urn, where $f(z) = z$, the ratio of red balls converges to a stable fixed point $z_* = f(z_*)$. Mathematically, the fixed points z_* are categorized as upcrossings and downcrossings, at which the graph $y = f(z)$ crosses the graph $y = z$ going upward and downward, respectively.

S. Mori (✉)
Faculty of Science and Technology, Department of Mathematics and Physics,
Hirosaki University, Hirosaki, Aomori, Japan

M. Hisakado
Nomura Holdings, Inc., Chiyoda-ku, Tokyo, Japan

A.-H. Sato (ed.), *Applications of Data-Centric Science to Social Design*,
Agent-Based Social Systems 14, https://doi.org/10.1007/978-981-10-7194-2_10

The downcrossing (upcrossing) fixed point is stable (unstable), and the probability that z converges to it is positive (zero).

If the number of stable fixed points changes as one changes the parameters of the function $f(z)$, the nonlinear Pólya urn shows continuous or discontinuous phase transitions (Mori and Hisakado 2015a,b). The order parameter is the limit value of the correlation between the first drawn ball and later drawn balls. If $f(z)$ is Z_2-symmetric and satisfies $f(z) = 1 - f(1 - z)$, the transition becomes continuous, and the order parameter satisfies a scaling relation in the nonequilibrium phase transition. One good candidate for experimental realization of the phase transition is the information cascade experiment (Bikhchandani et al. 1992; Devenow and Welch 1996; Anderson and Holt 1997). There, participants answer two-choice questions sequentially. In the canonical setting of the experiment, two urns, A and B, which are composed of different numbers of red and blue balls, are prepared. One of the two urns is chosen at random to be X, and the question is whether X is A or B. The participants can draw a ball from X and see which type of ball it is. This knowledge, which is called the private signal, provides some information about X. However, the private signal does not indicate the truth unequivocally, and participants have to decide under uncertainty. Participants are also provided with social information regarding how many prior participants have chosen each urn. The social information introduces network externality to the decision-making: as more participants choose A (B), later participants are more likely to identify X as A (B). The social interaction in which a participant tends to choose the majority choice even if it contradicts the private signal is called an information cascade or rational herding (Bikhchandani et al. 1992). In a simple model of information cascade, BHW's cascade model, if the difference in the numbers of subjects who have chosen A and B exceeds two, the social information overwhelms subjects' private signals. In the limit of many previous subjects, the decision is described by a threshold rule stating that a subject chooses an option if its ratio exceeds 1/2, $f(z) = \theta(z - 1/2)$ (Hisakado and Mori 2011). The function $f(z)$ that describes decisions under social information is called response function.

To detect the phase transition caused by the change in $f(z)$, we have proposed another information cascade experiment in which subjects answer two-choice general knowledge questions (Mori et al. 2012, 2013). If almost all of the subjects know the answer to a question, the probability of the correct choice is high, and $f(z)$ does not depend greatly on the social information. In this case, $f(z)$ has only one stable fixed point. However, when almost all the subjects do not know the answer, they show a strong tendency to choose the majority answer. Then $f(z)$ becomes S-shaped, and it could have multiple stable fixed points. In this chapter, we review the result of the information cascade experiment using the general knowledge quiz. We have performed four experiments up to now. We use all data to estimate the response function $f(z)$. Using $f(z)$, we explain the micro-macro features of the experiment.

2 Experimental Setup

The experiments reported here were conducted at the Information Science Laboratory at Kitasato University and at the Group Experiment Laboratory of the Center for Experimental Research in Social Sciences at Hokkaido University. The subjects included students from the two universities. We call the experiments as EXP1, EXP2, EXP3, and EXP4. EXP1 and EXP4 were performed at Kitasato University in 2010 and 2013. We performed EXP2 and EXP3 at Hokkaido University in 2011 and 2012.

In the experiments the subjects answer general knowledge two-choice questions one by one sequentially without referring to the previous subjects' choices in the first place. Afterward, they answer the same question by referring to the previous subjects' choices. We provide the number of subjects who have chosen each options. There are two ways to provide the information. One way is to provide it by gradually increasing the number of the previous subjects. We denote the number as r, and r increases in a set R, and at last the summary statistics of all previous subjects' choices is provided to the current subject. If the subject is $t+1$-th respondent, r increases in R until $r < t$ holds, and then the subject answers by referring to all $r = t$ previous subject choices. We adopted the procedure in EXP1 and EXP2. Another way is to provide the summary statistic of all $r = t$ previous subjects' choices. We adopted the procedure in EXP3 and EXP4. If we denote the answer without reference as $r = 0$ and the answer with reference to all previous subjects' choices as $r = t$, R should be $R = \{0, t\}$ in the latter case.

In EXP1, 62 subjects have participated and were divided into 2 groups of 31 subjects. They have answered 100 questions, and the length T of the subjects' sequence is 31. In EXP2 (3), 102 (120) students have participated and answered 120 question. We divide the subjects into two groups and the average length T is 50.8 (60). In EXP4, 125 students have participated and answered 120 questions. Each student answered 80 questions and $T = 83.15$. We denote the number of choice sequence and the (average) number of subject as I and $T(T_{\text{avg}})$, respectively. In Table 1, we summarize the data of the experiment. More tailed information about the experiments is found in Mori et al. (2012, 2013).

The subjects answered the quiz individually with and without information about the previous subjects' choices. This information, called social information, is given

Table 1 Experimental design. T is the length of the subjects' sequence and T_{avg} is the average value. R is the set of r. $r = t$ means that the subject $t + 1$ refers to all previous subjects' choices. I is the number of the sequence

Exp.	T, T_{avg}	$R = \{r\}$	I	Subject pool
EXP1	31	$\{0, 1, 2, 3, 5, 7, 9, t\}$	200	Kitasato Univ.
EXP2	50.8	$\{0, 1, 5, 11, 21, t\}$	240	Hokkaido Univ.
EXP3	60	$\{0, t\}$	240	Hokkaido Univ.
EXP4	83.2	$\{0, t\}$	120	Kitasato Univ.

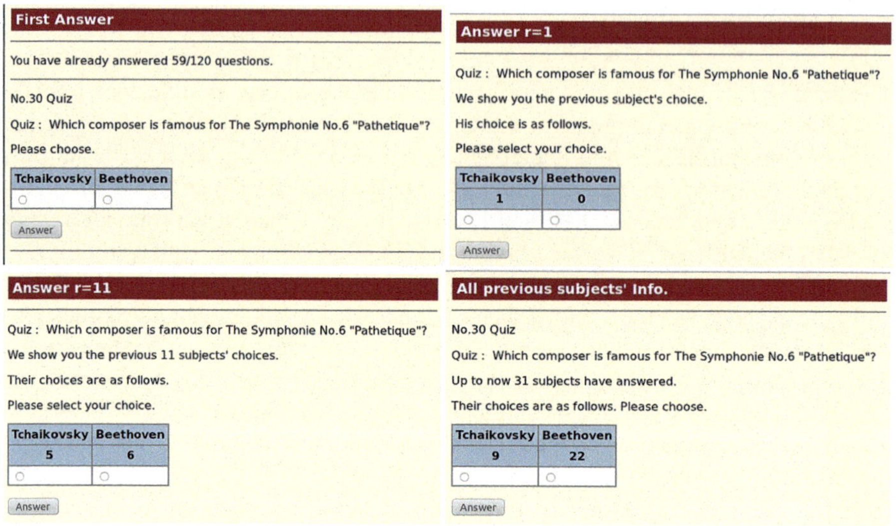

Fig. 1 Snapshot of the screen for $r = 0, 1, 11$ and t in EXP2. The summary statistics $\{N_0, N_1\}$ are given in the second row in the box in cases $r = 1, 11$ and t

as the summary statistics of the previous r subjects as $\{N_0, N_1\}$. $N_0 + N_1 = r$ holds. We denote the $t + 1$-th subject's answer to question i for case r by $X(i, r, t + 1)$, which takes the value 1 (0) if the choice is true (false). $\{N_0(i, r, t), N_1(i, r, t)\}$ are the numbers of subjects who choose each choice among the prior r subjects as $N_1(i, r, t) = \sum_{t'=t-r+1}^{t} X(i, r, t')$ and $N_0(i, r, t) = r - N_1(i, r, t)$.

Figure 1 shows the experience of the subjects in EXP2 more concretely. The subjects entered the laboratory and sat in the partitioned spaces. After listening to a brief explanation about the experiment and the reward, they logged into the experiment web site using their IDs and started to answer the questions. A question was chosen by the experiment server and displayed on the monitor. First, the subjects answered the question using their own knowledge only ($r = 0$). Later, the subjects received social information and answered the same question. Figure 1 shows the cases $r = 1, 11$ and $r = t$. Subjects could then use or ignore the social information in making decisions.

2.1 *Quiz Selection*

We explain the choice of the questions in the quiz. In the experiment, it was necessary to control the difficulty of the questions. If a question is too easy, all subjects know the answer. If the question is too difficult, no subjects know the answer. In order to study social influence by varying the ratio of people who do not know the answer, it is necessary to choose moderately difficult questions. In

Table 2 Five typical questions from the two-choice quiz in EXP2. q is the label of the questions in the quiz and $q \in \{0, 1, 2 \cdots, 119\}$

q	Question	Choice 0	Choice 1	Answer
0	Which insect's wings flap more in 1 minute?	Mosquito	Honeybee	0
1	During which period did the *Tyrannosaurus rex* live?	Jurassic	Cretaceous	1
3	Which animal has a horn at birth?	Rhinoceros	Giraffe	1
7	Which is forbidden during TV programs in Korea?	Commercials	Kissing scenes	0
8	Which instrument is in the same group as the marimba?	Vibraphone	Xylophone	1

EXP1, we selected 100 questions for which only 1 among the 5 experimenters knew the answer. This choice means that the ratio is estimated to be around 80% for the subjects. After EXP1, we calculated the correct answer ratio for each question. In general, the ratio should be greater than 0.5 for two-choice questions. A too small value of the ratio indicates some bias in the given choices of the question. We excluded questions with too small values of the ratio and prepared a new quiz with 120 questions in EXP2. Table 2 shows five typical questions from EXP2. In EXP3 and EXP4, we adopted the questions of EXP2 after slight replacement.

3 Data Analysis: Macroscopic Aspects

We denote the correct answer ratio up to respondent t for question i with reference to $r \in R$ as

$$Z(i, r, t) = \frac{1}{t} \sum_{s=1}^{t} X(i, r, s).$$

If $T(i)$ subjects answered to question i, the final value of $Z(i, r, t)$ is denoted as

$$Z(i, r) = Z(i, r, T(i)).$$

In order to see the social influence more pictorially, we show the scatter plots of $Z(i, r)$ vs. $Z(i, 0)$ in Fig. 2. The left figure shows the scatter plot of $Z(i, 0)$ and $Z(i, r)$ for $r \in \{1, 5, 21, t\}$ in EXP1 and EXP2. The right figure shows the plot of $Z(i, 0)$ and $Z(i, r = t)$ of all experiments. The x-axis shows $Z(i, 0)$ and the y-axis shows $Z(i, r)$ and $Z(i, r = t)$ in the figures.

If the subjects' answers are not affected by the social information, the data should distribute on the diagonal line. However, as the plots clearly indicate, this is not the case. As r increase in the left figure, the amount of social information increases. As r increases from $r = 1$ to $r = t$, the changes $|Z(i, r) - Z(i, 0)|$ increase, and the samples scatter more widely in the plane. In the right figure, as the EXP No. increases, the average length of the subjects' sequence increases. We

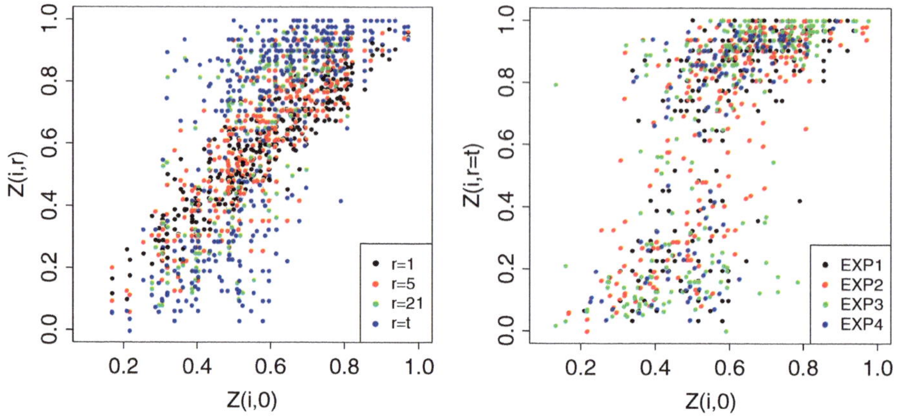

Fig. 2 (Left) Scatter plots of $Z(i, r)$ vs. $Z(i, 0)$ for $r \in \{1, 5, 21, t\}$ in EXP1 and EXP2. (Right) Scatter plots of $Z(i, r = t)$ vs. $Z(i, 0)$ in all experiments

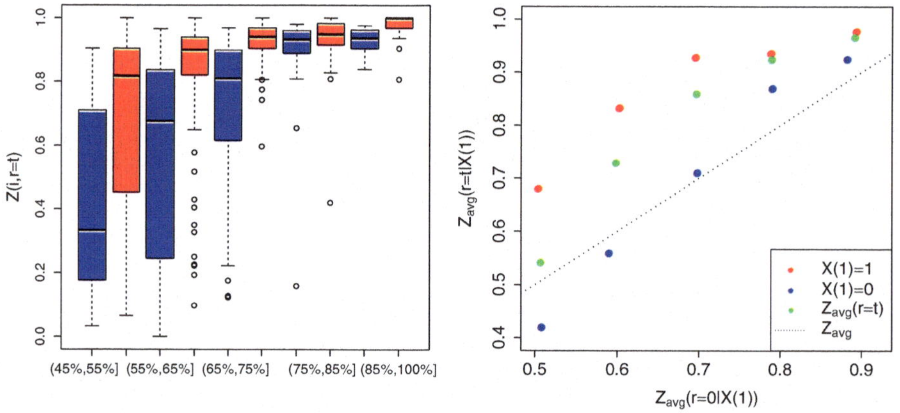

Fig. 3 (Top) Boxplot of $Z(i, t)$. (Bottom) Plot of $Z_{\text{avg}}(x)$ and Z_{avg} vs. q

do not notice clear difference between EXP1($T = 31$) and EXP4($T_{\text{avg}} = 83.2$). For the samples with $Z(i, 0) \geq 0.8$, the changes are almost positive, and $Z(i, r = t)$ takes a value greater than 0.9. The average performance improves by the social information there. On the other hand, for the samples with $0.4 \leq Z(i, 0) < 0.7$, the social information does not necessarily improve the average performance. There are many samples with negative change $Z(i, r = t) - Z(i, 0) < 0$. These samples are in the sub-optimal state which corresponds to the equilibrium q_- of nonlinear Pólya urn (See chapter "Domino Effect in Information Cascade").

Figure 3 shows boxplots and the average values of $Z(i, r = t)$. The data are stratified according to $Z(i, 0)$ and $X(i, 0, 1)$. From left to right, the bins for $Z(i, 0)$ are $Z(i, 0) \in \{(45\%, 55\%], (55\%, 65\%], (65\%, 75\%], (75\%, 85\%], (85\%, 100\%]\}$. Furthermore, the data are classified by $X(i, r = t, 1) = x \in \{0, 1\}$, and blue (red) boxplots (symbols) are the boxplots (symbols) for $x = 0(1)$.

The left figure shows the boxplots. When $Z(i, 0) \in (45\%, 55\%], (55\%, 65\%]$, the distribution of $Z(i, r = t)$ is very wide, and it greatly depends on x. For $Z(i, 0) \in (45\%, 55\%]$, the median of $Z(i, r = t)$ is below 40% for $x = 0$ and about 80% for $x = 1$, respectively. The results suggest that there are two equilibria. When $Z(i, 0) \in (75\%, 85\%], (85\%, 100\%]$, the difference between the median of $Z(i, r = t)$ for $x = 0$ and $x = 1$ becomes small. All $Z(i, r = t)$ except the two samples are greater than one-half. This suggests that there is only one equilibrium.

The right figure shows the average value of $Z(i, r = t)$. The average values are denoted as $Z_{\text{avg}}(r = t|x)$, and $Z_{\text{avg}}(r = t)$ is the unconditional average value. The difference between $Z(r = t|1)$ (red) and $Z(r = t|0)$ (blue) is larger when $Z_{\text{avg}}(0|x)$ is small. As $Z_{\text{avg}}(r = t)$ (green) is larger than q (thin dotted line), we observe the collective intelligence effect.

3.1 Correlation Function $C(t)$ and Related Quantities

As we have discussed in chapter "Domino Effect in Information Cascade", the order parameter c of the information cascade phase transition of nonlinear Pólya urn is defined as the limit value of the correlation function $C(t)$ (Mori and Hisakado 2015a). $C(t)$ is defined as

$$C(t) \equiv \text{Cor}(X(1), X(t+1))/\text{V}(X(1))$$

$$= \text{E}(X(t+1)|X(1) = 1) - \text{E}(X(t+1) = 1|X(1) = 0).$$

$C(t)$ behaves asymptotically with three parameters, c, c', and $l > 0$, as

$$C(t) \simeq c + c' \cdot t^{l-1}.$$

If there is one stable state at z_+, $f(z_+) = z_+$, $Z(t)$ converges to z_+. The memory of $X(1) = x$ disappears and $c = 0$. $C(t)$ decays to zero with power-law behavior as $C(t) \propto t^{l-1}$. The exponent l is given by the slope of $f(z)$ at the stable fixed point z_+ as $l = f'(z_+) < 1$. If there are two stable states at z_-, z_+ and $z_- < z_+$, the probability that $Z(t)$ converges to $z_\pm$ depends on $X(1) = x$. If c is subtracted from $C(t)$, the remaining terms also obey a power law as $C(t) - c \propto t^{l-1}$. The exponent l is given by the larger one of $\{f'(z_+), f'(z_-)\}$.

We estimate $C(t)$ using experimental data. We combine data of EXP1 and EXP2 and denote EXP1+EXP2. Likewise, we combine data of EXP3 and EXP4 and denote EXP3+EXP4. The reason to divide four experiments into two groups is that the way to provide social information is completely different between the groups. Figure 4 plots $C(t)$ vs. t. The left figure plots $C(t)$ of EXP1 + EXP2, and the right figure plots $C(t)$ of EXP3 + EXP4, respectively. We observe that $C(t)$ shows the tendency of the monotonic decrease apart from some fluctuation. For $Z(i, 0) \in (45, 55\%], (55\%, 65\%]$, $C(t)$ is about 0.2 at $t = 30$, which suggests that $c > 0$. For $Z(i, 0) \in (75, 85\%]$, $C(t)$ decays to zero. The right figure for EXP3 + EXP4 shares

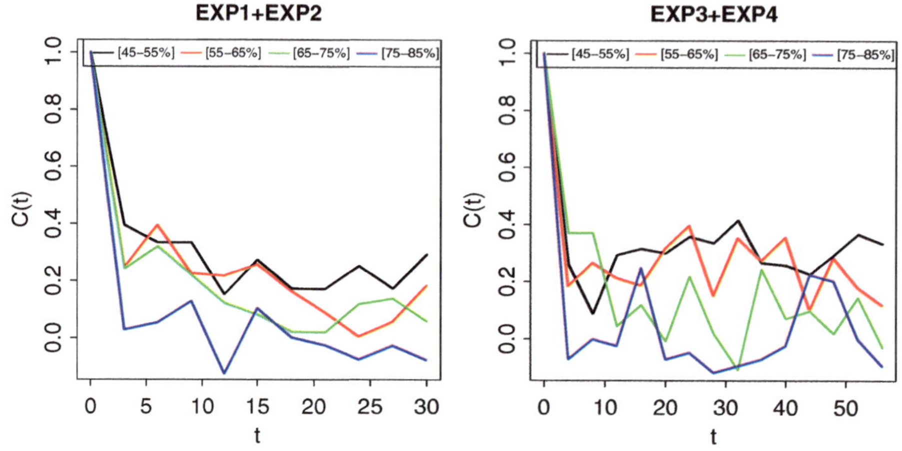

Fig. 4 Plot of $C(t)$ vs. t. The left (right) figure is the plot for EXP1 + EXP2 (EXP3 + EXP4). The data are stratified by $Z(i,0)$, and we show the results for the four bins, $Z(i,0) \in (45\%, 55\%]$, $(55\%, 65\%]$, $(65\%, 75\%]$, and $(75\%, 85\%]$

almost the same features with the left figure for EXP1 + EXP2. The results suggest the phase transition between the phase with $c > 0$ and the phase with $c = 0$.

In order to clarify the possibility of the phase transition, we estimate the relaxation time $\tau(t)$ and the second moment correlation time $\xi(t)$ using the n-th moment of $C(t)$ as (Mori and Hisakado 2015a,b)

$$M_n(t) \equiv \sum_{s=0}^{t-1} C(s)s^n$$

$$\tau(t) = M_0(t)$$

$$\xi(t) = \sqrt{\frac{M_2(t)}{M_0(t)}}$$

If we assume that $C(t) \propto t^{l-1}$, $M_n(t)$ behaves as

$$M_n(t) \propto \frac{1}{n+l} t^{n+l}$$

Using the asymptotic behavior of $M_n(t)$, we find $\tau(t)/t$ behaves as

$$\tau(t)/t \propto \frac{1}{l} t^{l-1},$$

and $\xi(t)/t$ behaves as

$$\xi(t)/t \propto \sqrt{\frac{l}{2+l}}.$$

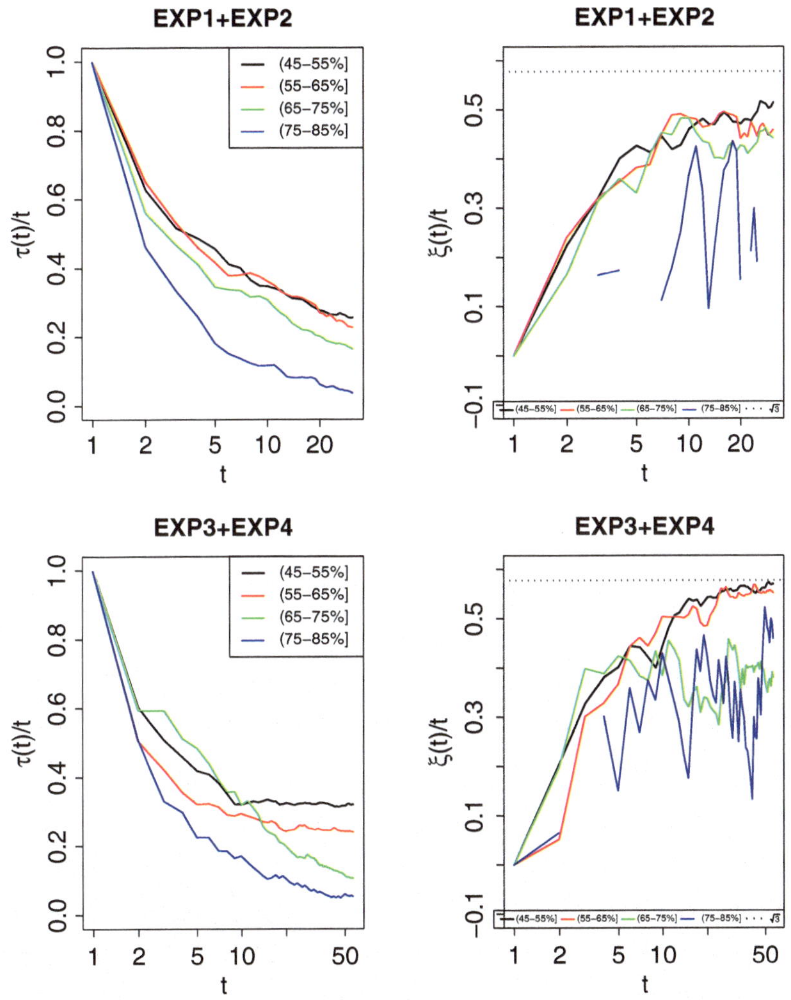

Fig. 5 Semilogarithmic plot of $\tau(t)/t$(top) and $\xi(t)/t$ (bottom) vs. t

When $c > 0$, we assume $C(t) \simeq c + c' t^{l-1}$, $c > 0$. We find $\tau(t)/t \simeq c + o(1/t)$ and $\xi(t)/t \simeq 1/\sqrt{3}$. If $\lim_{t\to\infty} \tau(t)/t > 0$ and $\lim_{t\to\infty} \xi(t)/t = 1/\sqrt{3}$, we can judge $c > 0$. If $\lim_{t\to\infty} \tau(t)/t = 0$ and $\lim_{t\to\infty} \xi(t)/t < 1/\sqrt{3}$, $c = 0$.

Figure 5 shows the semilogarithmic plot of $\tau(t)/t$ (top left and bottom left) and $\xi(t)/t$ vs. t(top right and bottom right). $\tau(t)/t$ converges to a positive value for $Z(i,0) \in (45\%, 55\%], (55\%, 65\%]$ for EXP3+EXP4. For $Z(i,0) \in (65\%, 75\%], (75\%, 85\%]$ in EXP3+EXP4 and all cases in EXP1+EXP2, it is difficult to judge whether $\tau(t)/t$ converges to zero or a positive value. The slopes at the right edge are negative; it suggests that $\tau(t)/t$ converges to zero. However, in EXP1+EXP2, $t \le 30$ and the judge about the limit $t \to \infty$ is impossible.

$\xi(t)/t$ converges to $1/\sqrt{3}$ for $Z(i,0) \in (45\%, 55\%], (55\%, 65\%]$ in EXP3+EXP4, which suggests that $c > 0$. For $Z(i,0) \in (65\%, 75\%], (75\%, 85\%]$, even though the fluctuation is large, $\xi(t)/t$ seems to converge to a value less than $1/\sqrt{3}$. In EXP1+EXP2, one observes almost the same feature with EXP3+EXP4; as $t \leq 30$, the converging value of $\xi(t)/t$ is unclear.

The behaviors of $C(t)$, $\tau(t)/t$, and $\xi(t)/t$ suggest that there occurs the information cascade phase transition in EXP3+EXP4. As the system size is very limited, we need to estimate the response function $f(z)$ of nonlinear Pólya urn. If the number of stable fixed points changes, it strongly supports the existence of the phase transition.

4 Data Analysis: Microscopic Aspects

The results of the macroscopic analysis in the previous section strongly suggest the existence of the information cascade phase transition. In this section, we study the microscopic behavior of the subjects. We study how the social information affects the subjects' choices. We estimate $f(z)$ of the nonlinear Pólya urn. $f(z)$ is defined as

$$P(X(t+1) = 1 | z(t) = z) = f(z).$$

Based on the results, we describe the system with nonlinear Pólya urn.

4.1 *Strength of Social Influence*

To measure how strongly the social information affects subjects' decision-making, we estimate the correlation coefficients between them. We estimate the correlation coefficients between $Z(i, r = t, t)$ and $X(i, r = t, t+1)$ for $t \geq 1$.

Figure 6 shows the plot of the correlation coefficients versus t. Overall, $\mathrm{Cor}(X(i, r = t, t+1), Z(i, r = t, t))$ increases with t and begins to fluctuate around 0.6 for $t \geq 10$. The result suggests that the microscopic behavior of the subjects becomes stationary for $t \geq 10$.

4.2 *Response Function $f(z)$*

As $\mathrm{Cor}(Z(i, r = t, t), X(i, r = t, t+1))$ increases with t and saturates at around $t = 10$, we use data for $t \geq 10$ to estimate $f(z)$. We prepared data set $\{Z(i, r = t, t), X(i, r = r, t+1)\}, t \geq 10$ for each bin $(45, 55\%], (55\%, 65\%], (65\%, 75\%], (75, 85\%]$. In each bin, we calculate the

Fig. 6 Plot of $\mathrm{Cor}(X(i, r = t, t + 1), Z(i, r = t, t))$ vs. t

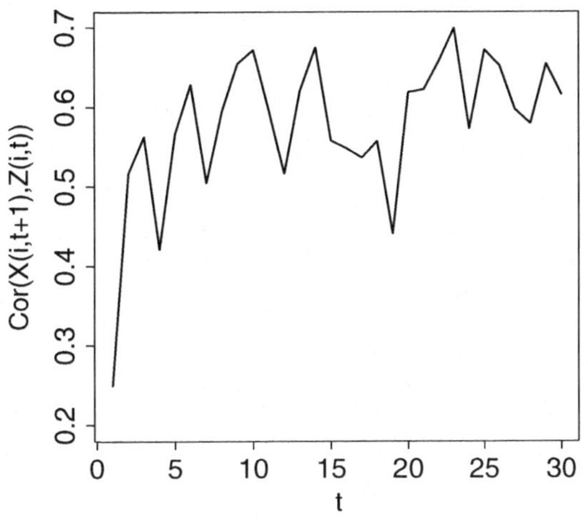

average value of $Z(i, r = t, t), X(i, r = t, t + 1)$ and obtain $f(z)$. We plot the results in Fig. 7. The top (bottom) figure shows the result for EXP1+EXP2 (EXP3+EXP4). It is clear that $f(z)$ is an almost monotonic increasing function of z. We also notice that $f(z)$ for bin $(45, 55\%]$ has two stable fixed points if one simply extrapolate $f(z)$ linearly. In other cases, there is only one stable fixed point. The results suggest that the system shows the information cascade phase transition by the change of the difficulty of the question.

The fitted curves in the figures adopt the next functional form

$$f(z) = (1 - p) \cdot 1 + p \cdot \mathrm{logistic}(\lambda(z - 1/2)). \tag{1}$$

Here logistic $(x) = 1/(1 + \exp(-x))$ is logistic function. The functional form assumes that there are two types of subjects, and $1 - p : p$ means the ratio of each type. The subjects in one type know the answer to the question. The probability that the subject choose the correct answer is 1. The subjects of another type do not know the answer. They have the tendency to follow the majority' choice, and we assume that the probability that the they choose the option with the vote share z is given by the logistic function. λ is the measure of the tendency to follow the majority's choice. If there is no social information, the probability to choose the correct option is $1 - p/2$. We estimate p by the formula $1 - p/2 = Z(i, 0)$. We estimate λ using the maximum likelihood method. In EXP1 + EXP2 (EXP3 + EXP4), $\lambda = 4.69(5.30)$.

Base on the form of $f(z)$, we can understand the behavior of $C(t)$. If the question is difficult and $Z(i, 0) \in (45, 55\%]$, there are two stable fixed point and $c > 0$. The first subject's choice affects the later subject's choices forever. If the question is easy and there is only one stable fixed point, $C(t)$ decays to zero, and the domino effect disappears. Even so, the decay of $C(t)$ obeys power law; the domino effect might remain for long time in the sequence.

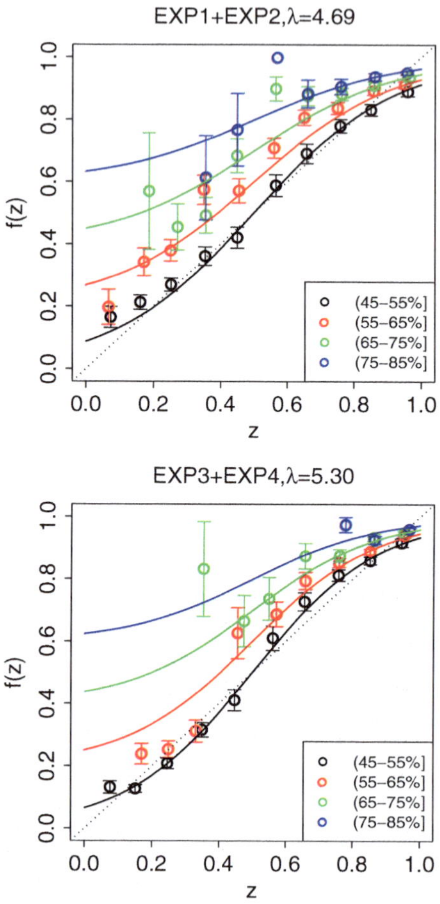

Fig. 7 Plot of $f(z)$ vs. z. The solid curves are fitted results with Eq. (1)

5 Conclusions

In this chapter, we explain the results of information cascade experiment that uses general knowledge two-choice quiz. Contrary to the conventional urn choice quiz, it is difficult to control and know the private signal in the experiment. We overcome the difficulty by letting subjects answer with and without referring to the previous subjects' choices. From the answer without reference, we estimate the difficulty of the questions. The distribution of the correct answer ratio depends on the difficulty of the question. If the questions are easy (difficult), the distribution is unimodal (bimodal). The change in the shape of the distribution suggests the phase transition of nonlinear P'olya urn. We estimate the response function $f(z)$ empirically and show that the system has one (two) stable fixed point(s) if the questions are easy (difficult). Based on $f(z)$, we understand the behavior of the correlation function $C(t)$ and verify the possibility of the phase transition macroscopically.

6 Data and R Script

One can download the data and the R script that is used to plot the figures in this chapter. One should visit https://sites.google.com/site/shintaromori/home/data and download AppDCSSDy_Chap10.tar.gz.

References

Anderson LR, Holt CA (1997) Information cascades in the laboratory. Am Econ Rev 87:847–862

Bikhchandani S, Hirshleifer D, Welch I (1992) A theory of fads, fashion, custom, and cultural changes as informational cascades. J Polit Econ 100:992–1026

Devenow A, Welch I (1996) Rational herding in financial economics. Euro Econ Rev 40:603–615

Hill B, Lane D, Sudderth W (1980) A strong law for some generalized urn processes. Ann Prob 8:214–226

Hisakado M, Mori S (2011) Digital herders and phase transition in a voting model. J Phys A Math Theor 44:275204–275220

Mori S, Hisakado M (2015a) Finite-size scaling analysis of binary stochastic processes and universality classes of information cascade phase transition. J Phys Soc Jpn 84:054001–054013

Mori S, Hisakado M (2015b) Correlation function for generalized Pólya urns: finite-size scaling analysis. Phys Rev E92:052112–052121

Mori S, Hisakado M, Takahashi T (2012) Phase transition to two-peaks phase in an information cascade voting experiment. Phys Rev E86:026109–026118

Mori S, Hisakado M, Takahashi T (2013) Collective adoption of max-min strategy in an information cascade voting experiment. J Phys Soc Jpn 82:0840004–0840013

Pemantle R (2007) A survey of random processes with reinforcement. Pobab Surv 4:1–79

Pólya G (1931) Sur quelques points de la théorie des probabilités. Ann Inst Henri Poincar 1:117–161

Information Cascade Experiment: Urn Quiz

Shintaro Mori and Masato Hisakado

1 Background

Many researchers have done the experimental studies of information cascade after the pioneering work by Anderson and Holt (1997). The experimental implementation of information cascade is based on the "basic" model (Bikhchandani et al. 1992), where two-choice question is the choice of one of two urns. In the experiment of Anderson and Holt, six subjects answer the two-choice quiz one by one after referring to all the previous subjects' choices. There are two urns, urn A and urn B, and urn $A(B)$ contains two (one) red and one (two) blue balls. Later, in addition to the original 2:1 and 1:2 compositions of red and blue balls, 5:4 and 4:5 compositions cases were also studied (Goeree et al. 2007). At the beginning, urn X is chosen at random from the two urns A and B, and the question is which urn is X. As the private signal, one draws a ball from X, and as public information, one knows the choices of all previous subjects. If the ball is red (blue), the posterior probability that urn X is A (B) is $2/3$ in the original 2:1 and 1:2 setting.

We denote the private signal for t-th subject in the sequential choices as $S(t)$. The probability that $S(t)$ is correct is q, q depends on the composition of balls in the urn. In the $m : n$ and $n : m$ composition case, where urn A (B) contains $m(n)$ red balls and $n(m)$ blue balls, the posterior probability that X is A is $q = m/(m+n)$ if one gets a red ball. If one gets a blue ball, the probability that X is B is $q = m/(m+n)$. In the original information cascade experiment, $t+1$-th subject answered the question based on his private signal and the history of the previous subjects' choices. In the

S. Mori (✉)
Faculty of Science and Technology, Department of Mathematics and Physics, Hirosaki University, Hirosaki, Aomori, Japan

M. Hisakado
Nomura Holdings, Inc., Chiyoda-ku, Tokyo, Japan

A.-H. Sato (ed.), *Applications of Data-Centric Science to Social Design*, Agent-Based Social Systems 14, https://doi.org/10.1007/978-981-10-7194-2_11

experiment reported here, the subject answered based on the private signal and the summary statistics of the previous subjects' choices (Hino et al. 2016). We model the sequential choice process as a nonlinear Pólya urn and estimate the response functions of the subjects. We study the possibility of the information cascade phase transition from the microscopic and macroscopic viewpoints.

2 Experimental Setup

The experiments reported here were conducted at Kitasato University. We recruited 307 students, mainly from the school of science. We prepared $I = 200$ questions for $q \in Q = \{5/9, 6/9, 7/9\}$ and $I = 400$ questions for $q = 8/9$. Subjects sequentially answered two-choice questions and received returns for each correct choice. The experiment was performed during three periods, $q \in \{5/9, 6/9\}$ in 2013, $q = 7/9$ in 2014, and $q = 8/9$ in 2015. We label the questions as $i = 1, 2, \cdots, I$. We obtained I sequences of answers of length $T = 63$ for $q = 5/9, 6/9$. The average length T is 54.0 for $q = 7/9$ and 60.5 for $q = 8/9$.

We prepared a question for $q = m/(m+n) \in Q$ by randomly choosing an urn X from two urns, A and B. For $q = m/(m+n) > 1/2$, urn A (B) contains m red (blue) balls and n blue (red) balls. Urn A (B) contains more red (blue) balls than blue (red) balls. We denote the answer to question $q \in Q, i \in \{1, \cdots, I\}$ as $U(q, i) \in \{A, B\}$. The subjects obtain information about X by knowing the color of a ball randomly drawn from it. The color of the ball is the private signal, as it is not shared with other subjects. If the ball is red (blue), X is more likely to be A (B). Further, q is the posterior probability that the randomly chosen ball suggests the correct urn. We prepared the private signal $S(q, i, t) \in \{A, B\}$ for T subjects and I questions in advance. Table 1 summarizes the design.

Subjects answered the questions sequentially using their respective private signals and information about the previous subjects' choices. This information, called social information, was given as the summary statistics of the previous subjects. When the subject answers question q, i after t subjects, the subject receives a private signal $S(q, i, t + 1)$ and social information $\{C_A(q, i, t), C_B(q, i, t)\}$ from the previous t subjects. Let $X(q, i, s) \in \{A, B\}$ be the s-th subject's choice; the social information $C_x(q, i, t), x \in \{A, B\}$ is written as

Table 1 Experimental design. $|ID|$ number of subjects, T_{avg} average length of subject sequence, $\{q\}$ precision of private signal, I number of questions

Date	$\|ID\|$	T_{avg}	$\{q\}$	I
2013.9∼ 2013.10	126	63	$\{5/9, 6/9\}$	200
2014.12	109	54.0	7/9	200
2015.9	121	60.5	8/9	400

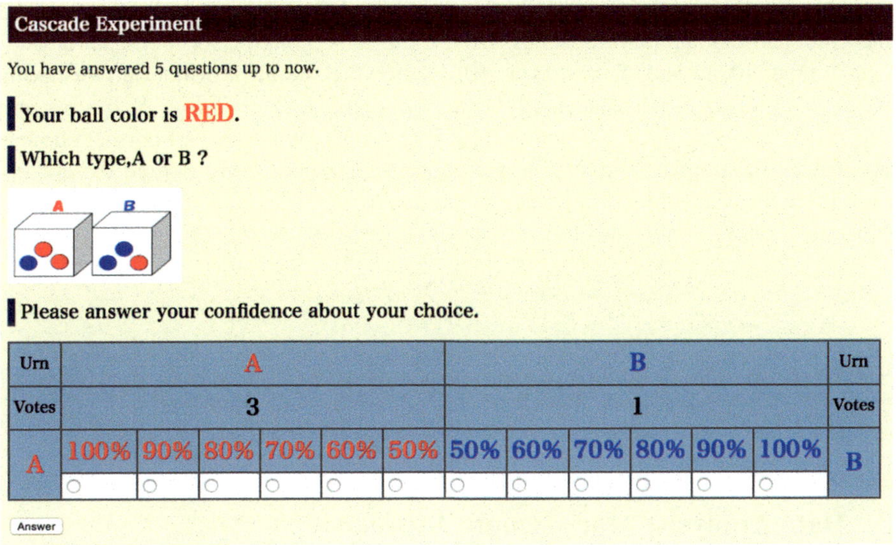

Fig. 1 Snapshot of the screen for $q = 6/9 = 2/3$. The private signal (RED) is shown on the second line. The summary statistics $\{C_A(t), C_B(t)\} = \{3, 1\}$ appear in the second row in the box. Subjects answer by choosing a radio button. In the experiment, subjects are asked to answer their confidence about their answers

$$C_x(q, i, t) = \sum_{s=1}^{t} \delta_{X(q,i,s),x},$$

where $C_A(q, i, t-1) + C_B(q, i, t-1) = t$ holds.

Figure 1 illustrates the experience of subjects more concretely. The second line shows the subject's private signal. The figure below the question shows the type of question, q. Before the experiment, the experimenter described the ball configuration in A and B and explained how the signal is related to the likelihood for each urn. The subjects can recall the question by looking at the figure. In the second row of the box, the social information $\{C_A, C_B\}$ is provided. In the screenshot shown in the figure, four subjects have already answered the question. Three of them have chosen A, and one has chosen B. The subject chooses A or B using the radio buttons in the last row of the box. They were asked to choose A or B by stating how confident they were about their answers, that is, to choose 100% if they were certain about their choices and to choose 50% if they were not at all confident about their choices. The reward for the correct choice does not depend on the confidence level. Irrespective of the degree of confidence, subjects receive a fixed return for the correct choice. After they chose an option and put answer button below the box, they know whether their choices are correct or not in the next screen. For more details about the experimental procedures, please refer to Hino et al. (2016).

Hereafter, instead of A and B, we use 1 and 0 to describe the correct and incorrect choices and private signal. We use the same notation for them, as follows: $S(q,i,t) \in \{0,1\}$ and $X(q,i,t) \in \{0,1\}$. For the social information, we define $\{C_1(q,i,t), C_0(q,i,t)\}$ as $C_1(q,i,t) \equiv C_{U(q,i)}(q,i,t) = \sum_{s=1}^{t} X(q,i,s)$ and $C_0(q,i,t) \equiv t - C_1(q,i,t)$. Here, $C_1(q,i,t)$ shows the number of correct choices up to the t-th subject for question $q \in Q, i \in \{1,\cdots,I\}$. We denote the correct ratio up to the t-th subject for question q,i as $Z(q,i,t)$:

$$Z(q,i,t) = \frac{1}{t}\sum_{s=1}^{t} X(q,i,s).$$

We denote the length of the subjects' sequence as $T(q,i)$ and the final value of $Z(q,i,T(q,i)))$ as $Z(q,i)$.

3 Data Analysis: Macroscopic Aspects

In this section, we show the macroscopic aspects of the results of the analysis of the experimental data. We show how the difficulty of quiz q and the first subject's answer $X(q,i,1)$ affect the $Z(q,i)$.

3.1 Distribution of $Z(q,i)$

We study the relationship between the precision of the private signal q and $Z(q,i)$. As we are interested in the dependence of $Z(q,i)$ on $X(q,i,1)$, we divide the samples based on $X(q,i,1) = x$. We denote the average value of $Z(q,i)$ for $x \in \{0,1\}$ and $q \in Q$ as $Z_{\mathrm{avg}}(q|x)$, respectively.

$$Z_{\mathrm{avg}}(q|x) = \frac{\sum_{i=1}^{I} Z(q,i)\delta_{X(q,i,1),x}}{\sum_{i=1}^{I} \delta_{X(q,i,1),x}}.$$

The unconditional average value of $Z_{\mathrm{avg}}(q|x)$ is then given as

$$Z_{\mathrm{avg}}(q) = q \cdot Z_{\mathrm{avg}}(q|1) + (1-q)\cdot Z_{\mathrm{avg}}(q|1).$$

The deviation of $Z_{\mathrm{avg}}(q)$ from q is a measure of the collective intelligence.

Figure 2 shows boxplots of $Z(q,i)$ for the samples with $X(q,i,1) = x \in \{0,1\}$ and $q \in Q$ and plot of $Z_{\mathrm{avg}}(q|x)$ vs. q and x. From left to right, q increases. Blue (red) boxplots (symbols) are for $x = 0(1)$. When q is small, the distribution of $Z(q,i)$ greatly depends on x. For $q = 5/9$, the median is about 50% for $x = 0$ and 80% for $x = 1$, respectively. For $q = 8/9$, the difference becomes small. All $Z(q,i)$

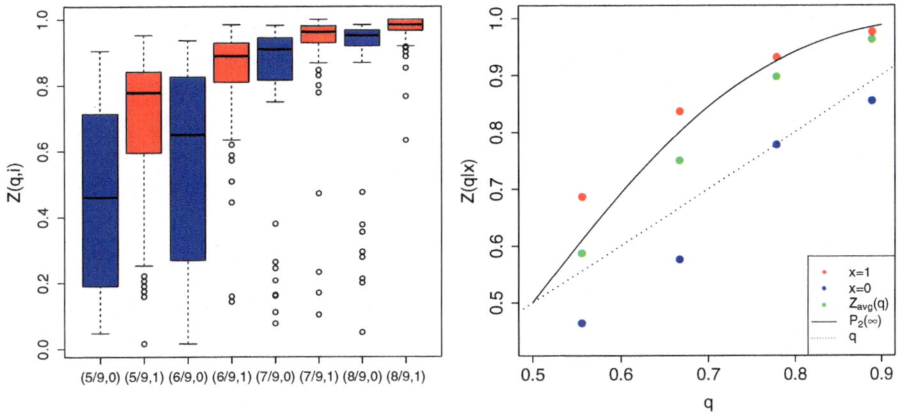

Fig. 2 (Left) Boxplot of $Z(q, i)$. (Right) Plot of $Z_{\rm avg}(q|x)$ and $Z_{\rm avg}(q)$ vs. q. The thin solid line shows $P_2(\infty)$ given by Eq. (1) in chapter "Domino Effect in Information Cascade"

are larger than one-half for $x = 1$. This suggests that $Z(q, i, t)$ converges to almost 1 as t increases. On the other hand, if $x = 0$ for $q = 8/9$, there are some samples with $Z(q, i) < 1/2$. We cannot judge whether all $Z(q, i, t)$ converge to almost 1 in the limit $t \to \infty$. If $x = 0$ with $q \in \{5/9, 6/9\}$, the distribution of $Z(q, i)$ is wide, suggesting the existence of two fixed points where $Z(q, i, t)$ converges.

From the right figure, one notices that $Z(q|x)$ monotonically increases with q. The difference between $Z(q|1)$(red) and $Z(q|0)$(blue) is larger when q is small. As $Z_{\rm avg}(q)$(green) is larger than q(thin dotted line), there is the collective intelligence effect. The thick solid line plots $P_2(\infty)$ in Eq. (1) in chapter "Domino Effect in Information Cascade" as a function of q. One sees that $P_2(\infty)$ describes $Z_{\rm avg}(q)$ relatively well. However, it does not mean that the experiment should be described by the BHW's simple cascade model.

3.2 Correlation Function $C(t)$ and Related Quantities

As we have discussed in chapter "Domino Effect in Information Cascade", the order parameter of the information cascade phase transition of nonlinear Pólya urn with $f(z)$ is defined as the limit value of the correlation function $C(t)$ (Mori and Hisakado 2015a). $C(t)$ is defined as

$$C(t) \equiv \mathrm{Cor}(X(1), X(t+1))/\mathrm{V}(X(1))$$

$$= \mathrm{E}(X(t+1)|X(1)=1) - \mathrm{E}(X(t+1)=1|X(1)=0).$$

$C(t)$ behaves asymptotically with three parameters, c, c' and $l > 0$, as

$$C(t) \simeq c + c' \cdot t^{l-1}. \tag{1}$$

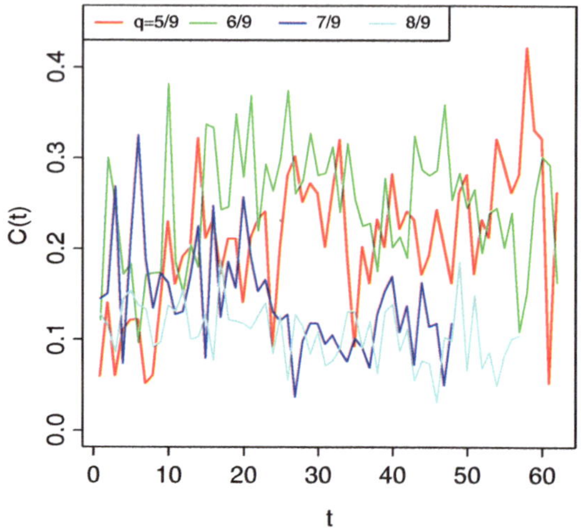

Fig. 3 Plot of $C(t)$ vs. t for $q = 5/9, 6/9, 7/9$ and $8/9$

If there is one stable state at z_+, $f(z_+) = z_+$, $Z(t)$ converges to z_+. The memory of $X(1) = x$ disappears and $c = 0$. $C(t)$ decreases to zero with power-law behavior as $C(t) \propto t^{l-1}$. The exponent l is given by the slope of $f(z)$ at the stable fixed point z_+ as $l = f'(z_+) < 1$. If there are two stable states at z_-, z_+ and $z_- < z_+$, the probability that $Z(t)$ converges to $z_\pm$ depends on $X(1) = x$. If c is subtracted from $C(t)$, the remaining terms also obey a power law as $C(t) - c \propto t^{l-1}$. The exponent l is given by the larger one of $\{f'(z_+), f'(z_-)\}$, as the term with the larger value governs the asymptotic behavior of $C(t) - c$.

Figure 3 plots $C(t)$ as a function of t. $C(t)$ fluctuates around 0.25 for $q \in \{5/9, 6/9\}$. For $q \in \{7/9, 8/9\}$, $C(t)$ decreases slowly and takes small values for large t. The results suggest that there occurs the phase transition between $q = 7/9, 8/9$ with $c = 0$ and $q = 5/9, 6/9$ with $c > 0$. However, it is difficult to judge whether $C(t)$ decreases to zero or fluctuates around some positive values in the limit $t \to \infty$.

In order to clarify the possibility of the phase transition, we estimate the relaxation time $\tau(t)$ and the second-moment correlation time $\xi(t)$ using the n-th moment of $C(t)$ as (Mori and Hisakado 2015a,b),

$$M_n(t) \equiv \sum_{s=0}^{t-1} C(s)s^n$$

$$\tau(t) = M_0(t)$$

$$\xi(t) = \sqrt{\frac{M_2(t)}{M_0(t)}} \tag{2}$$

If we assume that $C(t) \propto t^{l-1}$ and $C(0) = 1$, $M_n(t)$ behaves as

$$M_n(t) \propto \frac{1}{n+l} t^{n+l}$$

Using the asymptotic behavior of $M_n(t)$, we find $\tau(t)/t$ behaves as

$$\tau(t)/t \propto \frac{1}{l} t^{l-1},$$

and $\xi(t)/t$ behaves as

$$\xi(t)/t \propto \sqrt{\frac{l}{2+l}}.$$

When $c > 0$, we assume $C(t) \simeq c + c' t^{l-1}$, $c > 0$. We find $\tau(t)/t \simeq c + o(1/t)$ and $\xi(t)/t \simeq 1/\sqrt{3}$. If $\lim_{t\to\infty} \tau(t)/t > 0$ and $\lim_{t\to\infty} \xi(t)/t = 1/\sqrt{3}$, we can judge $c > 0$. If $\lim_{t\to\infty} \tau(t)/t = 0$ and $\lim_{t\to\infty} \xi(t)/t < 1/\sqrt{3}$, $c = 0$.

Figure 4 shows the semilogarithmic plot of $\tau(t)/t$ and $\xi(t)/t$ vs. t. $\tau(t)/t$ converges to a positive value for $q = 5/9, 6/9$. For $q = 7/9, 8/9$, it is difficult to judge whether $\tau(t)/t$ converges to zero or a positive value. The slope at the right edge is negative; it suggests that $\tau(t)/t$ converges to zero. $\xi(t)/t$ converges to $1/\sqrt{3}$ for $q = 5/9, 6/9$, which suggests that $c > 0$. For $q = 7/9, 8/9$, the slope of $\xi(t)/t$ on the right edge is almost zero and takes about 0.5. If $c = 0$, $\xi(t)/t \to \sqrt{1/(l+2)} < \sqrt{1/3}$. The result suggests that $c = 0$ for $q = 7/9, 8/9$.

The behaviors of $C(t)$, $\tau(t)/t$, and $\xi(t)/t$ suggest that there occurs the information cascade phase transition at $q_c \in (6/9, 7/9)$. As the system size is limited in the experiment, we need to estimate the response function $f(z)$ of nonlinear Pólya urn. If the number of stable fixed points changes between $q = 6/9$ and $7/9$, it strongly supports the existence of the phase transition.

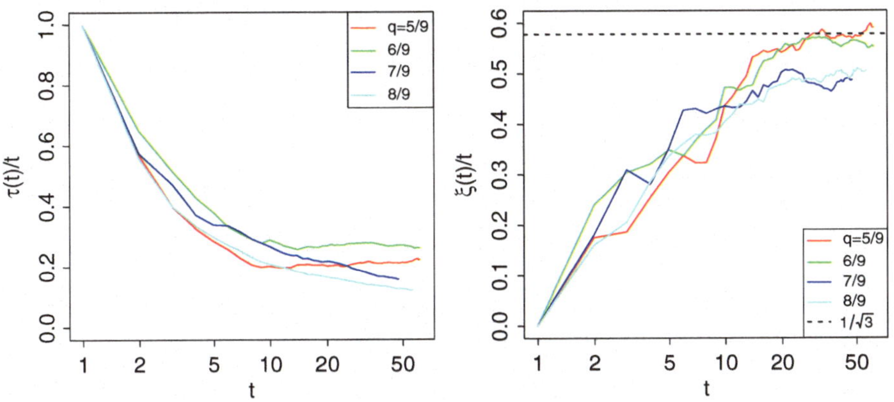

Fig. 4 Semilogarithmic plot of $\tau(t)/t$ (left) and $\xi(t)/t$ (right) vs. t

4 Data Analysis: Microscopic Aspects

The results of the macroscopic analysis in the previous section strongly suggest the existence of the information cascade phase transition. In this section, we study the microscopic behavior of the system. We study how the subjects choose based on the private signal and the social influence.

4.1 Strength of Social Influence and Private Signal

To measure how strongly the social information and private signal affect subjects' decision-making, we compare the correlation coefficients between them and the subjects' decisions. We estimate the correlation coefficients between $S(q,i,t)$ and $X(q,i,t)$ and between $Z(q,i,t)$ and $X(q,i,t+1)$ for $t \geq 1$.

Figure 5 shows the plots of the correlation coefficients vs. t. Overall, $\text{Cor}(S(q,i,t), X(q,i,t))$ decreases and $\text{Cor}(Z(q,i,t), X(q,i,t+1))$ increases with t. For $q = 5/9$, $\text{Cor}(S(q,i,t), X(q,i,t))$ starts at very small values. We think that subjects were confused with small q, and they could not trust their private signals. However, $\text{Cor}(S(q,i,t), X(q,i,t))$ rapidly increases and behaves similarly with other cases. At around $t = 20$, the correlation coefficients fluctuate around certain values. The results suggest that the subjects' behaviors becomes stationary for $t \geq 20$. $\text{Cor}(S(q,i,t), X(q,i,t))$ and $\text{Cor}(Z(q,i,t), X(q,i,t+1))$ fluctuate around 0.3 and 0.6, respectively. The result indicates that the social influence is stronger than the private signal.

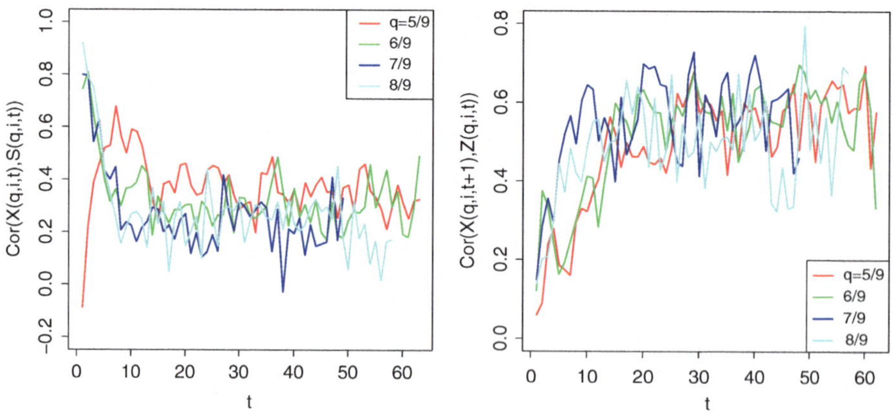

Fig. 5 Plot of $\text{Cor}(S(q,i,t), X(q,i,t))$(left) and $\text{Cor}(Z(q,i,t), X(q,i,t+1))$ (right) vs. t

4.2 Response Function $f(z, s)$ and $f(z)$

We study how subjects' decisions are affected by the social information and private signal. We study the probabilities that $X(t+1)$ takes 1 under the condition that $Z(t) = z$ and $S(t+1) = s$. We denote them as

$$f(z, s) \equiv \Pr(X(t+1) = 1 | Z(t) = z, S(t+1) = s).$$

By the symmetry under the transformations $S \leftrightarrow 1-S$, $X \leftrightarrow 1-X$, and $Z \leftrightarrow 1-Z$, $f(z, s)$ has the Z_2 symmetry

$$1 - f(1-z, 0) = f(z, 1).$$

In the estimation of $f(z, s)$ using experimental data $\{S(q, i, t), Z(q, i, t-1), X(q, i, t)\}$, we exploit the symmetry. As we are interested in the stationary behavior of $f(z, s)$, and $\mathrm{Cor}(S(q, i, t), X(q, i, t))$ and $\mathrm{Cor}(Z(q, i, t), X(q, i, t+1))$ reach their stationary values at $t = 20$, we use data $\{S(q, i, t), Z(q, i, t-1), X(q, i, t)\}$ for $t \geq 20$.

Figure 6 shows plots of $f(z, 1)$ vs. z. It is clear that $f(z, 1)$ are monotonically increasing functions of z. For $q = 5/9, 6/9$, the behaviors are almost the same. For $q = 7/9, 8/9$, few samples appear in the middle bins, and the error bars are large.

We fit the experimental data $\{S(q, i, t), Z(q, i, t-1), X(q, i, t)\}, q \in Q, 20 \leq t \leq T(q, i)$ with logistic function. By the symmetry of $f(z, s)$, we assume

$$f(z, s) = \frac{1}{1 + \exp(-\beta_1(s - 1/2) - \beta_2(z - 1/2))}.$$

By the maximum likelihood estimate, we obtain $\beta_1 = 2.92$ and $\beta_2 = 7.64$. The solid lines in the top figure show the fitted result.

As the private signal takes 1 with probability q, the probability that the $t+1$-th subject chooses the correct option under the social influence $Z(t) = z$ is estimated as

$$f(z) \equiv \Pr(X(t+1) = 1 | Z(t) = z) = q \cdot f(z, 1) + (1-q) \cdot f(z, 0).$$

We denote the averaged response function as $f(z)$. Then the voting process $\{X(t)\}, t = 1, 2, \cdots$ becomes a nonlinear Pólya urn process with $f(z)$. The bottom figure of Fig. 6 shows $f(z)$ vs. z for $q \in Q$.

For $q = 5/9$ (red), $f(z)$ crosses the diagonal (black broken line) at three points. The left and right fixed points are stable, and the middle one is unstable. $Z(q, i, t)$ converges to the two stable fixed points with positive probability, and the order parameter c is positive, $c > 0$. For $q = 6/9$ (green), $f(z)$ touches and crosses the diagonal. The touchpoint and the fixed points are both stable and $c > 0$. For $q = 7/9$ (blue) and $q = 8/9$ (light blue), $f(z)$ have only one stable fixed point.

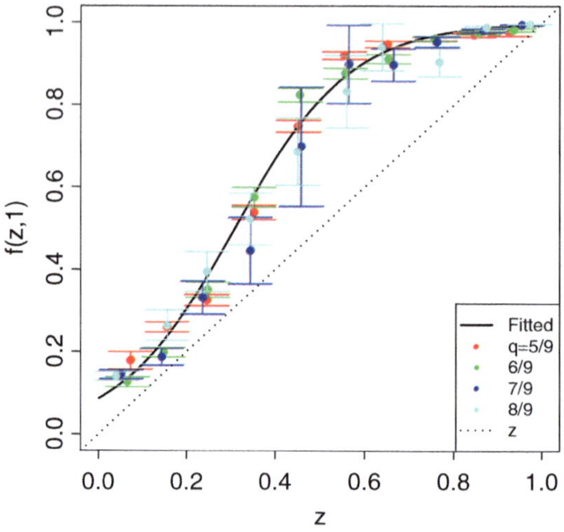

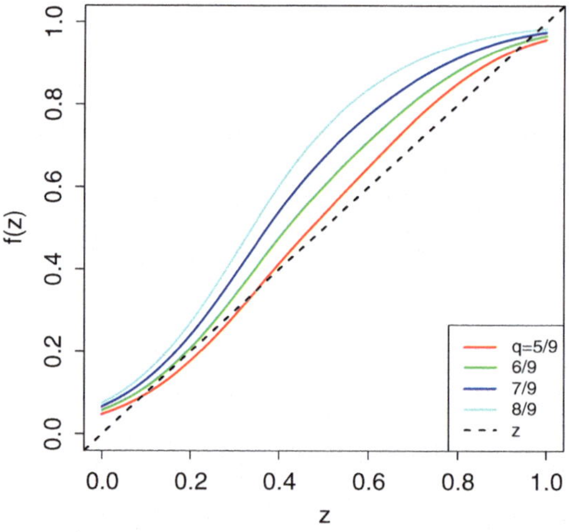

Fig. 6 (Top) Response functions $f(z, 1)$ for $q \in Q$. The solid line shows the fitted result with logistic $(\beta_1(z - 0.5) + \beta_2(s - 0.5)$ with $\beta_1 = 2.92$ and $\beta_2 = 7.64$. (Bottom) Plot of $f(z)$ vs. z for $q \in Q$ and $f(z) = q \cdot f(z, 1) + (1 - q) \cdot f(z, 0)$. $f(z)$ is the fitted result with logistic function

$Z(q, i, t)$ converges to the unique fixed point and $c = 0$. However, for $q = 7/9$ the departure from the diagonal is small, and it is difficult to judge whether there is only one stable fixed point or there are two stable fixed points.

5 Summary

We explain the information cascade experiment in the canonical setup. Two-choice urn quiz is adopted, and we can control the precision of the private signal by changing the composition of red and blue balls in two urns. We show the results of the analysis of the experimental data. Both the macroscopic and microscopic analyses strongly suggest the existence of the information cascade phase transition. If the private signal q is $5/9, 6/9$, there are two stable states in the system, and the limit value of the correlation function becomes positive. The domino effect continues forever, and the first subject's choice affects infinitely later subjects' choices. If $q = 7/9, 8/9$, there is only one stable state in the system. The domino effects decay to zero with power-law behavior, and the first subject's choice does not affect infinitely later subjects' choices.

However, the system and sample size of the experiment are very limited. In order to realize larger system size, web-based online experiment should be adopted in the future experiment.

6 Data and R Script

One can download the data and the R script that is used to plot the figures in this chapter. One should visit https://sites.google.com/site/shintaromori/home/data and download AppDCSSD_Chap11.tar.gz.

References

Anderson LR, Holt CA (1997) Information cascades in the laboratory. Am Econ Rev 87:847–862

Bikhchandani S, Hirshleifer D, Welch I (1992) A theory of fads, fashion, custom, and cultural changes as informational cascades. J Polit Econ 100:992–1026

Goeree JK, Palfrey TR, Rogers BW, McKelvey RD (2007) Self-correcting information cascades. Rev Econ Stud 74:733–762

Hino M, Irie Y, Hisakado M, Takahashi T, Mori S (2016) Detection of phase transition in generalized Polya urn in Information cascade experiment. J Phys Soc Jpn 85:034002–034013

Mori S, Hisakado M (2015a) Finite-size scaling analysis of binary stochastic processes and universality classes of information cascade phase transition. J Phys Soc Jpn 84:054001–054013

Mori S, Hisakado M (2015b) Correlation function for generalized Pólya urns: finite-size scaling analysis. Phys Rev E92:052112–052121

Information Cascade and Bayes Formula

Masato Hisakado and Shintaro Mori

1 Introduction

Human beings estimate public perception by observing the actions of other individuals, following which they exercise a choice similar to that of others. This phenomenon is also referred to as social learning or imitation and studied several fields (Galam 1990). Because it is usually sensible to do what other people are doing, collective herding behavior is assumed to be the result of a rational choice according to public perception. In ordinary situations this is the correct strategy and sometimes erroneous decisions. As a macro phenomenon, we can see that large social movement start is the absence of central control or public communications. A well-known studied example is the bank run on the Toyokawa Credit Union in 1973. The incident was caused by a false rumor, the process of which was subsequently analyzed by Ito et al. (1974a,b). These phenomena are known as an information cascade (Bikhchandani et al. 1992). We can define information cascade is that the few persons' personal information is introduced without correction. In other words the correlation function is not zero in the large time limit.

Herders' behavior is known as the influence response function. Threshold rules have been derived for a variety of relevant theoretical scenarios as the response function. Some empirical and experimental evidence has confirmed the assumptions that individuals follow threshold rules when making decisions in the presence of social influence (Watts and Dodds 2007; Watts 2002). This rule posits that individuals will switch between two choices only when a sufficient number of other

M. Hisakado (✉)
Nomura Holdings, Inc., Chiyoda-ku, Tokyo, Japan

S. Mori
Faculty of Science and Technology, Department of Mathematics and Physics,
Hirosaki University, Hirosaki, Aomori, Japan

A.-H. Sato (ed.), *Applications of Data-Centric Science to Social Design*,
Agent-Based Social Systems 14, https://doi.org/10.1007/978-981-10-7194-2_12

persons have adopted the choice. We refer to individuals such as these as digital herders (Hisakado and Mori 2011). From our experiments, we observed that human beings exhibit behavior between that of digital and analog herders (Mori et al. 2012). Analog herders vote for each candidate with probabilities that are proportional to candidates' votes (Hisakado and Mori 2010). In the previous paper Hisakado and Mori (2016), we discuss the digital herders who have threshold in half.

To investigate the phenomena, there are several experiments. Here, participants answer two-choice questions sequentially (Anderson and Holt 1997; Mori et al. 2012). In the canonical setting of the experiment, two urns, A and B, with different configurations of red and blue balls are prepared. One of the two urns is chosen at random to be urn X, and the question is whether urn X is A or B. The participants can draw a ball from urn X and see which type of ball it is. This knowledge, which is called the private information, provides some information about X. However, the private information indicate the true situation equivocally, and participants have to decide under uncertainty.

Participants are also provided with social information regarding how many prior participants have chosen each urn. The social information introduces externally to the decision-making; as more participants choose urn A (B), later participants are more likely to identify urn X as urn A (B). The social interaction in which a participant tends to choose the majority choice even if it contradicts the private signal is called an information cascade or rational herding. In simple experiments of information cascade, if the difference in the numbers of subjects who have chosen each urn exceeds two, the social information overwhelms subjects private information (Anderson and Holt 1997). In the limit of many previous subjects, the decision is described by a threshold rule stating that a subject chooses an option if its ratio exceeds 1/2, $f(z) = \theta(z - 1/2)$. The function $f(z)$ that describes decisions under social information is a response function.

To detect the phase transition caused by the change in $f(z)$, we have proposed another information cascade experiment in which subjects answer two-choice general knowledge questions. If almost all of the subjects know the answer to a question, the probability of the correct choice is high, and $f(z)$ does not depend greatly on the social information. In this case, $f(z)$ has only one stable fixed point. However, when almost all the subjects do not know the answer, they show a strong tendency to choose the majority answer. Then $f(z)$ becomes S shaped, and it could have multiple stable fixed points. We have shown that when the difficulty of the questions is changed, the number of stable fixed points of the experimentally derived $f(z)$ changes.

If the questions are easy, there is only one stable fixed point, z+, and the ratio of the correct choice z converges to that value. If the questions are difficult, two stable fixed points, z+ and z−, appear. The stable fixed point to which z converges becomes random. To detect the randomness using experimental data, we study how the variance of z changes as more subjects answer questions of fixed difficulty. We showed that the variance converges to zero in the limit of many subjects for easy questions. For difficult questions, it converges to a finite and positive value, which suggests the existence of multiple stable states in the system.

Since the phase transition, we sometimes did wrong conclusion. We cannot choose the correct pod. In this chapter we are interested in whether we can repair the voting process by Bayesian statistics. We study the method to recalculate the votes by weighting each vote and to conclude the correct decisions.

The remainder of this chapter is organized as follows. In Sect. 2, we introduce our pod model and discuss the information cascade transition. In Sect. 3, we discuss the Bayes formula to estimate the ratio of red and blue balls. Finally, the conclusions are presented in Sect. 4.

2 Model

Voters sequentially answered a two-choice question. We prepared questions, urn X, by randomly choosing an urn from two different urns, urn A and urn B, which contain ball A (red) and ball B (blue) in different proportions. For $q = n/m > 1/2$, urn A (B) contains n A (B) balls and $m - n$ B (A) balls. Urn A (B) contains more A (B) balls than B (A) balls. The voters obtain information about urn X by knowing the color of a ball randomly drawn from it. The color of the ball is the private signal, as it is not shared with other voters. Here we assume X is Urn A.

If the ball is ball A (B), urn X is more likely to be A (B). We prepared the private signal $S(t) \in \{A, B\}$. Voters answered the questions individually using their respective private signals and information about the previous voters' choices. This information, called social information, was given as the summary statistics of the previous voters. If the voter answers question, after $t - 1$ voters, the voter receives a private signal $S(t)$ and social information $c_A(t-1)$ and $c_B(t-1)$ from the previous $t - 1$ voters. Let $X(t) \in \{A, B\}$ be the t-th subject's choice; the social information $c_i(t-1), i \in \{A, B\}$ is written as

$$c_i(t-1) = \sum_{t=1}^{t-1} \delta_{X(t),i},$$

where $c_A(t-1) + c_B(t-1) = t - 1$holds.

Hereafter, instead of A and B, we use 1 and 0 to describe the correct and incorrect choices and private signal. We use the same notation for them, as follows: $S(t) \in \{0, 1\}$ and $X(t) \in \{0, 1\}$.

$c_1(t)$ shows the number of correct choices up to the t-th voter. We denote the percentage up to the t-th voter as

$$Z(t) = \frac{1}{t} \sum_{i=1}^{t} X(i).$$

We are interested in how voters' decisions are affected by the social information and private signal. We study the probabilities that $X(t+1)$ takes 1 under the condition that $Z(t) = z$ and $S(t+1) = s$. We denote them as $f(z, s) = Pr(X(t+1) = 1|Z(t) = z, S(t+1) = s)$. By the symmetry we have the hypothesis there is the relation

$$1 - f(1-z, 0) = f(z, 1). \tag{1}$$

We can write the process as

$$\begin{aligned} c_1(t) = k \rightarrow k+1 : P_{k,t} &= qf(k/t, 1) + (1-q)f(k/t, 0), \\ c_1(t) = k \rightarrow k : Q_{k,t} &= 1 - P_{k,t}, \end{aligned} \tag{2}$$

We define a new variable Δ_t such that

$$\Delta_t = 2c_1(t) - t = c_1(t) - c_0(t). \tag{3}$$

We change the notation from k to Δ_t for convenience. Then, we have $|\Delta_t| = |2k - t| < t$. Thus, Δ_t holds within $\{-t, t\}$. Given $\Delta_t = u$, we obtain a random walk model:

$$\begin{aligned} \Delta_t = u \rightarrow u+1 : P_{u,t} &= qf(\frac{1}{2} + \frac{u}{2t}, 1) + (1-q)f(\frac{1}{2} + \frac{u}{2t}, 0), \\ \Delta_t = u \rightarrow u-1 : Q_{u,t} &= 1 - P_{u,t}. \end{aligned}$$

We now consider the continuous limit $\epsilon \rightarrow 0$,

$$X_\tau = \epsilon \Delta_{[\tau/\epsilon]}, \tag{4}$$

where $\tau = t\epsilon$. Approaching the continuous limit, we can obtain the stochastic differential equation:

$$\mathrm{d}X_\tau = \left[2qf\left(\frac{1}{2} + \frac{X_\tau}{2\tau}, 1\right) + 2(1-q)f\left(\frac{1}{2} + \frac{X_\tau}{2\tau}, 0\right) - 1\right] d\tau + \sqrt{\epsilon}. \tag{5}$$

We are interested in the behavior at the limit $\tau \rightarrow \infty$. The relation between X_∞ and the voting ratio to 1 is $2Z - 1 = X_\infty/\tau$. We consider the solution $X_\infty \sim \tau^\alpha$, where $\alpha \leq 1$, since the maximum speed is τ when $q = 1$. The slow solution is $X_\infty \sim \tau^\alpha$, where $\alpha < 1$ is hidden by the fast solution $\alpha = 1$ in the upper limit of τ. Hence, we can assume a stationary solution as $X_\infty = \bar{v}\tau$, where $\bar{v}$ is constant. We can obtain the self-consistent equation:

$$\bar{v} = 2qf\left(\frac{1}{2} + \frac{\bar{v}}{2}, 1\right) + 2(1-q)f\left(\frac{1}{2} + \frac{\bar{v}}{2}, 0\right) - 1. \tag{6}$$

The number of solutions depends on the function $f(z, s)$. The number of solution is important. If there is only one solution, the voting converges the solution. On the other hand, if there are three solutions, the middle solution is unstable. The other solutions become good and bad equilibria. Hence there is phase transition we call information cascade transition. The response function which is obtained from the urn experiments is in Chap. 11.

3 Bayes Formula

3.1 Bayesian Updating

In this section we consider how to estimate the urn X using Bayes formula. We observe the final voting history without personal information. Sometimes the final votes do not show the correct answers. We recalculate the final votes by Bayes formula. Table 1 is estimated conditional probabilities. $\hat{f}(z, i)$ is the estimated probabilities that the voter takes 1 under the condition the social information $Z = z$ and personal signal $S = i; i = 1, 0$. We estimate the Table 1 from the training data.

When the final vote is 1(0) and prior probability of personal signal $S = i$ is $\hat{q}_i$, the posterior probabilities of personal signal are $i, ; i = 1, 0$ is $\hat{q}_i \hat{f}(z, i)/(\sum_j \hat{q}_j \hat{f}(z, j))$ $(\hat{q}_i(1 - \hat{f}(z, i))/(\sum_j \hat{q}_j(1 - \hat{f}(z, j)))$. There is the relation $\sum_j \hat{q}_j = 1$. In standard case, we take the prior probability $\hat{q}_1 = 1/2$ at $t = 1$. We consider the process of $\hat{q}_1(t)$, Bayesian updating.

$$X(t) = 1 : \hat{q}_1(t) = g(\hat{q}_1(t-1)) = \frac{\hat{q}_1(t-1)\hat{f}(z, 1)}{(\sum_j \hat{q}_j(t-1)\hat{f}(z, j))},$$

$$X(t) = 0 : \hat{q}_1(t) = g(\hat{q}_1(t-1)) = \frac{\hat{q}_1(t-1)(1 - \hat{f}(z, 1))}{\sum_j (t-1)\hat{q}_j(1 - \hat{f}(z, j))}, \tag{7}$$

where g is the updating function. We consider the equilibrium of this process. Here we confirm the notations. $\hat{x}$ means estimated parameter x. The expected value of posterior probability $g(\hat{q}_1)$, $E[g(\hat{q}_1)]$ is

Table 1 Personal signal and final votes

Personal signal \ final votes	1	0
1	$\hat{f}(z, 1)$	$1 - \hat{f}(z, 1)$
0	$\hat{f}(z, 0)$	$1 - \hat{f}(z, 0)$

$$E[g(\hat{q}_1)] - \hat{q}_1 = [qf(Z,1) + (1-q)f(z,0)]\frac{\hat{q}_1\hat{f}(z,1)}{(\sum_j \hat{q}_j\hat{f}(z,j))}$$
$$+[q(1-f(Z,1)) + (1-q)(1-f(z,0))]\frac{\hat{q}_1(1-\hat{f}(z,1))}{\sum_j \hat{q}_j(1-\hat{f}(z,j))}$$
$$-\hat{q}_1. \quad (8)$$

Here we consider the case $\hat{f}(z,i) = f(z,i)$. It means the estimated response function is adequate good approximation of real response function.

We can calculate (8)

$$E[\hat{q}_1] - \hat{q}_1 = \frac{(q-\hat{q}_1)\hat{q}_1(1-\hat{q}_1)(\hat{f}(z,1)-\hat{f}(z,0))^2}{(\sum_j \hat{q}_j\hat{f}(z,j))(\sum_j \hat{q}_j(1-\hat{f}(z,j)))}. \quad (9)$$

When the prior probability $\hat{q}_1$ is larger than real q, the posterior probability $\hat{q}_1$ increases. When the prior probability $\hat{q}_1$ is smaller than real q, the posterior probability $\hat{q}_1$ decreases. The data set of voting is enough; the posterior probability $\hat{q}_1$ converges to q necessary. It does not depend on z.

We consider the case $\hat{f}(z,1) \neq \hat{f}(z,0)$. It means that the personal signal is effective for voting. In Fig. 1, we show the prior probability $\hat{p}_1$ and the expected posterior probability $E[g(\hat{q}_1)]$. There are three intersections, $\hat{q}_1 = 0$, 1, and q. The solutions 1 and 0 are not stable solutions. The middle intersection $\hat{q}_1 = q$ is the stable solution of (9). We can conclude that the Bayesian update converges to the correct q.

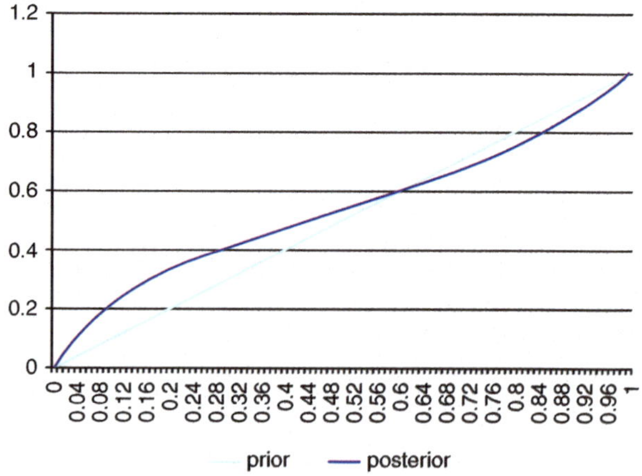

Fig. 1 We show the relation between the prior probability and the posterior probability. The X-axis is the prior probability, and Y-axis is the posterior probability. The middle intersection is the stable point

3.2 Fixed Prior Probability

Using Bayes formula which is introduced in the previous subsection, we can estimate the ratio of red ones to blue ones. If the data is adequate and the response function is collected, the ratio converges to the real q.

But we do not have adequate data and the response function has errors. In this section we study the pod problem to estimate whether the number of blue balls in pod X is larger than red balls or not.

Bayesian updating (7) has volatility; we set the prior probability as $1/2$.

$$X(t) = 1 : \hat{q}_1(t) = g(1/2) = \frac{\hat{f}(z,1)}{(\sum_j \hat{f}(z,j))},$$

$$X(t) = 0 : \hat{q}_1(t) = g(1/2) = \frac{(1-\hat{f}(z,1))}{(\sum_j (1-\hat{f}(z,j)))}. \tag{10}$$

We define the sum of the posterior probability for the pods 1 and 0 as Q_1 and Q_0,

$$Q_i = \sum_t \hat{q}_i(t), \tag{11}$$

where $i = 1, 0$. It is the sum of the divided votes for pods 1 and 0, when the prior probability is $1/2$. There is the constraint $Q_1 + Q_2 = t$.

When $Q_1 > Q_0$, we can estimate the pod is A. When $Q_0 > Q_1$, we can estimate the pod is B. By setting the prior probability $1/2$, we can make the influence of the prior probability small. Equation (9) becomes

$$E\left(\frac{Q_1 - Q_0}{2t}\right) = E[g(1/2)] - 1/2 = \frac{(2q_1 - 1)(\hat{f}(z,1) - \hat{f}(z,0))^2}{(\sum_j \hat{f}(z,j))(\sum_j (1-\hat{f}(z,j)))} > 0. \tag{12}$$

It does not depend on z, the path of the voting. When there is the adequate data, if we select the larger one of Q_i, we can estimate correctly whether X is A or B.

4 Concluding Remarks

We consider a voting experiment using two-choice questions. An urn X is chosen at random from two urns A and B, which contain red and blue balls in different configurations. Subjects sequentially guess whether X is A or B using information about the prior subjects' choices and the color of a ball randomly drawn from X. The color tells the subject which is X with probability q. We describe the sequential voting process by a nonlinear Pólya urn model. The model suggests the possibility of a phase transition when q changes. When there is not the phase transition, in the limit $t \rightarrow \infty$, we can choose the correct pod. When there is the phase transition, the votes sometimes converge to the wrong equilibrium. We consider the method to estimate the ratio of red and blue balls using the Bayes formula. It is the corrections of the voting conclusions. If the estimated response function is correct, we can estimate the correct ratio of red and blue balls. The method will not be affected whether there is the phase transition or not.

We can see ratings in many Internet sites which are related to the consumers. It is the important information to choose an option for consumers and useful as the social system. Using these ratings, we can choose the option without adequate information. But we sometimes have questions whether the rating is correct or not after we have tried the option. The reason is the ratings may be affected by the ratings of the other persons.

The rating is the sequential voting process which we discussed in this chapter. The consumers have the personal information and social information. For example, we consider the gourmet Internet sites; the personal information is the stand-alone rating. The social information is the rating which are represented in the site before his/her rating. If we can estimate the correct response function by observation and experiments, we can estimate the true rating using the method which we explained in this chapter. The true rating is the rating which gathered independent ratings. In this mean we explained the noise reduction system by the Bayes formula in this chapter.

Appendix A Multi-choice Case

In this appendix we consider how to estimate the answer using Bayes formula for multi-choice case. As the Sect. 3.1, we observe the final voting history without personal information. We recalculate the final votes by Bayes formula.

Table 2 is the estimated conditional probabilities. $\hat{f}_j(z, i)$ is the estimated probabilities that the voter takes j, $j = 0, 1, 2 \cdots r$ under the condition the social information $Z = z$ and personal signal $S = i; i = 0, 1 \cdots r$. There is the constraint $\sum_k \hat{f}_k(z, j) = 1$.

When the final vote is j and prior probability of personal signal $S = i$ is $\hat{q}_i$, the posterior probabilities of personal signal are i is $\hat{q}_i \hat{f}_j(z, i)/(\sum_k \hat{q}_k \hat{f}_j(z, k))$. There is the normalization relation $\sum_k \hat{q}_k = 1$. In standard case we take the prior probability $\hat{q}_1 = 1/(r + 1)$ at $t = 1$. We consider the process of $\hat{q}_1(t)$, Bayesian updating.

Table 2 Personal signal and final votes of multi-choice

Personal signal \ final votes	0	1	$\cdots$	r
0	$\hat{f}_0(z, 0)$	$\hat{f}_1(z, 0)$	$\cdots$	$\hat{f}_r(z, 0)$
1	$\hat{f}_0(z, 1)$	$\hat{f}_1(z, 1)$	$\cdots$	$\hat{f}_r(z, 1)$
$\cdots$	$\cdots$	$\cdots$	$\cdots$	$\cdots$
r	$\hat{f}_0(z, r)$	$\hat{f}_1(z, r)$	$\cdots$	$\hat{f}_r(z, r)$

$$X(t) = j : \hat{q}_i(t) = g(\hat{q}_i(t-1)) = \frac{\hat{q}_i(t-1)\hat{f}_i(z,j)}{(\sum_j \hat{q}_k(t-1)\hat{f}_j(z,k))}, \tag{13}$$

a where g is the updating function. We consider the equilibrium of this process. Here we confirm the notations. $\hat{x}$ means estimated parameter x.

The expected value of posterior probability $g(\hat{q}_1)$, $E[g(\hat{q}_1)]$ is

$$E[g(\hat{q}_i)] - \hat{q}_i = \sum_j [(\sum_k q_k f_j(z,k)) \frac{\hat{q}_i \hat{f}_j(z,i)}{(\sum_l \hat{q}_l \hat{f}_j(z,l))}] - \hat{q}_i. \tag{14}$$

Here we consider the case $\hat{f}_j(z,i) = f_j(z,i)$. It means the estimated response function is adequate good approximation of real response function. We consider the case $\hat{f}_k(z,i) \neq \hat{f}_k(z,j)$, when $i \neq j$.

We consider the equilibrium solutions of $E[g(\hat{q}_i)] - \hat{q}_i = 0$. The equation is $r+2$th rank equation. There are $(r+2)$ solutions. $r+1$ solutions are $\hat{q}_i = 1, \hat{q}_j = 0$ where $j \neq i$. The other solution is $\hat{q}_i = q_i$.

We investigate the former solutions. We consider $\hat{q}_0 = 1 - r\delta$, $\hat{q}_i = \delta$ where $i \neq 0$ and $\delta << 1$.

$$\begin{aligned} E[g(\hat{q}_0)] - \hat{q}_0 &= \sum_j \left[\left(\sum_k q_k f_j(z,k) \right) \frac{(1-r\delta)\hat{f}_j(z,0)}{((1-r\delta)\hat{f}_j(z,0) + \delta \sum_{i\neq 0} \hat{f}_j(z,i))} \right] - 1 \\ &< \sum_j \left[\left(\sum_k q_k f_j(z,k) \right) \right] - 1 = 0 \end{aligned} \tag{15}$$

$\hat{q}_0$ leaves from 1. Hence, these solutions are unstable. The later solution, $\hat{q}_i = q_i$, correct solution becomes stable. Then we can conclude the process converges to the correct solution.

Appendix B Response Function of Humans

Recently, the response function $f(z)$ of foreign exchange (FX) traders is estimated. The traders decide their bid-ask prices referring to the price in the market. In our model, the price is the social information. According to the social information and private information, the traders decide the mid-price of bid-ask spread. The response function is tanh type function, which is nonlinear function. The response function depends on the trader, but it can be scaled by adjusting parameters. In Table 3 we summarize the response function of humans. It depends on the circumstances whether the response function is linear or nonlinear.

Table 3 Response functions of humans

No.	Experiments and observation	Linear or Non-linear	Reference
1	Two choice question I	Non-linear	Chapter 10 (Mori et al. 2012)
2	Two choice question II	Non-linear	Chapter 11 (Hino et al. 2016)
3	Horse race betting market	Linear	Chapter 13
4	BBS contributors	Linear	Chapter 8 (Hisakado et al. 2018)
5	FX trader	Non-linear	Kanazawa et al. (2018)

References

Anderson LR, Holt CA (1997) Information cascades in the laboratory. Am Eco Rev 87(5):847–862

Bikhchandani S, Hirshleifer D, Welch I (1992) A theory of fads, fashion, custom, and cultural change as information cascades. J Polit Econ 100:992–1026

Galam G (1990) Social paradoxes of majority rule voting and renormalization group. Stat Phys 61:943–951

Hino M, Irie Y, Hisakado M, Takahashi T, Mori S (2016) Detection of phase transition in generalized Póla urn in information cascade experiment. J Phys Soc Jpn 85(3):034002–034013

Hisakado M, Mori S (2010) Phase transition and information cascade in a voting model. J Phys A 43:315027

Hisakado M, Mori S (2011) Digital herders and phase transition in a voting model. J Phys A 22:275204

Hisakado M, Mori S (2016) Phase transition of Information cascade on network. Physica A 450:570–584

Hisakado M, Sano F, Mori S (2018) Pitman-Yor process and empirical study of choice behavior. J Phys Soc Jpn 87(2):024002–024019

Ito Y, Ogawa K, Sakakibara H (1974a) Stu Jour 11(3):70 (in Japaneese)

Ito Y, Ogawa K, Sakakibara H (1974b) Stu Jour 11(4):100 (in Japaneese)

Kanazawa K, Sueshige T, Takayasu H, Takayasu M (2018) Derivation of Boltzmann equation for financial Browinian motion: direct observation of the collective motion of high frequency traders. Phys Rev Lett 120:138301

Mori S, Hisakado M, Takahashi T (2012) Phase transition to two-peaks phase in an information cascade voting experiment. Phys Rev E 86:26109–026118

Watts DJ (2002) A simple model of global cascades on random networks. Proc Natl Acad Sci USA 99(9):5766–5771

Watts DJ, Dodds PS (2007) Influentials, networks, and public opinion formation. J Consum Res 34:441–458

Part III
Applications of Data Analysis to Social Design

How Betters Vote in Horse Race Betting Market

Shintaro Mori and Masato Hisakado

1 Introduction

Racetrack betting is a simple exercise of gaining a profit or losing one's wager. However, one needs to make a decision in the face of uncertainty, and a closer inspection reveals great complexity and scope. The field has attracted many academics from a wide variety of disciplines and has become a subject of wider importance (Hausch et al. 2008). Compared to the stock or currency exchange markets, racetrack betting is a short-lived and repeated market. It is possible to obtain starker views of aggregated better behavior and study the market efficiency. One of the main findings of the previous studies is the favorite-longshot bias (FL bias) in the racetrack betting market (Griffith 1949; Hausch et al. 2008). Final odds are, on average, accurate measures of winning probability, and short-odds horses are systematically undervalued, and long-odds horses are systematically overvalued. The bias becomes smaller as time proceeds. Long ago, it was possible to make money by using the bias; it is no more possible as the bias becomes small (Hausch et al. 2008). As the market can predict the future well, the betters' decision mechanism is an interesting research theme (Ali 1977). Recently, the prediction market emerges where the betters' information is collected efficiently as in the horse race betting market and the price of the market can predict the future (Wolfers et al. 2004).

From an econophysical viewpoint, horse race betting market is an interesting subject. Park and Dommany have done an analysis of the distribution of final odds (dividends) of the races organized by the Korean Racing Association (Park and

S. Mori (✉)
Faculty of Science and Technology, Department of Mathematics and Physics,
Hirosaki University, Hirosaki, Aomori, Japan

M. Hisakado
Nomura Holdings, Inc., Chiyoda-ku, Tokyo, Japan

A.-H. Sato (ed.), *Applications of Data-Centric Science to Social Design*,
Agent-Based Social Systems 14, https://doi.org/10.1007/978-981-10-7194-2_13

Dommany 2001). They found power law behavior in the distribution and proposed a simple betting model. Ichinomiya also found the power law in the races of the Japan Racing Association (JRA) (Ichinomiya 2006) and proposed another betting model. We have studied the relationship between the rank of a racehorse in JRA and the result of victory or defeat (Mori and Hisakado 2010). Horses are ranked according to the win bet fractions and we studied the ROC curve. ROC curve is a 2D curve which measures the accuracy of a classification algorithm, and the coordinates (x, y) of the curve are the false positive rate (FPR) and the true positive rate (TPR) of the predictions. The area under the ROC curve (AUROC) or the normalized index AR (accuracy ratio) is used to estimate the accuracy of the classification. In the horse race betting, we study the the ROC curve $(x(v), y(v))$ for the win bet fraction v. In particular, we focus on the long-odds region, where the win bet fraction v is small and the corresponding horses are considered to be weak in the market (Mori and Hisakado 2010). The double logarithmic plot of the ROC curve for small v becomes straight; that means the power law relation holds as $(1-y(v) \propto (1-x(v))^{\alpha}$ for small v. We show that in a Pólya model, where betters are pure analog herder and vote on the horses according to the probabilities that are proportional to the votes shares, the ROC curve is given by the incomplete gamma functions whose cumulative functions show the scale invariance. Furthermore, the exact scale invariant relation $(1-y(v)) = (1-x(v))^{\alpha}$ holds exactly over the entire range $0 \leq x(v), y(v) \leq 1$ in a limit of the model.

In this chapter, we start with the review of the efficiency of the racetrack betting market by studying the JRA winning bet data in Sect. 2. We study the relation of the win betting fractions and the probabilities of win using about one million win bet fraction data of JRA from 1986 to 2008. We show the FL bias is negligibly small and the market is efficient. We introduce the expected ROC (EROC) curve, whose coordinates are the cumulative sum of $1 - v$ and v. By studying the discrepancy between ROC and EROC, or the areas under the curves, we measure the efficiency of the market. In Sect. 3, we study the time series data of the odds of the winning bet of JRA in 2008. We estimate how the accuracy and efficiency change as betting proceeds. After that, we study the betting behavior in Sect. 4. We classify the horses according to the final win bet fraction v_f and study the response function of the betters. The crossing point of the response function with the diagonal indicates the equilibrium value of the win bet fraction. In the initial stage of the betting, the equilibrium value is far from the final win bet fraction. However, betters adjust the response function as betting proceeds. We explain the typical response functions of noisy, herding, and arbitrage betters and interpret the empirically observed response function as the combination of them.

2 Efficiency and Accuracy

We study all the data on horse race betting obtained from the Japan Racing Association (JRA) for the period 1986–2008. There were 100,0825 horses and 78,564 winning horses. We study the efficiency of the market by estimating the

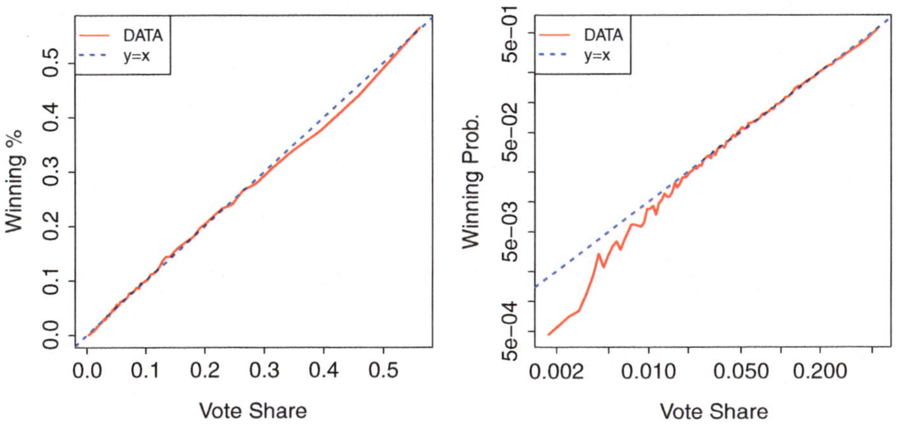

Fig. 1 Plot of the winning horse rate vs. win bet fraction v. The right figure is the double logarithmic plot

winning probability of the horses with win bet fraction v. If the market is efficient and there is no bias, the winning probability should coincide with v. We bin the horses according to the value of v, and each bin contains 10^4 horses, except for the last bin which contains 10,825 horses. We calculate the rate of the winning horses and the average values of the win bet fractions. Figure 1 plots the results.

One sees that the curve almost coincides with the diagonal and the market is almost efficient. From the left figure, we see small discrepancy around $v \simeq 0.4$. The curve is lower than the diagonal and the strength of the horses is overestimated. The right figure is the double logarithmic plot of the relationship in order to see the longshot bias. The curve is also below the diagonal for $v < 0.01$ and the longshot horses are overestimated. There is a small favorite bias for $v > 0.5$, which cannot be seen. It is impossible to utilize the inefficiency of the market to get profit by buying the favorite horses winning bet even if their strength is underestimated, as the discrepancy is too small.

We measure the accuracy of win bet fraction as the classification of the winning and losing horses. As the estimate of the accuracy, we use the ROC curve. For a threshold win bet fraction v, we measure the rates of losing horses $x(v)$ among all losing horses (false positive rate, FPR) and the rate of the winning horse $y(v)$ among all winning horses (true positive rate, TPR) whose win bet fractions are larger than v. For $v > 1$, there is no horse with such a value of v and $x(v)$, $y(v)$ are zeros. The curve starts from the origin. As v decreases from 1 and v is large, the winning probability is high and the curve rapidly rises. As v becomes smaller, the direction of the curve bents downward. At $v = 0$, all the losing and winning horses are counted and $(x(0), y(0) = (1, 1)$. If the classification is random, the ROC curve goes on the diagonal. If the classification is perfect, the TPR reaches $(0, 1)$ for some v and then ROC goes to $(1, 1)$. As the estimate of the accuracy of the prediction, the area under the ROC curve (AUROC) is often used. As the diagonal ROC curve means random classification, the area of the diagonal 1/2 set the standards for the accuracy of the classification. The AUROC for the perfect classification is 1. Accuracy ratio (AR) is a normalized estimate of the accuracy of the classification, defined as

$$\text{AR} = 2\left(\text{AUROC} - \frac{1}{2}\right).$$

AR takes 0 (1) for the random (perfect) classification.

AR measures how different is the ranking of the horses according to the win bet fraction from the complete case where all the winning horse appears first and then losing horse follows. In order to estimate the efficiency of the win bet fraction v as the winning probability, we estimate the expected ROC (EROC) curve. Here, instead of FPR and TPR, we calculate the cumulative sum of $1-v_i$ and v_i divided by the number of losing and winning horses for $v_i > v$. If the v_i matches the winning probability, EROC curve matches ROC curves. The discrepancy between the ROC curve and the EROC curves means the inefficiency of the market. We estimate AR of the EROC curve and denote it as EAR (expected accuracy ratio).

Figure 2 shows the ROC and EROC curve for the win bet fractions. The discrepancy is small. AR is 0.734 and EAR is 0.734; they are almost the same. In order to see the longshot cases, we show the double logarithmic plot of $1 - x(v)$ and $1 - y(v)$. As $x(v)$, $y(v)$ are almost 1 for small v, we can observe the ROC curve for the longshot cases at the lower left corner. The ROC curve becomes straight at the corner, which means the relation $(1 - y(v)) \propto (1 - x(v))^{\alpha}$ holds with some α. From the figure, $\alpha \simeq 1.5$. The scale invariance of the ROC curve with $\alpha > 1$ in the longshot regime means that the better can sort the winning horses even if the winning probability is less than 1%. If the better cannot sort horses for small

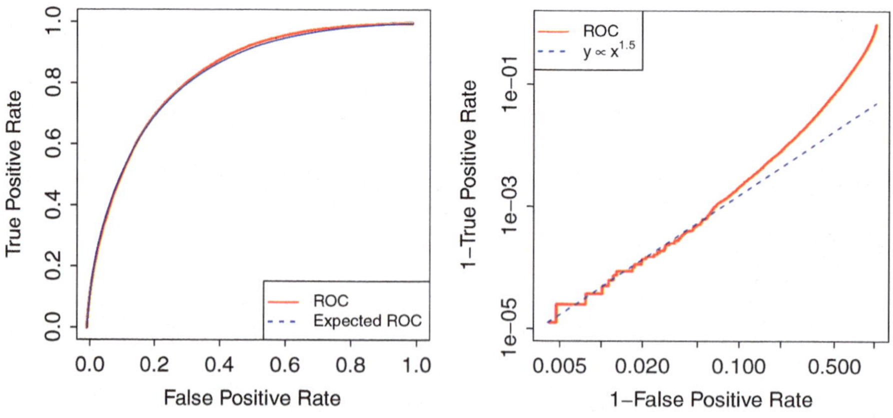

Fig. 2 (Left) ROC curve and expected ROC curve. AR and EAR is the normalized area under the curves. (Right) Double-logarithmic plot of $(1 - x(v), 1 - y(v))$ of the ROC curve. The fitted functions $(1 - y(v)) \propto (1 - x(v))^{1.5}$ are also plotted

probability region, the winning horses appear at random and $\alpha = 1$. $\alpha \simeq 1.5$ indicates the classification ability of the market participants for the rare events.

3 Time Series Data Analysis

We study the time series data of the odds in win betting market of JRA in 2008 (Mori and Hisakado 2009). A win bet requires one to name the winner of the race. The data set contains times series of total votes (public win pool) and odds of 3450 races. There are 16 horses on average in a race. The time interval of the records of odds is several minutes, and the number of records in each race K is in the range $13 \sim 262$. The total number of records in the data set is 307,964. A typical sample from the data is shown in Table 1.

We transform the odds to the vote shares and the number of votes. $O(k, h)$ denotes the odds of horse h in the k-th record, and $V(k)$ denotes the total number of votes up to the k-th records. We write the vote and the win bet fraction of horse h in k-th record as $V(k, h)$ and $v(k, h)$, respectively. As the final total votes $V(K)$ differ from race to race, we introduce a normalized time variable t which is defined as $t(k) = V(k)/V(K)$. As betting proceeds, $t(k)$ increase from 0 to 1. We write the final value of the win bet fraction of horse h as $v_f(h) = v(K, h)$.

In order to see the betting process pictorially, we pick up 100 winning and 100 losing horses in the data set. We arrange the 200 horses in the decreasing order of the win bet fraction from left to right. On the left-hand side of the sequence, more popular horses exist. On the right-hand side, there are less popular horses.

Figure 3 shows the time evolution of the ranking of the win bet fraction of the horses. The bottom line shows the initial configuration of the horses. The top line shows the final configuration of the horses. Initially, the horses are arranged in the sequence at random. The win bet fractions do not contain much information about the strength of the horses, and the winning and the losing horses are mixed up. As the betting progresses from the bottom to the top, the phase separation between the two categories of the horses does occur. Winning (losing) horses move to the left (right) in general. In the end, to the left (right) are more winning (losing) horses. In the betting process, betters have succeeded in choosing the winning horses to some extent. We also note that there remain winning horses in the right. This means that we can find winning horses with very small win bet fractions.

In order to quantify the improvement of the accuracy and the efficiency of the predictions of the racetrack betters, we use AR and EAR as in the previous section. For t, we choose the minimum k such that $t(k) = V(k)/V(K) > t$. As t increases from 0 to 1, k increase from 1 to K. With the choice of k for t, we calculate the ROC and EROC curves and estimate AR and EAR.

Figure 4 shows AR and EAR as the functions of t. As the betting progresses, AR increases monotonically and it almost reaches its maximum at $t = 0.2$. Afterward, the increase in AR is very slow, and the following bets do not increase the accuracy of the prediction as to which horse wins the race. Or, the accuracy of the ranking by

Table 1 Time series of odds and pool for a race is transformed in the time series of votes and vote shares. There are 16 horses and 53 records in the race. We show the data only for the first three horses. The second horse wins in the race

k	$V(k)$	$O(k,1)$	$O(k,2)$	$O(k,3)$	$\cdots$	$V(k,1)$	$V(k,2)$	$V(k,3)$	$\cdots$	$v(k,1)$	$v(k,2)$	$v(k,3)$	$\cdots$
1	39	3.1	2.8	0.0	$\cdots$	10	11	0	$\cdots$	10/39	11/39	0/39	$\cdots$
2	561	44.3	2.9	221.1	$\cdots$	10	155	2	$\cdots$	10/561	155/561	2/561	$\cdots$
3	775	47.0	3.9	203.6	$\cdots$	13	155	2	$\cdots$	13/775	155/775	2/775	$\cdots$
4	1039	54.6	4.5	163.8	$\cdots$	15	185	5	$\cdots$	15/1039	185/1039	5/1039	$\cdots$
5	1269	55.6	4.6	142.9	$\cdots$	18	219	7	$\cdots$	18/1269	219/1269	7/1269	$\cdots$
6	1481	55.6	5.2	145.9	$\cdots$	21	274	8	$\cdots$	21/1481	274/1481	8/1481	$\cdots$
$\vdots$	$\vdots$	$\vdots$	$\vdots$	$\vdots$	$\vdots$	$\vdots$	$\vdots$	$\vdots$	$\vdots$				
51	78,982	44.6	4.1	82.9	$\cdots$	1387	15,428	745	$\cdots$	0.0176	0.1953	0.0094	$\cdots$
52	143,569	50.0	4.1	86.7	$\cdots$	2246	28,014	1294	$\cdots$	0.0156	0.1951	0.0090	$\cdots$
53	246,315	36.6	3.7	106.8	$\cdots$	5303	53,764	1814	$\cdots$	0.0215	0.2183	0.0074	$\cdots$

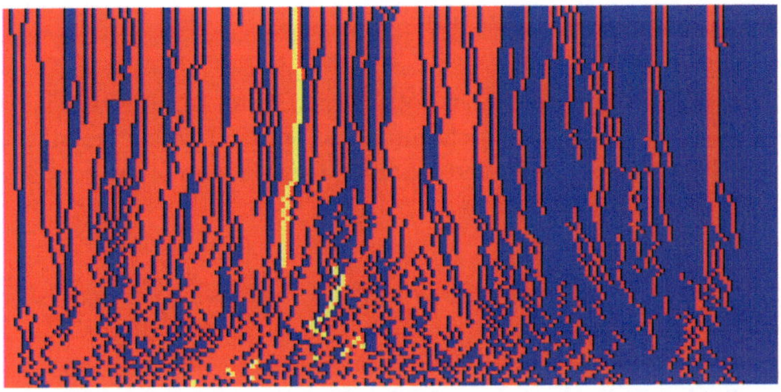

Fig. 3 Pictorial presentation of the betting process. We choose 100 winning (red) and 100 losing (blue) horses randomly from the data set and follow their ranking of the win bet fraction as the betting proceeds. One losing horse is tagged by yellow dots

Fig. 4 Plot of AR and EAR as the functions of t

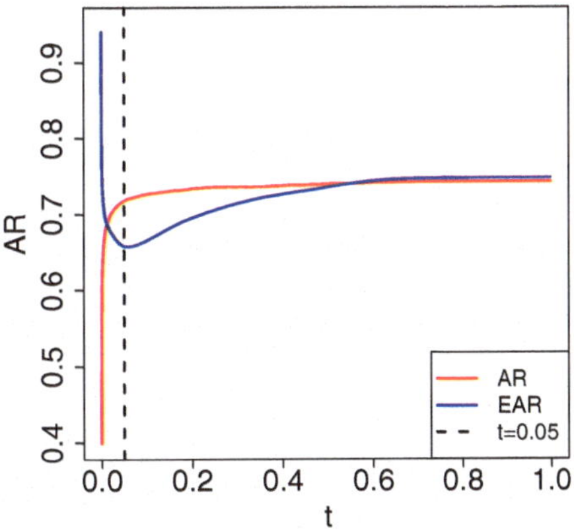

the win bet fraction does not increase anymore. More interesting behavior can be found in the time evolution of EAR. At $t = 10^{-6} (k = 1)$, EAR is very large and is 0.934. Almost all votes are concentrated on a small number of horses. However, AR at $t = 10^{-6} (k = 1)$ is small and the horses with large win bet fractions (popular) are not so strong. There are many winning horses which do not obtain votes. Afterward, EAR decreases rapidly and at $t = 0.05$ EAR reaches its minimum. The bets are scattered among many horses, and this implies the rich variety of the betters' predictions as to which horse wins the race. This also means that the strong horses are undervalued and the weak horses are overvalued, that is, the FL bias state. The discrepancy between AR and EAR decreases afterward, and at $t = 0.5$, they almost coincide as EAR increases faster than AR. The bets begin to be concentrated on stronger horses and the FL bias decreases. The concentration of votes to stronger horses continues, and we observe a small discrepancy between AR and EAR for $t > 0.5$.

4 Better's Response Function

We estimate $f(z)$ of the nonlinear Pólya urn (Hill et al. 1980) using the time series data $\{V(k,h)\}$. Here, $f(z)$ determines the probability that the binary random variable $X \in \{0,1\}$ takes 1 under the condition that the ratio of the previous random variables that takes 1 is z.

$$P(x(t+1)=1|v(t)=v)=f(v)\ ,\ v(t)=\frac{1}{t}\sum_{s=1}^{t}X(s).$$

As $v(k,h)$ evolves to $v_f(h)$ at $k=K$ and the final value $v_f(h)$ almost coincides with the winning probability of horse h. It suggests that the betters know the winning probability of the horses at the collective level and $f(v)$ should depend on v_f. In addition, the betting behavior changes as betting proceeds as we observe in the behavior of AR and EAR in the previous section. We divide the time series data $v(k,h)$ according to the values of $t=k/K$ and v_f as $t<5\%, t>5\%$ and $v_f \in \{[0-1\%],[1-4\%],[4-8\%],[8-12\%],[12-20\%],[20-50\%]\}$. We estimate the ratio of the votes that are cast to horse h as

$$\frac{V(k+1,h)-V(k,h)}{V(k+1)-V(k)}.$$

We estimate the average value of the ratio as function of the deviation of $v(k,h)$ from $v_f(h)$.

$$f(v_f+v)=\overline{\frac{V(k+1,h)-V(k,h)}{V(k+1)-V(k)}}|_{v(k,h)=v_f(h)+v}.$$

$\overline{A}|_B$ indicates the average of A over the samples with condition B. We bin data according to the values $v(k,h)-v_f(h), k/K$, and we estimate the average value of $(V(k+1,h)-V(k,h))/(V(k+1)-V(k))$ and $v(k,h)-v_f$ in each bin.

Figure 5 shows the results. As $f(v_f+v)$ should be v_f at $v=0$ if $v(k,h)$ converges to v_f, we plot $f(v_f+v)-v_f$ as function of v. If $f(v_f+v)=v_f$ at $v=0$, the graph $(v, f(v_f+v)-v_f)$ should cross the origin. In the figures, we plot the symbol • to show the place of the origin. The red (blue) curve corresponds to the result for $t<5\%(>5\%)$.

At first, we check the position of the crossing point of the curve with the diagonal. If the crossing point is larger than 0, the converging value of v is greater than v_f, and it corresponds to the overestimate of the strength of the horse. If the crossing point is smaller than zero, it means the underestimate. One can clearly see the crossing point depends on the stage of the betting $t<5\%, t>5\%$ and the final value v_f. In general, the position of the crossing points approaches to the origin from the initial stage of the betting ($t<5\%$) to the latter stage ($t>5\%$). The market participants modify the odds and reduce the bias. It is consistent with the behavior

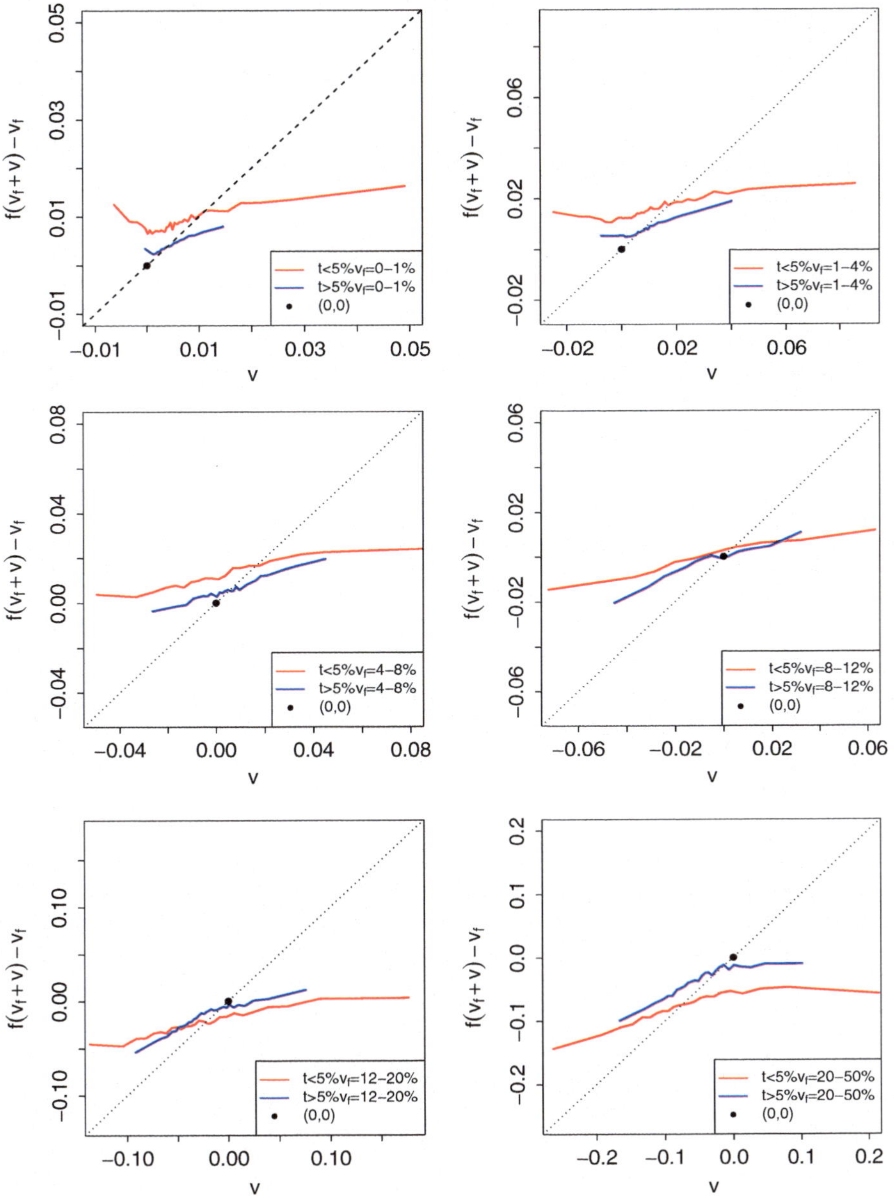

Fig. 5 Plots of $f(v - v_f)$ vs. $v - v_f$ for $v_f \in [0 - 1\%], [1 - 4\%], [4 - 8\%], [8 - 12\%], [12 - 20\%], [20 - 50\%]$. The symbol • shows the position of the origin

of the discrepancy of AR and EAR for $t > 5\%$. As the crossing points approach to the origin, the bias becomes small and the efficiency of the market improves. We also observe that the crossing points is larger than zero for $v_f \in \{[0 - 1\%],$

$[1-4\%], [4-8\%]\}$ and smaller than zero for $v_f \in \{[12-20\%], [20-50\%]\}$. About the case $v_f \in [8-12\%]$, the crossing points almost coincide with the origin for both initial and later stage of the betting. The results indicate that the weaker (stronger) horses are over(under)estimated.

Next, we study the shape of the curves $(v, f(v_f+v)-v_f)$. Before that, we explain $f(v_f+v)$ for three types of voters.

1. Noisy voter
   The first type is the noisy voter who casts his vote randomly with some probability. $f(v_f+v)$ is constant function and the curve becomes a horizontal line. One can regard the noisy voter as the independent voter who casts his vote independently from the choice of other betters. Independent voter bring some information about the strength of the horses by the choice of the betting probability. If there are only independent voters in the market and they vote with probability v_f, the win bet fraction of the horse converges rapidly to v_f.
2. Herder
   The second type is herder who has the tendency to choose the popular choice. The simplest herder is analog herder whose probability coincides with the win bet fraction v (Mori and Hisakado 2010; Hisakado and Mori 2010). The digital herder is a voter who has a threshold value v_{th}, and if $v > v_h$, he casts his votes and otherwise not (Hisakado and Mori 2011). The nonlinear herder has a monotonic increasing function $f(v_h+v)$ (Mori and Hisakado 2015). In general, the curve is ever-increasing for the herders. For simplicity, we only consider the analog herder here. As $f(v_f+v)-v_f = v$ for the analog herder, it is the diagonal line.
3. Arbitrager
   The third type is arbitrager who has some estimate about the winning probability of the horse v_* (Ali 1977; Manski 2006; Wolfers et al. 2004). If the win bet fraction is smaller than v_*, the odds of the horses provide a good chance of making money for the better, and he casts his vote. In a sense, arbitrager prefers the minority choice. The estimate v_* varies among arbitragers and the distribution of v_* determines $f(v_f+v)$. If the distribution of v_* is described by $g(v)$ and we denote its cumulative function as $G(v) = \int_v^1 g(u)du$, the probability that arbitrager votes at v_f+v is given by $P(v_* > v_f+v) = G(v_f+v)$.

   $$f(v_f+v) = G(v_f+v).$$

   If we assume that the arbitragers realize the efficiency of the market, $G(v_f) = v_f$ should hold (Ali 1977). In general, the curve $(v, f(v_f+v)-v_f) = (v, G(v_f+v)-v_f)$ of the arbitrager is ever-decreasing and crosses the origin if the equality $G(v_f) = v_f$ holds.

The plots of the empirically estimated $f(v_f+v)$ show that it decreases for small $v < 0$; it is ever-increasing or becomes flat for large $v - v_f$. In particular, in all six cases, $f(v_f+v)$ shows the increase from the left edge of the curve to the right edge of it, and it suggests the existence of herders. For the initial stage, the switch point from decreasing to increasing exits for the cases $v_f \in [0 - 1\%], [1 - 4\%]$ and the position is almost zero. The arbitrager's $g(v)$ might notice v_f and $G(v_f + v)$ decreases sharply at $v = 0$. The crossing point with the diagonal is larger than zero, and the betters overestimate the strength of the horse. For other cases, we do not notice such sharp switching point from decreasing to increasing. For the latter stage of the betting ($t > 5\%$), we notice the switching point at $v = 0$. For the cases $v_f \in [0 - 1\%], [1 - 4\%], [4 - 8\%]$, $f(v_f + v)$ changes from decreasing to increasing. It suggests that the arbitragers completely disappear and there remain herders and noise voters in the market. If we assume there remains some arbitrager in the market, $g(v_f + v)$ should decrease very slowly with v, and $G(v_f + v)$ does not decrease rapidly with v. In the range, we cannot distinguish noisy voters from the arbitragers. For the cases $v_f \in [8 - 12\%], [12 - 20\%], [20 - 50\%]$, the slope of the curve decreases at $v = 0$. Furthermore, we observe a hinge for $v_f \in [8 - 12\%]$, and it suggests that the arbitragers notice v_f and $G(v_f + v)$ drops at $v = 0$ for $v_f \in [8 - 12\%]$. We can interpret the slope change at $v = 0$ as the decreasing rate of $g(v_f + v)$ becomes larger for $v > 0$. For $v_f \in [20 - 50\%]$, the slope for $v > v_f$ is almost zero; we can interpret it as there is only noisy voter after the disappearance of the arbitrager.

5 Conclusion

We study the accuracy and efficiency of the horse race betting market in Japan. We observe a small discrepancy between the vote share of a win bet and the winning probability; it is negligibly small. The favorite-longshot (FL) bias does also exist; one cannot make money by using the bias. The accuracy of the predictions is excellent and the accuracy ratio (AR) is about 73.4%. In order to understand the emergence of the accuracy and the efficiency of the market, we study the time series data of odds. We estimate the time evolution of AR and EAR. Initially, the odds does not predict the winning of the horses so well. As time goes on and the 20% betting has finished, the sorting of the horses by the odds has completed, and AR almost reaches its maximum value. After 5% of betting, the adjustment of the odds starts, and at 50% of betting, EAR coincides with AR and the efficiency of the market recovers. In addition, we derive the probabilistic rule of the betters. We assume three types of voters, noisy voter, analog herder, and arbitrager. We interpret the probabilistic rule as the combination of the response function of the three types of voters. In the initial stage of the betting ($t < 5\%$), the equilibrium values of the vote share are rather different from their final values. In most cases, the probability of the betting becomes an increasing function of the previous vote shares which suggests the existence of herders. At the latter stage of the betting ($t > 5\%$), the equilibrium values approach to the final value, which means that the betters do fine-tuning of the odds and realize the efficiency of the market. The decomposition of the probabilistic rules by the combination of the three types of voters depends on the strength of the horse. When the horses are weak, the slope of the response function becomes steeper at the final value, which suggests the disappearance of the arbitrager there. If the horses are strong, the slope becomes smaller there. This suggests that the arbitrager whose estimate of the winning probability exceeds the final value of the win bet fraction begins to decrease faster there. Even so, the existence of the herder is necessary to explain the probabilistic rule, because the herder's response function is the unique one that increases monotonically.

6 Data and R Script

One can download the data and the R script that is used to plot the figures in this chapter. One should visit https://sites.google.com/site/shintaromori/home/data and download AppDCSSD_Chap13.tar.gz.

References

Ali M (1977) Probability and utility estimates for racetrack bettors. J Polit Econ 85:803–815
Griffith RM (1949) Odds adjustments by American horse race bettors. Am J Psychol 62:290–294
Hausch DB, Lo VSY, Ziemba T (2008) Efficiency of racetrack betting markets, 2008 edn. World Scientific, Singapore
Hill B, Lane D, Sudderth W (1980) A strong law for some generalized urn processes. Ann Prob 8:214–226
Hisakado M, Mori S (2010) Phase transition and information cascade in voting model. J Phys A Math Theor 43:315207–315219
Hisakado M, Mori S (2011) Digital herders and phase transition in a voting model. J Phys A Math Theor 44:275204–275220
Ichinomiya T (2006) Power-law distribution in Japanese racetrack betting. Physica A 368:207–214
Manski C (2006) Interpreting the predictions of prediction markets. Econ Lett 91:425–429
Mori S, Hisakado M (2009) Emergence of scale invariance and efficiency in racetrack betting market. In: Matsushita M, Aruka Y, Namatame A, Sato H (eds) Proceedings of the 9th Asia-Pacific complex systems conference complex 09, pp 258–267. Available via arXiv https://arxiv.org/pdf/0911.3249
Mori S, Hisakado M (2010) Exact scale invariance in mixing of binary candidates in voting model. J Phys Soc Jpn 79:034001–034008
Mori S, Hisakado M (2015) Correlation function for generalized Pólya urns: finite-size scaling analysis. Phys Rev E92:052112–052121
Park K, Dommany E (2001) Power law distribution of dividends in horse races. Europhys Lett 53:419–425
Wolfers J, Zitzewitz E (2004) Prediction markets. J Econ Persp 18:107–126

Smart Micro-sensing: Reaching Sustainability in Agriculture via Distributed Sensors in the Food Chain

R. Dolci and L. Boschis

1 Introduction

The last two decades has witnessed the growth of the industrial approach to agriculture, with biotech and infotech becoming the buzzwords in farming. This also led to a consolidation in two parts of the market: the companies producing farming machinery and the ones developing new seeds and treatments. Today less than ten multinationals control most processing and distribution of agricultural products and in so doing leave sustainability to the smaller and weaker players, the farmers (Ikerd 1996).

The difference in level of sophistication between large multinationals and small farmers suggests that some technologies and products could be misused and lead to safety issues. For example, glyphosate has made news worldwide for its possible role in causing acute myeloid leukemia (AML) or other pathologies (Andreotti et al. 2018). Regardless of continued research in the toxicity of this chemical, there is general consensus that frequency and quantity of its application are paramount to keeping crop safe.

Food safety is therefore the top priority issue of the whole production chain, from crops and cattle breeding to processing facilities. Foodborne disease, indeed, is currently a critical public health concern worldwide. Being able to accurately assess the presence of contaminants, pathogens, or allergens in foods is crucial, and there is a constant research for new performing analytical methods and data management tools.

R. Dolci (✉)
Aizoon USA Inc., Lewiston, ME, USA
e-mail: rob.dolci@aizoon.us

L. Boschis
Trustech s.r.l., Torino, Italy
e-mail: laura.boschis@trustech.it

A.-H. Sato (ed.), *Applications of Data-Centric Science to Social Design*,
Agent-Based Social Systems 14, https://doi.org/10.1007/978-981-10-7194-2_14

Quality by design (QbD) is simultaneously the new leading paradigm in the food industry and is directly inspired by the process analytical technology (PAT) concept, originally introduced for the pharmaceutical industry in 2004 by the US Food and Drug Administration (FDA, United States Food and Drug Administration 2004). This paradigm bases its shift on the hypothesis that the quality of the (food) products can and should be incorporated by process design and not by postproduction quality testing. It is evident that farmers need to be part of this all-encompassing effort to produce safe products, and the ability to monitor the variables of their fields is crucial in such respect.

Currently, food safety is assessed by standard techniques, such as ELISA, PCR, chromatography, and spectroscopy analyses, carried out in central laboratories. This approach is not time-effective and exhaustive, because only a limited number of representative samples are tested. The development of point-of-use tests (POUT) logically allows for a timely control on the farming process, by setting up local extensive monitoring programs and rapid screenings. This requires high sensibility, high selectivity, and rapid and reliable biosensors, integrated with an automated control system, something Aizoon works on often exploiting machine learning algorithms to develop a twin digital process of what happens in the field. In fact, available physical sensors such as for temperature, relative humidity, PH, and nitrogen allow for some tuning of the farming processes but do not indicate the definite presence and concentration of pathogens such as aflatoxins and others.

By coupling real-time or near-real-time process monitoring capabilities with advanced sensors, the industry is gradually moving from inferential monitoring toward continuous measurement of core quality parameters. Research-based development and implementation of PAT in farming is focused on improved control, rapid final product quality evaluation, and increased productivity. A positive side effect can be industrial innovation and enhanced competitiveness. In fact, the interest toward "smart" analytical approaches based on the use of micro- and nanosensors is continuously growing worldwide.

Portable and automated devices for the detection and characterization of biotic and abiotic contaminant during the whole food supply chain are pivotal in this context and constitute a dynamic area in food processing, which is experiencing important developments coming mainly from the standpoint of food safety. Several sensing systems are described in the literature (Bhardwaj and McGoron 2014). Commonly they are based on an optical or electrochemical transducer functionalized with a biological recognizing element (enzymes, cells, or affinity molecules) (Narsaiah et al. 2012; Thakur and Ragavan 2013; Dong et al. 2014; Kim et al. 2016). Several new commercially available tests are issued every year, but no one performs all the requirements needed for food analysis protocols, and it is still difficult to satisfy both accuracy and on-field reliability.

A new analytic platform named EliChip has been developed by Trustech (www.trustech.it) and combines capillary data collection to elaboration during the entire farming and food processing. A corn farmer can tune their seeding, irrigation, treatment, harvesting, and storage processes if relevant data points inform about the efficiency and safety of the produce. While sensing relative humidity, airflow, and temperature will help optimize irrigation, the sensing of molds or toxins such as aflatoxins will guide the storage and processing of corn.

The EliChip device is based on a microfluidic ready-to-use disposable card (lab-on-a-chip, LOC) and a small portable related reader. EliChip has been designed to be automated and highly integrated with the whole control system software of food smart factories. Following this vision, new portable and connected biosensors like EliChip are very important to monitor chemical, physical, and microbiological parameters, in addition to data and information collected during the whole food production chain, which will allow the capillary control of industry production processes. This centralized monitoring system will allow food companies to control the entire production and distribution chain by detecting occurring anomalies through diagrams that will quicken the identification of the problem. The dashboard will therefore enable the real-time and near real-time intervention directly in the production stages, thus improving the efficiency and limiting the random variability, the rebalancing, and the product losses. Tools like EliChip are crucial to empower this concept of "Smart Factory."

In this section we describe the EliChip smart sensor and present the experimental data regarding food allergens and mycotoxins detected. EliChip is a microfluidic point-of-care biosensor able to detect biotic and abiotic contaminants (e.g., allergens and mycotoxins) during the food supply chain. EliChip probes are based on affinity biomolecules; both antibodies and aptamers can be used. In this paper, two significant case studies are reported: lysozyme as food allergen and aflatoxin B1 as food contaminant.

Lysozyme, like gluten and lactose, increases allergy-related risks in susceptible individuals. It is naturally present in eggs and egg derivates and is often added as preservative in various foods (such as cheese and wine), medications, and vaccines. The residues of this protein could represent a serious risk for allergic subjects. According to the EU Directive on labeling (2003/89/EU), if one or more of the ingredients listed in Annex IIIa of Directive 2000/13/EC (and following amendments) are present in wine, they must be indicated on the labeling of the product. Egg proteins like lysozyme used during the production need to be present at a detection level lower than 0.25 mg/L.

Aflatoxins are hepatotoxic and carcinogenic secondary metabolic products, synthetized by fungi belonging in particular to the *Aspergillus flavus* and *A. parasiticus* species (Council for Agricultural Science and Technology 1989; Kensler et al. 2010). Aflatoxins are found as natural contaminants in many feedstuffs of plant origin, especially in cereals but also in fruits, hazelnuts, almonds, and their products, intended for human or animal consumption (Binder et al. 2007). It is well known that AB1 can cause chronic diseases in humans and animals and can determine several adverse effects, such as hepatotoxicity, genotoxicity, and immunotoxicity (Liu and Felicia 2010). European legislation sets a maximum permitted concentration in human food (2–12 μg/kg), infant foods (0.1 μg/kg), and animal feeds (5–50 μg/kg) (Liu and Felicia 2010).

2 System Overview and Description of the Analytical Approach

For farming and the entire food chain to be sustainable, especially in consideration of the sophistication brought to the field by large multinationals and the importance of safety, we argue the relevance of gathering and sharing data at farmer level, some kind of data democratization. It is farmers who decide how much pesticide (e.g., glyphosate) to apply and how frequently, so no matter how innovative is the latest product from the chemical providers, knowledge needs to be with the farmer. When considering Internet of Things (IoT) benefits to sustainable agriculture, one can see the advantages of sensors and big data analytics in informing operators of the best way to farm (Popovic et al. 2017).

In this specific work, a portable and automated diagnostic platform able to perform miniaturized ELISA assays has been set up, developed, and applied to some case studies. The system is suitable for point-of-use (POU) analysis in agrifood quality control applications and can be a remote node or a larger network connected to the control software of a "Smart Factory."

The EliChip device is composed of:

- Disposable ready-to-use LOC cards.
- A light and automated reader.

The LOC has been designed to perform direct, indirect, competitive, and noncompetitive assays based on antibodies as well as aptamers. The microfluidic architecture is made of microchambers devoted to reagent reservoirs, reaction chambers, optical detection areas, and waste collection chambers, all connected through microchannels, as described in Fig. 1.

Four analytical lines are present in the LOC architecture, in order to perform four tests in parallel. This configuration allows choosing between a semiquantitative multi-target measurement, being able to perform either four different analyses in

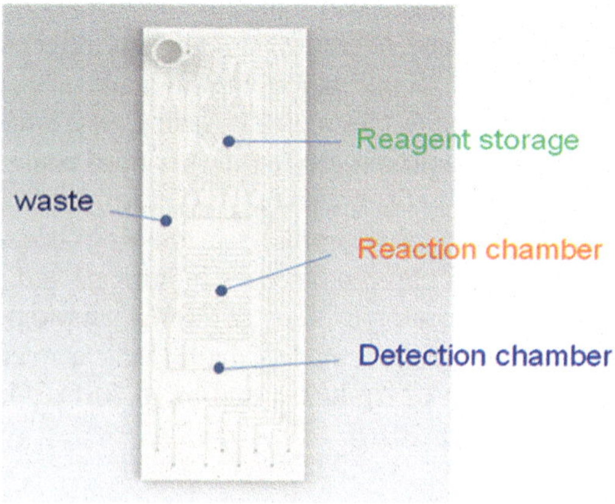

Fig. 1 Microfluidic architecture of EliChip LOC

parallel, and a single quantitative target measurement, being able to perform a calibration curve and one target quantitative analysis at the same time.

To perform a calibration curve at the same time of the analysis enables the accurate quantification of the target, and this is a main advantage of EliChip that allows an accurate and precise detection even outside the lab. Most of currently available screening devices do not give this possibility, thus lacking in accuracy and precision, especially for low-concentrated targets like aflatoxins.

In this configuration, the first line is dedicated to the analyte analysis, while the second is dedicated to the blank (or max signal control, depending whether the ELISA is competitive or not), and the last two lines are measures of known reference standard analytes. The LOC, realized in a biocompatible thermoplastic polymer such to be producible in industrial scale with injection molding, is designed to have all reagents preloaded; antibodies or aptamers are bound to the surface of the reaction region. In this configuration, the device is ready to use. The only operational step required is loading the sample and inserting the LOC into the instrument, which can be done by either a user or a robotic system.

Once inserted in the instrument, the LOC card is automatically connected to micropumps and electrovalves inside the reader, and the software can properly manage the loading of reagents. Appropriate ELISA protocols are executed automatically pumping reagents and samples through the microfluidic channels architecture. When the final reaction occurs, absorbance readings are performed. The software processes the data acquired by interpolating them with the calibration curve and calculating the concentration of the target analyte. Then, the result of analysis can be sent to the Smart Factory central control software.

The colorimetric detection was chosen as detection system. This is composed of a light source, a light-emitting diode (LED), and a photodiode used as collector. Experimental tests have shown that the method is selective and very sensitive, reaching detection limits in the order of parts per trillion (ppt).

The method was tested for aflatoxin M1 in milk and B1 in corn and wheat and to discover traces of allergens like lysozyme in wine.

The molecular probes used were antibodies, properly immobilized inside the reaction chambers.

EliChip microfluidic architecture is designed to allow the performance both of competitive and not competitive assays. In the reactive region, antibodies are bound on the surface. The HPR, linked to the secondary antibody, and TMB in the solution react producing a change of color directly related to the concentration of the target, which can be evaluated with absorbency (Fig. 2).

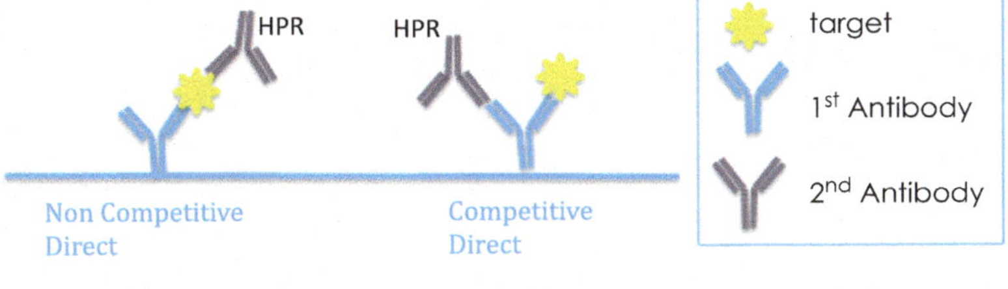

Fig. 2 Typical ELISA protocol

3 Experimental Results and Preliminary Outcomes of the Research

The EliChip LOC device and system have been set up to execute both competitive and noncompetitive assays, in order to detect small molecules as well as macromolecules. Regarding small molecules, like aflatoxins, a competitive ELISA design is preferred, since it allows the detection of very low concentration of the target. As for macromolecules, like gluten and lysozyme, a noncompetitive ELISA design is preferred, in order to reach high specificity. Experiments were carried out to compare standard laboratory ELISA and EliChip LOC measurement of different concentrations of target molecules. As molecular model for competitive ELISA assay design, different aflatoxin B1 concentrations (0.1 ng/mL, 0.6 ng/mL, 1.2 ng/mL) have been tested. As molecular model for noncompetitive ELISA assay design, different lysozyme concentrations (1.5 ng/mL, 6.25 ng/mL, 12.5 ng/mL) have been tested. The resulting curves overlapped with standard ELISA, as showed in Figs. 3 and 4.

In addition, in order to test the reproducibility of analytical results, different experimental sessions have been planned to calculate the coefficient of variation (CV). The CV is defined as the ratio of the standard deviation σ to the mean μ, and it is a very important parameter to evaluate the reliability of the measurements. Usually, for commercial ELISA assays, the CV is <10%. The EliChip LOC device has been optimized to reach this requirement, and the data collected are shown in Table 1.

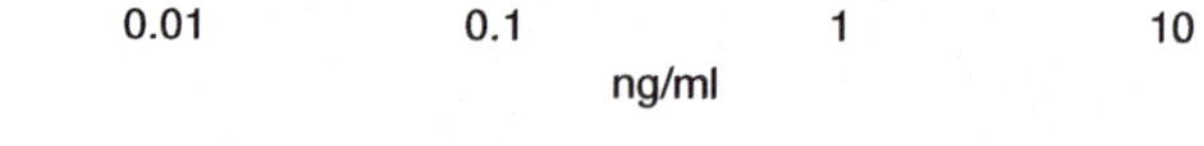

Fig. 3 Comparison between the performances of EliChip (blue line) and a laboratory ELISA test for aflatoxin B1 (purple line)

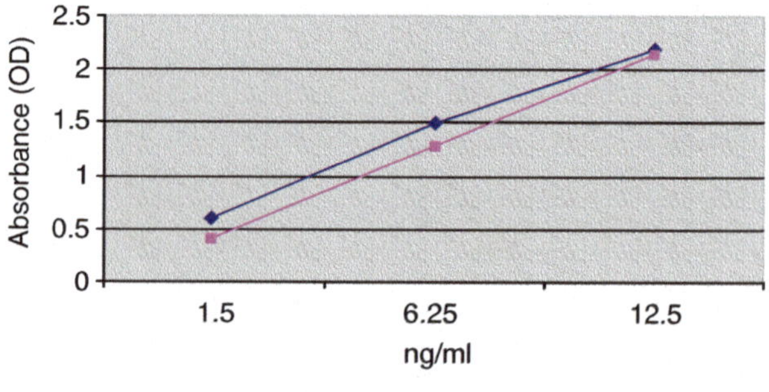

Fig. 4 Comparison between the performances of EliChip (blue line) and a laboratory ELISA test for lysozyme (purple line)

Table 1 EliChip optimization

Aflatoxin B1	
Session	CV%
1	7
2	8.1
3	8.5
Lysozyme	
Session	CV%
1	6.4
2	10
3	9

These results are referred to EliChip LOC devices fabricated by lab-scale prototyping techniques and operator-depended assembly. Thus, the reproducibility of the fabrication and of the bio-functionalization needs to be achieved using industrial-grade equipment.

Preliminary outcomes confirm that the developed portable device shows similar test quality compared to conventional laboratory ELISA, opening new interesting challenges in the setting up of detection systems to be exploited during the food supply and production chain.

4 Conclusion and Future Work

In conclusion, it is evident how the fragmented world of agriculture, where a multitude of farmers operate under much economic pressure from large multinationals in the machinery and chemical sectors and from markets that are based on commodity, can benefit from a more sophisticated approach to data analytics. At present the use of third-party laboratories to monitor the presence of toxins and other pathogens causes delays and costs to farmers who would need a much more real-time indication of problems they need to act upon swiftly.

In this regard, the availability of a portable and economic lab-on-a-chip (LOC) that allows for immunoenzymatic reactions to test for the presence of toxins and pathogens is a welcome development.

This chapter has presented EliChip as an example of one such portable sensor. Future work will be focused on the development of virtual sensors or the application of machine learning algorithms to the estimation of variables from the big data analysis of the other known physical and biological conditions.

Acknowledgment The work was part of the project "Food Digital Monitoring," financed by the Piedmont Region with European Funds for Regional Development (Call: Smart Factory Platform).

References

Andreotti G et al (2018) Glyphosate use and cancer incidence in the agricultural health study. JNCI 110(5):509–516. https://doi.org/10.1093/jnci/djx233

Bhardwaj V, McGoron AJ (2014) Biosensor technology for chemical and biological toxins: progress and prospects. Photon J Biomed Eng 112:380–392

Binder EM et al (2007) Worldwide occurrence of mycotoxins in commodities, feeds and feed ingredients. Anim Feed Sci Technol 137(3):265–282

Council for Agricultural Science and Technology (1989) Mycotoxins: economic and health risks. CAST, Ames, p 99

Dong Y et al (2014) Aptamer and its potential applications for food safety. Crit Rev Food Sci Nutr 54(12):1548–1561

FDA, United States Food and Drug Administration (2004) Guidance for industry: PAT – a framework for innovative pharmaceutical development, manufacturing and quality assurance. U.S. Department of Health and Human Services, Rockville

Ikerd J (1996) Sustainable agriculture: a positive alternative to industrial agriculture, University of Missouri. Presented at the Heartland Roundup in Manhattan, KS – December 7

Kensler TW et al (2010) Aflatoxin: a 50-year odyssey of mechanistic and translational toxicology. Toxicol Sci 120(Suppl 1):S28–S48

Kim SG et al (2016) Ultrasensitive bisphenol a field-effect transistor sensor using an aptamer-modified multichannel carbon nanofiber transducer. ACS Appl Mater Interfaces 8(10):6602–6610

Liu Y, Felicia W (2010) Global burden of aflatoxin-induced hepatocellular carcinoma: a risk assessment. Environ Health Perspect 118(6):818

Narsaiah K et al (2012) Optical biosensors for food quality and safety assurance—a review. J Food Sci Technol 49(4):383–406

Popovic T et al (2017) Architecting an IoT-enabled platform for precision agriculture and ecological monitoring: a case study. In: Computers and electronics in agriculture, Elsevier

Thakur MS, Ragavan KV (2013) Biosensors in food processing. J Food Sci Technol 50(4):625–641

High-Frequency Data Analysis of Foreign Exchange Markets

Aki-Hiro Sato

1 Introduction

A challenging new field called *econophysics* has been established during the past quinquennium. This movement was derived from the fact that recently, physicists have begun writing research papers on finance and economics. Econophysics is a movement to apply statistical physics to economically motivated problems, and it is attracting many physicists. The reason is that they can be considered exciting subjects of "complex systems," indeed, the definition of complex systems has not yet been properly formalized. However, it is pointed out that complex systems include a self-organization mechanism (Haken 1983). In other words, they should be understood as circular causal systems including information feedback loops (Wiener 1961). Namely, cause and effect look like a reversal. For example, a cause produces an effect and the effect leads to the next cause.

Economical systems (of course, similarly, social systems) obviously present "complexity" derived from interactions between humans. Such interactions allow communication with each other and common understandings to converge into a range or diverge. Namely, information flow through interactions can produce both negative and positive feedback. For example, a common understanding is established at a certain point when the negative feedback appears (Helbing 2001). On the other hand, a misunderstanding is more amplified as deeper communication when positive feedback exists.

A.-H. Sato (✉)
Yokohama City University, Kanazawa-ku, Yokohama-shi, Kanagawa, Japan

Japan Science and Technology Agency PRESTO, Kawaguchi-shi, Saitama, Japan

Statistical Research and Training Institute, Ministry of Internal Affairs and Communications, Shinjuku-ku, Tokyo, Japan
e-mail: ahsato@yokohama-cu.ac.jp

A.-H. Sato (ed.), *Applications of Data-Centric Science to Social Design*,
Agent-Based Social Systems 14, https://doi.org/10.1007/978-981-10-7194-2_15

Recently, investigations of foreign exchange rates have been conducted intensively by several researchers (Gworek et al. 2010; Kaltwasser 2010; Alfarano et al. 2006; Liu et al. 2010). Moreover, it became possible to examine the foreign exchange market using tick-by-tick data (Sazuka et al. 2009; Inoue and Sazuka 2010; Ohnishi et al. 2004; Sato and Takayasu 1998; Hashimoto and Ito 2010). Ohnishi et al. (2004) proposed a weighted-moving-average analysis for the tick-by-tick data of yen–dollar exchange rates. Meanwhile, Hashimoto and Ito examined the market impact of Japanese macroeconomic statistics news within minutes of their announcements on the yen–dollar exchange rates (Hashimoto and Ito 2010). In general, after participants perceive information, they determine their investment attitude. However, the impact of news on trading decisions is an open question (Kyle 1985).

Nevertheless, it may include meaningful information to investigate both quotation and transaction activities. In this chapter, we focus on the numbers of quotations and transactions for each currency pair in the foreign exchange market as a test bed of collective human behavior that can be empirically observed, and we attempt to develop a method to characterize them as a multivariate time series. In fact, Bonanno et al. investigated the spectral density of both the logarithm of prices and the daily number of trades of a set of stocks traded on the New York Stock Exchange (Bonanno et al. 2000). They detected a $1/f$-like behavior for the spectral density of the daily number of trades. It shows that the number of transactions is generated by a long memory process similar to volatility. In such studies on financial time series, several researchers take a stance that the traded volume or number of transactions (quotations) is a proxy variable for unobservable information arrivals perceived by market participants.

The statistical properties of price fluctuations are important in understanding how real markets behave. A large amount of high-frequency financial data are available owing to high-performance computers and huge data storage. Mantegna et al. investigated fluctuations in the stock market price index of the S&P 500 (Mantegna and Stanley 1995). Meanwhile, Mandelbrot analyzed the market price changes in an open market and proposed the multi-fractal model to describe them (Mandelbrot 1997).

The price fluctuations were treated as stochastic processes. Traditionally, stochastic differential equations are often used to understand the characteristics of option pricing (Black and Scholes 1973; Merton 1973). Specifically, they assume that a stock price follows a geometric Brownian motion,

$$dY_t = \nu Y_t dt + \sigma Y_t dW_t, \tag{1}$$

where ν is the expected return per unit time, σ^2 is the variance per unit time, and W_t is a Wiener process. More recently, stochastic processes based on time-dependent volatility (so-called clustered volatility) have been proposed (Engle 1982; Bollerslev 1986; Nelson 1990). Especially, ARCH-type stochastic processes are well-known. In fact, this approach has clarified the properties of market price fluctuations, but it does not answer why the market prices fluctuate, because it is a phenomenological approach.

An approach using agent-based market models aims to construct an artificial market model having dealers who trade virtual securities in computers and to calculate numerically market prices that result from dealers' interactions (Takayasu et al. 1992; Hirabayashi et al. 1993; Palmer et al. 1994; Brock and LeBaron 1996; Bak et al. 1997; Johnson et al. 1998; Egenter et al. 1999; Lux and Marchesi 1999; Matassini and Franci 2001; Jefferies et al. 2001). These models have dealers interacting with market prices and some even include learning dealers. Advantages of this approach are that one can follow market prices easily and infer data unknown in the real market through numerical simulations.

2 Statistical Property of Foreign Exchange Rates

2.1 Fundamental Properties

Mantegna et al. investigated time series of the stock market index of the S&P500 and reported in their famous paper (Mantegna and Stanley 1995) that a probability density function of changes obeys a power law distribution. They indicated that the power law exponent is estimated as 1.4. However, it is clarified that the exponent is not universal in other research. Besides stock market prices, it is known that the probability density function of foreign exchange rates also follows a power law distribution (Takayasu et al. 2000).

Here, we deal with high-frequency financial data, the so-called tick data of yen–dollar exchange rates. They are collected by Bloomberg from representative dealers, and the dealers report their transaction prices on which they have traded to Bloomberg whenever they trade; then, Bloomberg announces the prices to the rest of the world right away. These data are not sampled at a constant interval but at every report. As a result, the interval of their time stamp is not constant. Data from 6 October 1998 to 4 December 1998 with 477,060 ticks except Saturday and Sunday are available. The foreign currency market is usually open on weekdays, and the trades are successive all over the globe.

Figure 1 displays a typical example of time series of the yen–dollar currency rate. Blank spaces represent weekends. Let ticks of the yen–dollar exchange rate be numbered in order, the rate of the s-th tick be denoted as r_s, and the corresponding change be $\Delta r_s = r_s - r_{s-1}$. Figure 2 shows typical examples of yen–dollar exchange rates and their changes. The sampling term is from 6 October 1998 to 9 October 1998 with 36,758 data points (36,787 data points in the changes). The time series of Δr_s has a continuation of changes, which is called volatility clustering. Volatility means variance/standard deviation in physics.

We focus on a frequency distribution of changes at s, its autocorrelation functions (defined in detail below), and a distribution of time intervals between trades as statistical properties of the sequence. Figure 3 exhibits a probability density function estimated from rate changes Δr_s and the corresponding cumulative distribution function, which is defined by

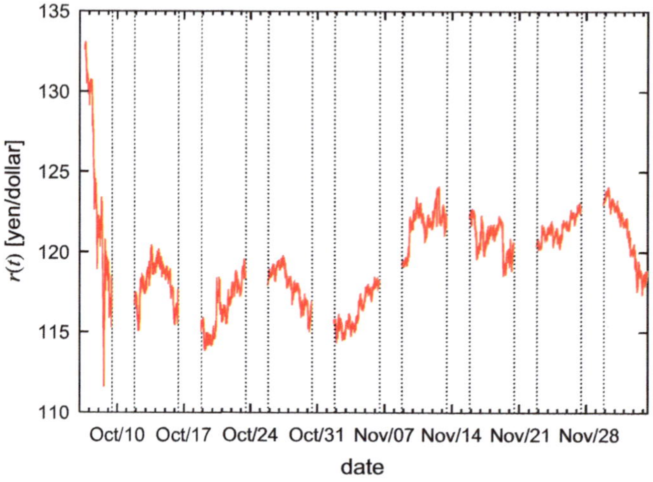

Fig. 1 Typical examples of time series of yen–dollar exchange rates from 06 October 1999 to 04 December 1999. $r(t)$ represents the rate at time t. Blank spaces represent weekends

$$P(\geq |x|) = \int_{-\infty}^{-|x|} p(x')dx' + \int_{|x|}^{\infty} p(x')dx', \tag{2}$$

where $p(x)$ denotes the probability density function. It is found that the probability density function has fatter tails than a Gaussian distribution with the same variance (0.006375043). A cumulative distribution function has a linear slope with a quick decay at 1 yen in log–log plots. It is known that if the cumulative distribution function has a linear slope in log–log plots, it follows the power law distribution,

$$P(\geq |x|) \propto |x|^{-\beta}, \tag{3}$$

where β is a power law exponent and $0 < \beta < 2$. Fitted to the cumulative distribution function numerically, we get $\beta = 1.8$.

Figure 4 shows the autocorrelation coefficient $R_{s'}^{(1)}$ and volatility correlation coefficient $R_{s'}^{(2)}$, defined as

$$R_{s'}^{(1)} = \frac{\langle r_s r_{s+s'} \rangle - \langle r_s \rangle \langle r_{s+s'} \rangle}{\langle r^2 \rangle - \langle r \rangle^2}, \tag{4}$$

$$R_{s'}^{(2)} = \frac{\langle r_s^2 r_{s+s'}^2 \rangle - \langle r_s^2 \rangle \langle r_{s+s'}^2 \rangle}{\langle r^4 \rangle - \langle r^2 \rangle^2}. \tag{5}$$

The autocorrelation coefficient $R_{s'}^{(1)}$ displays a short-time correlation with a negative value at $s' = 1$. On the other hand, the volatility correlation coefficient $R_{s'}^{(2)}$ shows

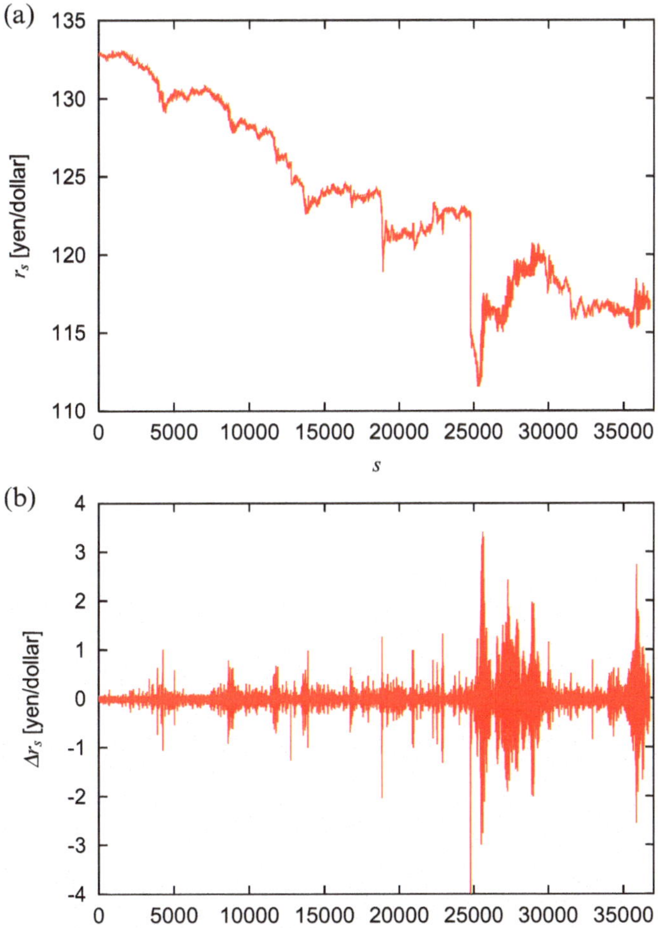

Fig. 2 Typical examples of sequences of the yen–dollar exchange rates numbered in order (**a**) and their changes (**b**) from 06 October 1999 to 09 October 1999, where r_s represents the rate at the s-th tick and Δr_s a corresponding change

a far longer correlation than $R_{s'}^{(1)}$. As mentioned above, this means that a high-volatility regime sustains for long ticks (volatility clustering).

Consider time stamps of time series in Fig. 1. As shown in Table 1 an interval of their transaction time is not constant and appears random. We define an interval between the $(s-1)$-th trade and the s-th trade as $\tau_s = t_s - t_{s-1}$, where t_s is the time when the s-th transaction occurs. We show averaging intervals over days at the time t in Fig. 5. It is obviously found that intervals between successive trades depend fully on transaction time. This implies that trading activity is related to human activity and that activity averages differ from each other.

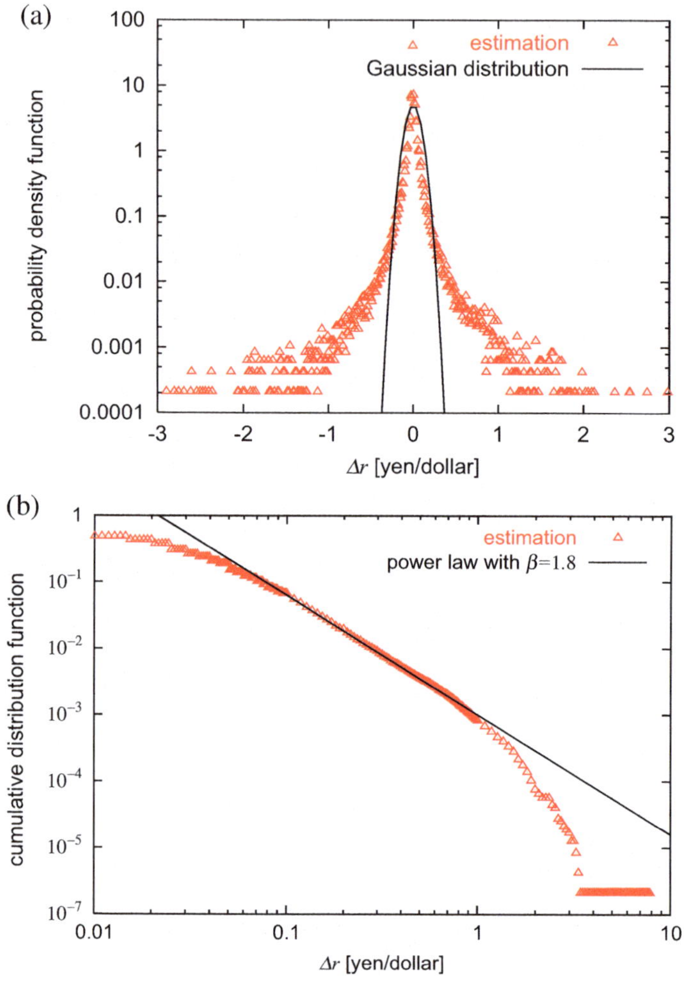

Fig. 3 (**a**) Semi-log plots of a probability density function of changes in yen–dollar exchange rates together with a Gaussian distribution having the same variance as the data (variance = 0.006375043). Triangles represent estimated distributions and a solid-line Gaussian. (**b**) Log–log plots of a corresponding cumulative distribution function. A solid line represents the power law with $\beta = 1.8$

Here, we focus on the mean interval averaged over T seconds from the observing point (Takayasu et al. 2002), and we introduce a normalized time interval $\mu_{T,s}$, which is defined as follows:

$$\langle \tau_s \rangle_T = \frac{1}{N_s(T)} \sum_{i=n-N_s(T)}^{s-1} \tau_i, \tag{6}$$

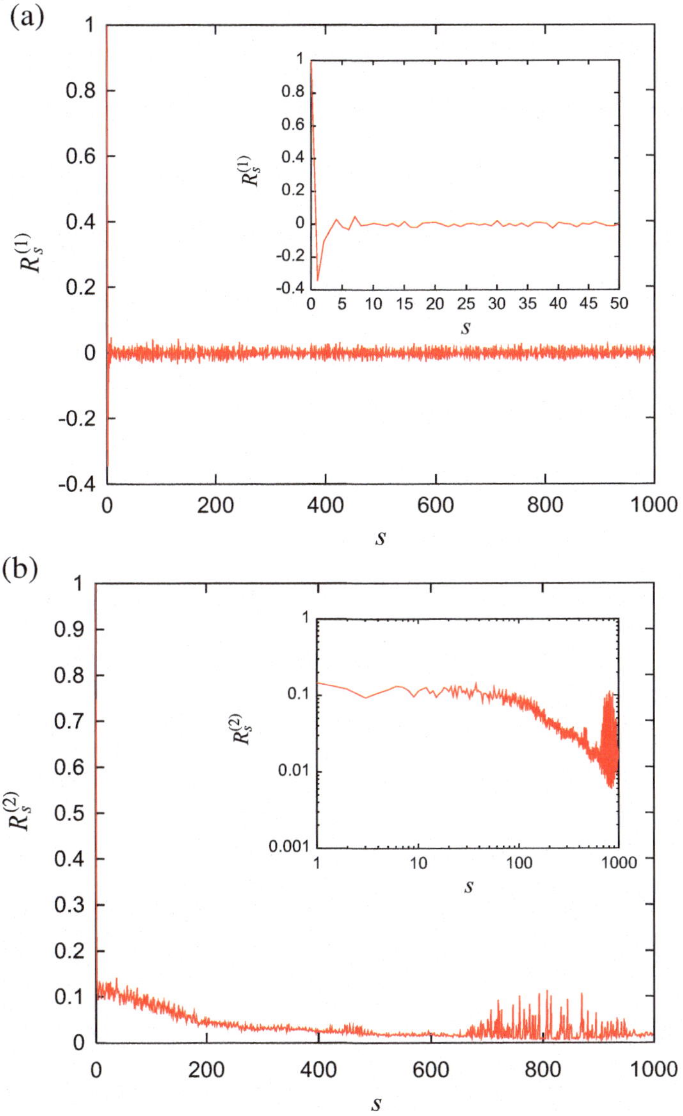

Fig. 4 (**a**) Typical examples of an autocorrelation coefficient of Δp_s. A small panel in the graph represents the focusing snap near zero. (**b**) Corresponding volatility correlation coefficient. A small panel into the graph represents log–log plots of it

Table 1 Typical examples of time series of yen–dollar currency rates. The transaction time resolution is a second. The time stamps are US Central Time (Chicago time)

Trade day	Transaction time	Transaction price
1998/10/06	13:00:21	132.780
1998/10/06	13:00:21	132.780
1998/10/06	13:00:21	132.780
1998/10/06	13:00:23	132.780
1998/10/06	13:00:32	132.780
1998/10/06	13:00:44	132.780
1998/10/06	13:00:48	132.780
1998/10/06	13:01:00	132.810
1998/10/06	13:01:00	132.810

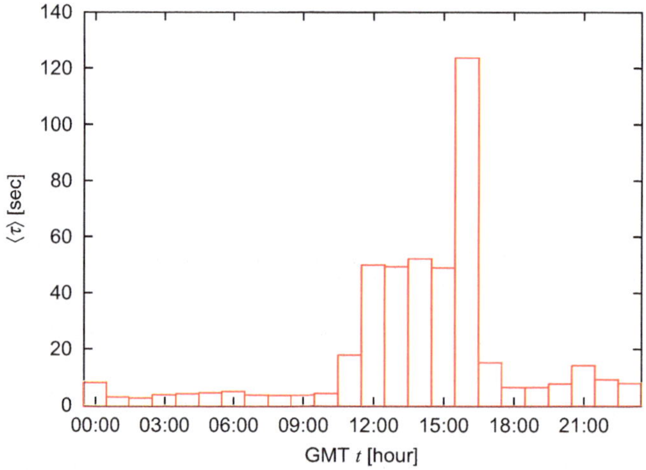

Fig. 5 Mean values of intervals between two successive trades over an hour, where t represents the time and $\langle\tau\rangle$ an average of intervals over the sampling day. European activity runs from 00:00 to 07:00. American activity runs through from 07:00 to 17:00. Asian activity runs through from 17:00 to 24:00. The time stamp is US Central Time (Chicago time)

$$\langle\mu_{T,s}\rangle = \frac{\tau_s}{\langle\tau_s\rangle_T}, \tag{7}$$

where $N_s(T)$ represents the number of transactions in the preceding T seconds, and $\langle\tau_s\rangle_T$ is the corresponding mean interval. For example, if the $(s-1)$-th tick occurs more than T seconds before the s-th tick, then $N_s(T)$ is set to unity.

Figure 6 shows cumulative distribution functions of normalized transaction intervals at bin size T. The normalized interval distributions are plotted for two typical sizes of T. The thick line is normalized using a bin size of 150 s, which almost overlaps with the theoretical line of exponential decay, $P(\geq \mu_{T,s}) = \exp(-\mu_{T,s})$, shown by a broken line. The thin line shows the distribution normalized using a bin size of 400 s. Therefore, this means that we can tune the mean interval of the normalized time to be close to unity, where $\langle\mu_{T,s}\rangle = 1$ if we choose an

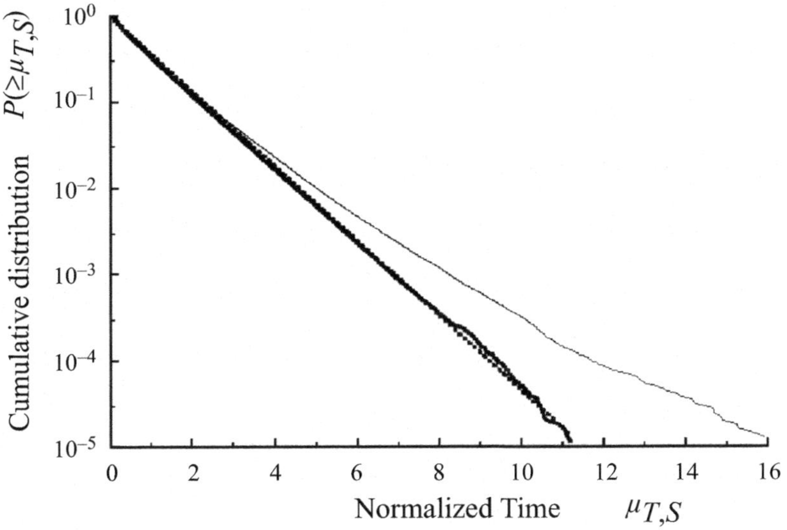

Fig. 6 Semi-log plots of cumulative distributions of a normalized transaction interval. The thick line is normalized using a 150 s bin size, and the thin line is normalized using a 400 s bin size. The broken line shows the exponential, $e^{-\mu_{T,s}}$

appropriate bin size T. It has longer intervals compared with the case of $T = 150$ s. We found that for any bin size smaller than 150 s, we always obtained a clear exponential decay. On the other hand, if we chose a bin size larger than 150 s, we obtained a longer tail that deviated from the exponential decay. Thus, the deviation from the exponential decay in our overall data set can be viewed as an artifact of changes in the mean interval with a time scale larger than 150 s. In practice, we conclude from these results that the transaction intervals are a quasi-stationary Poisson process within the range of 150 s and local mean interval changes in the time order of several hundred seconds.

This process can be modeled as the autoregressive conditional duration (ACD) model in the context of econometrics (Engle and Russell 1998). Suppose that τ_s can be represented as

$$\tau_s = \psi_s \epsilon_s, \tag{8}$$

where ϵ_s is an *i.i.d.* nonnegative random number and ψ_s is assumed to be a linear relationship.

$$\psi_s = \omega + \sum_{j=1}^{p} \alpha_j \tau_{s-j} + \sum_{j=1}^{q} \beta_{s-j} \psi_{s-1}, \tag{9}$$

If we assume ϵ_s is sampled from a standard exponential distribution

$$f(\epsilon) = e^{-\epsilon}, \tag{10}$$

then this is called EACD(p,q). In contrast, if we observe the number of transactions X in every T, then X is modeled as a Poisson process

$$P(X;T) = e^{-\lambda(t)T}\frac{(\lambda(t)T)^X}{X!}, \tag{11}$$

where $\lambda(t)$ is an intensity described as

$$\lambda(t) = \frac{1}{\psi_{N(t)-1}}. \tag{12}$$

However, a multidimensional viewpoint is needed to understand states of the foreign exchange market. In the next subsection, we will introduce a way to deal with the tick-by-tick data of the foreign exchange market from a comprehensive perspective.

2.2 *Fluctuation Scaling*

Fluctuation scaling refers to the fact that there exists a scaling relationship between the mean of constituent flows at the i-th observation point and their standard deviation on a system. The origin of Taylor's power law has been successively discussed in the literature of ecology (Kilpatrick and Ives 2003; Samaniego et al. 2011; Kendal 2004) and physics (Fronczak et al. 2010; Eisler and Kertész 2006; Eisler et al. 2008; Sato et al. 2010; Kendal and Jorgensen 2011). Taylor's law has been recently hypothesized to result from the second law of thermodynamics and the behavior of the density of states. This hypothesis is predicated on physical quantities, such as free energy and an external field, though it remains unproven. The Tweedie models can express power variance functions over a wide range of measurement scales (Kendal and Jorgensen 2011). However, there are few studies on the relationship between cross-correlation and Taylor's law for multivariate counting processes.

Multivariate Poisson distributions with correlations have been considered to model multivariate counting data (Teicher 1954). Guerrero considers an entropy of a multivariate Poisson distribution (Guerrero-Cusumano 1995). While the multivariate Poisson model is the most important among discrete multivariate distributions, it has several shortcomings in its application. The main drawback of the multivariate Poisson model is its complicated form of the joint probability function. As a result, multivariate Poisson distributions are defined as characteristic functions (Lukacs and Beer 1977). Karlis and Meligkotsidou studied mixtures of multivariate Poisson distributions and their parameter estimation procedure based on the maximum likelihood (ML) estimator and the expectation–maximization (EM) algorithm (Karlis and Meligkotsidou 2007). However, the convergence of parameter estimates of both the ML and EM estimators strongly depends on starting points of computation. In this way, we need much computational power to find adequate parameter estimations. A simpler way to generate a correlation among variables is to use a linear transformation to independent univariate Poisson random variables. However, it is not sufficient to consider Taylor's scaling relationship.

Here, we attempt to consider an infinite mixture of the multivariate Poisson distribution in a simpler way with fluctuations in common mode. Let the market activity X of the foreign exchange market be described by a Poisson distribution with a parameter μ. Let K_i be further a fraction that is devoted to an asset i, where $\sum_{i=1}^{N} K_i = 1$. Observing the market during the period T, we have time series where X is sampled from

$$P(X;T) = e^{-\mu T} \frac{(\mu T)^X}{X!}, \tag{13}$$

and activity X_i for the asset i from

$$P_i(X_i;T) = e^{-\mu K_i T} \frac{(\mu K_i T)^{X_i}}{X_i!}. \tag{14}$$

Their fluctuations in activity X_i are equal to the square root of mean value.

$$\sigma_i = \sqrt{\mu K_i T} = \langle X_i \rangle^{1/2}. \tag{15}$$

Let us assume that the whole market is equally described by a set of Poisson processes with intensities described by $G(\mu)$, $\quad \mu \in (0, +\infty)$. It follows that now

$$P(X;T) = \int_0^\infty e^{-\mu T} \frac{(\mu T)^X}{X!} G(\mu) d\mu, \tag{16}$$

and

$$P_i(X_i;T) = \int_0^\infty e^{-\mu K_i T} \frac{(\mu K_i T)^{X_i}}{X_i!} G(\mu) d\mu. \tag{17}$$

Then, we have

$$\langle X_i \rangle(T) = \int_0^\infty e^{-\mu K_i T} \frac{(\mu K_i T)^{X_i}}{X_i!} X_i G(\mu) d\mu = K_i T \langle \mu \rangle, \tag{18}$$

and

$$\begin{aligned}\langle X_i^2\rangle(T) &= \langle K_i^2T^2\mu^2\rangle + \langle K_iT\mu\rangle \\ &= K_i^2T^2\langle\mu^2\rangle + K_iT\langle\mu\rangle \\ &= K_i^2T^2\int_0^\infty \mu^2G(\mu)d\mu + K_iT\int_0^\infty \mu G(\mu)d\mu \\ &= K_i^2T^2\int_0^\infty \mu^2G(\mu)d\mu + \langle X_i\rangle. \end{aligned} \tag{19}$$

Because we have

$$\begin{aligned}\langle\mu^2\rangle &= \int_0^\infty (\mu-\langle\mu\rangle)^2G(\mu)d\mu - \langle\mu\rangle^2 + 2\int_0^\infty \mu\langle\mu\rangle G(\mu)d\mu \\ &= \int_0^\infty (\mu-\langle\mu\rangle)^2G(\mu)d\mu + \langle\mu\rangle^2, \end{aligned} \tag{20}$$

$$\begin{aligned}\sigma_i^2(T) &= \langle X_i^2\rangle(T) - \langle X_i\rangle^2(T) \\ &= K_i^2T^2\langle\mu^2\rangle + \langle X_i\rangle - K_i^2T^2\langle\mu\rangle^2 \\ &= K_I^2T^2(\langle\mu^2\rangle - \langle\mu\rangle^2) + \langle X_i\rangle, \end{aligned} \tag{21}$$

we got

$$\begin{aligned}\sigma_i^2(T) &= K_i^2T^2\int_0^\infty (\mu-\langle\mu\rangle)^2G(\mu)d\mu + K_iT\langle\mu\rangle \\ &= \langle X_i\rangle^2(T)\frac{\int_0^\infty(\mu-\langle\mu\rangle)^2G(\mu)d\mu}{\langle\mu\rangle^2} + \langle X_i\rangle(T) \\ &= \langle X_i\rangle(T)\Big[1 + \langle X_i\rangle(T)\frac{\int_0^\infty(\mu-\langle\mu\rangle)^2G(\mu)d\mu}{\langle\mu\rangle^2}\Big]. \end{aligned} \tag{22}$$

Thus, we have

$$\sigma_i^2(T) = \langle X_i\rangle(T)[1 + \langle X_i\rangle(T)\frac{\sigma_\mu^2}{\langle\mu\rangle^2}], \tag{23}$$

where $\langle\mu\rangle = \int_0^\infty \mu G(\mu)d\mu$ and $\sigma_\mu^2 = \int_0^\infty(\mu-\langle\mu\rangle)^2G(\mu)d\mu$, but because $\langle X_i\rangle(T) = K_iT\langle\mu\rangle$, we also have

$$\sigma_i^2(T) = \langle X_i\rangle(T)\Big[1 + K_iT\frac{\sigma_\mu^2}{\langle\mu\rangle}\Big]. \tag{24}$$

Because from Eq. (13), we have $\langle X\rangle(T) = \langle\mu\rangle T$ and $\langle X^2\rangle(T) = T^2\int_0^\infty \mu^2 G(\mu)d\mu + \langle\mu\rangle T$, we get

$$\begin{cases} \langle\mu\rangle = \frac{\langle X\rangle(T)}{T}, \\ \sigma_\mu^2 = \frac{\langle X^2\rangle(T)-\langle X\rangle(T)-\langle X\rangle^2(T)}{T^2}. \end{cases} \quad (25)$$

Hence, we have

$$\frac{\sigma_i^2(T)}{\sigma^2(T)} = \frac{\langle X_i\rangle(T)\left[1+\langle X_i\rangle(T)\frac{\sigma_\mu^2}{\langle\mu\rangle^2}\right]}{T^2\sigma_\mu^2 + T\langle\mu\rangle}$$

$$= \frac{K_i^2 T\sigma_\mu^2 + K_i\langle\mu\rangle}{T\sigma_\mu^2+\langle\mu\rangle}. \quad (26)$$

From the high-resolution data of the foreign exchange market, we extract the number of quotations (transactions) $X_{i,T}(t,s)$ $(i=1,\ldots,N; t=0,\ldots,Q_s-1)$ in the interval $[tT,(t+1)T]$ for the s-th observation period.

$$\langle X_i\rangle(T,s) = \frac{1}{Q_s}\sum_{t=0}^{Q_s-1} X_{i,T}(t,s), \quad (27)$$

$$\sigma_i^2(T,s) = \frac{1}{Q_s}\sum_{t=0}^{Q_s-1}\Big(X_{i,T}(t,s)-\langle X_i\rangle(T,s)\Big)^2, \quad (28)$$

$$\langle X\rangle(T,s) = \frac{1}{Q_s}\sum_{t=0}^{Q_s-1}\sum_{i=1}^{N} X_{i,T}(t,s), \quad (29)$$

$$\sigma^2(T,s) = \frac{1}{Q}\sum_{t=0}^{Q_s-1}\Big(\sum_{i=1}^{N} X_{i,T}(t,s)-\langle X\rangle(T,s)\Big)^2. \quad (30)$$

Because we can empirically obtain these means and the variance as functions in terms of T, the parameter of the common mode $\langle\mu\rangle(s)$ and $\sigma_\mu^2(s)$ within s should be computed from these values at a different T. From Eq. (25), we may determine these values by minimizing the squared errors

$$e^2 = \sum_{T=1}^{T_{\max}}\Big(\langle\mu\rangle(s) - \frac{\langle X\rangle(T,s)}{T}\Big)^2$$
$$+\lambda\sum_{T=1}^{T_{\max}}\Big(\sigma_\mu^2(s) - \frac{\langle X^2\rangle(T,s)-\langle X\rangle(T,s)-\langle X\rangle^2(T,s)}{T^2}\Big)^2,$$

where λ is an arbitrary constant. Partially differentiating it in terms of $\langle\mu\rangle(s)$ and $\sigma_\mu^2(s)$ and setting them to zero lead to

$$\langle\mu\rangle(s) = \frac{1}{T_{\max}} \sum_{T=1}^{T_{\max}} \frac{\langle X\rangle(T,s)}{T}, \tag{31}$$

$$\sigma_\mu^2(s) = \frac{1}{T_{\max}} \sum_{T=1}^{T_{\max}} \frac{\langle X^2\rangle(T,s) - \langle X\rangle(T,s) - \langle X\rangle^2(T,s)}{T^2}. \tag{32}$$

By using the bootstrap method, we computed $\langle\mu\rangle(s)$ and $\sigma_\mu^2(s)$ from the number of quotations and transactions for each 1-week observation period by using Eqs. (27), (28), (29), and (30). Figure 7 shows the fluctuation coefficients $\langle\mu\rangle(s)/\sigma_\mu(s)$ for each week, respectively. The financial crises are labeled as the (I) Paribas shock (August 2007), (II) Bear Stearns shock (February 2008), (III) subprime crisis driven by Lehman shock (September 2008 to March 2009), (IV) Euro crisis (April to May 2010), and (V) 3/11 earthquake and tsunami disaster of Japan. From Fig. 7, it is found that two phases exist before and after August 2009 in both graphs. Both the coefficients of variation have highly similar tendencies. Moreover, during the significant shocks and crises, they increase.

Furthermore, from the actual data, we confirm whether Eq. (23) holds. Figure 8 (top) shows the relationship between the mean of the number of quotations and their variance during different intervals T for several weeks. We found that these plots fit well with Eq. (23). We further confirmed whether Eq. (24) holds. Figure 8 (bottom) shows $\sigma_i^2(T,s)/\langle X_i\rangle(T,s) - 1$ versus T computed from the number of quotations. From these graphs, we found that the linear tendency of $\sigma_i^2(T)/\langle X_i\rangle(T)$ on $\langle X_i\rangle(T)$ exists for any i. The slope of this relationship is further proportional to K_i.

Moreover, the least squares method is applicable to finding an adequate coefficient $\sigma_\mu^2(s)/\langle\mu\rangle^2(s)$ in Eq. (23). The squared error is defined as

$$e^2 = \sum_{T=1}^{T_{\max}} \sum_{i=1}^{N} \Big[\sigma_i^2(s) - \langle X_i\rangle(T,s)\big\{1 + \langle X_i\rangle(T,s)/\kappa^2(s)\big\}\Big]^2, \tag{33}$$

where $\kappa(s) = \frac{\langle\mu\rangle(s)}{\sigma_\mu(s)}$. Partially differentiating e^2 in terms of $\kappa(s)$ and setting it zero lead to

$$\frac{\partial e^2}{\partial \kappa} = 4 \sum_{T=1}^{T_{\max}} \sum_{i=1}^{N} \Big[\sigma_i^2 - \langle X_i\rangle(T)\big\{1 + \langle X_i\rangle(T)/\kappa^2\big\}\Big]\langle X_i\rangle^2(T)\kappa^{-3} = 0. \tag{34}$$

Hence,

$$\sum_{T=1}^{T_{\max}} \sum_{i=1}^{N} \sigma_i^2 \langle X_i\rangle^2(T) - \sum_{T=1}^{T_{\max}} \sum_{i=1}^{N} \langle X_i\rangle^3(T)\big\{1 + \langle X_i\rangle(T)/\kappa^2\big\} = 0.$$

$$\sum_{T=1}^{T_{\max}} \sum_{i=1}^{N} \sigma_i^2 \langle X_i\rangle^2(T) - \sum_{T=1}^{T_{\max}} \sum_{i=1}^{N} \langle X_i\rangle^3(T) - \kappa^{-2} \sum_{T=1}^{T_{\max}} \sum_{i=1}^{N} \langle x_i\rangle^4(T) = 0.$$

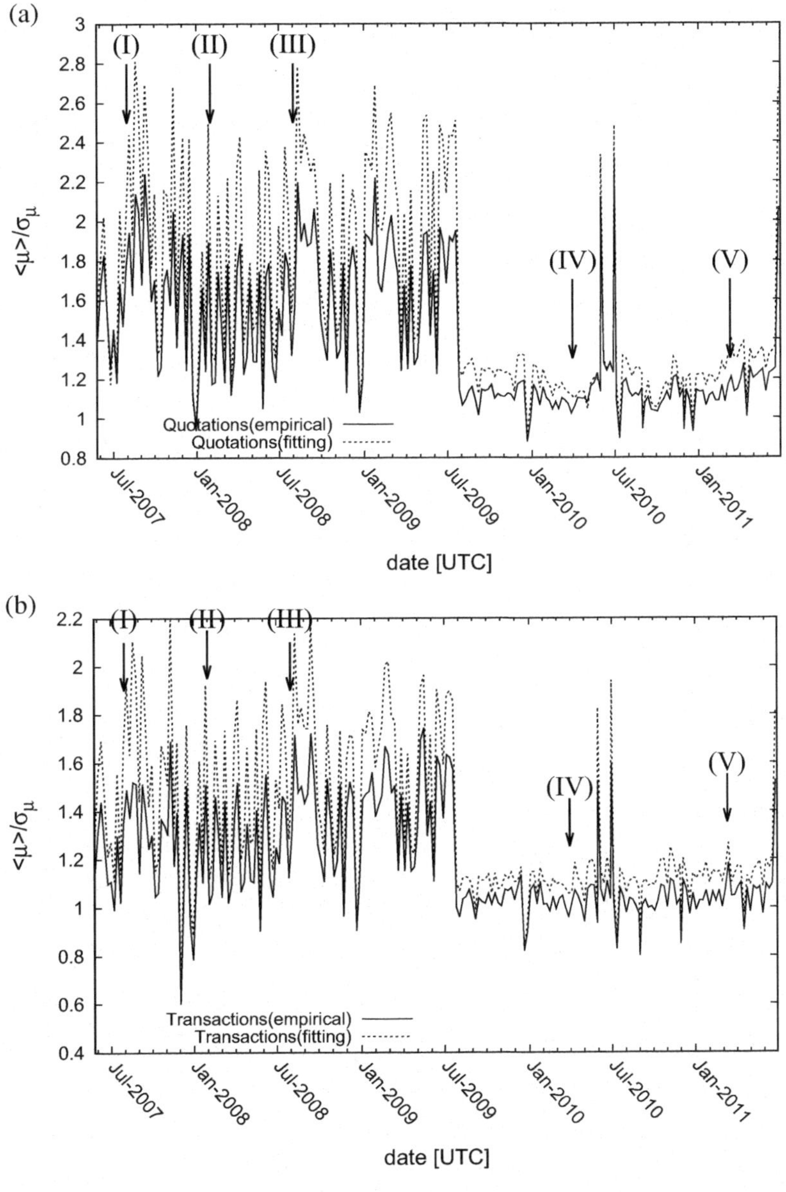

Fig. 7 Fluctuation coefficients of the total number of (**a**) quotations and (**b**) transactions per minute for each week. Values of $\langle\mu\rangle(s)/\sigma_{\mu}(s)$ for the period from 28 May 2007 to 31 June 2011. The solid curves are obtained from empirical means and standard deviations, and dashed curves are from by the least squares method for empirical values of $\langle X_i\rangle(T, s)$ and $\sigma_i^2(T, s)$

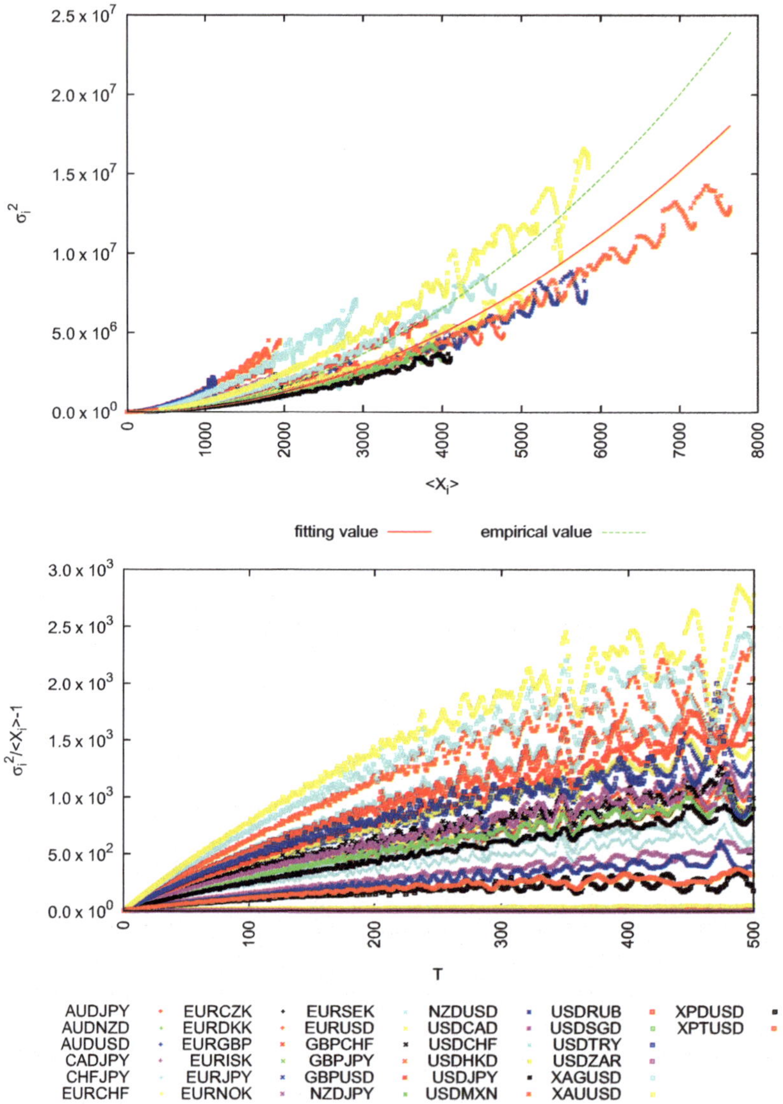

Fig. 8 (top) Scatter plots of the mean number of quotations during different intervals T (1[min] $\leq$ T $\leq$ 500[min]) during the period from 4 to 10 August 2008. (bottom) Scatter plots of $\sigma_i^2(T, s)/\langle X_i \rangle(T, s)$ of quotations in terms of T (1[min] $\leq$ T $\leq$ 500[min]) during the period from 4 to 10 August 2008

$$\frac{\langle\mu\rangle(s)}{\sigma_\mu(s)} = \sqrt{\frac{\sum_{T=1}^{T_{\max}}\sum_{i=1}^{N}\langle X_i\rangle^4(T,s)}{\sum_{T=1}^{T_{\max}}\sum_{i=1}^{N}\sigma_i^2(T,s)\langle X_i\rangle^2(T,s) - \sum_{T=1}^{T_{\max}}\sum_{i=1}^{N}\langle X_i\rangle^3(T,s)}}. \tag{35}$$

It is confirmed that the values of $\langle\mu\rangle(s)/\sigma_\mu(s)$ obtained from Eqs. (31) and (32) are significantly close to fitting the values computed from Eq. (35), which is calculated with the least squares method to Eq. (23). As shown in Fig. 7, we find that these values are close to each other for all the observation weeks. Because the fluctuation coefficients of μ correspond to the common mode, it is concluded that comprehensive behavior of the foreign exchange market may be captured by the total number of quotations and transactions in the market.

Moreover, we may calibrate K_i by using an empirical variance–covariance matrix. The squared error between an empirical variance–covariance matrix and a theoretical one should be minimized. Because we have the empirical variance–covariance matrix

$$\begin{aligned}\mathrm{Cov}_{ij}^{\mathrm{emp}}(T,s) = \frac{1}{Q_s}\sum_{t=0}^{Q_s-1}\Big(X_{i,T}(t,s) - \langle X_i\rangle(T,s)\Big)\\ \times\Big(X_{j,T}(t,s) - \langle X_j\rangle(T,s)\Big),\end{aligned} \tag{36}$$

and a theoretical one

$$\mathrm{Cov}(X_i,X_j)(T,s) = \begin{cases} K_i(s)K_j(s)\sigma_\mu^2(s)T^2 & (i\neq j)\\ K_i^2(s)\sigma_\mu^2(s)T^2 + K_i(s)\langle\mu\rangle(s)T & (i=j)\end{cases}, \tag{37}$$

we further define their squared error as

$$\begin{aligned}e^2 &= \sum_{T=1}^{T_{\max}}\sum_{l=1}^{N}\sum_{m=1}^{N}\Big(\mathrm{Cov}_{lm}^{\mathrm{emp}}(T,s) - \mathrm{Cov}(X_l,X_m)(T,s)\Big)^2\\ &= 2\sum_{l=1}^{N-1}\sum_{m=l+1}^{N}\Big(\mathrm{Cov}_{lm}^{\mathrm{emp}} - K_lK_m\sigma_\mu^2T^2\Big)^2 + \sum_{l=1}^{N}\Big(\mathrm{Cov}_{ll}^{\mathrm{emp}} - \big(K_l^2\sigma_\mu^2T^2 + K_l\langle\mu\rangle T\big)\Big)^2.\end{aligned} \tag{38}$$

Partially differentiating Eq. (38) in terms of $K_i(s)$ and setting it zero imply

$$\begin{aligned}\frac{\partial e^2}{\partial K_i} &= -4\sum_{m=l+1}^{N}\Big(\mathrm{Cov}_{lm}^{\mathrm{emp}} - K_lK_m\sigma_\mu^2T^2\Big)K_m\sigma_\mu^2T^2\\ &- 2\Big(\mathrm{Cov}_{ii}^{\mathrm{emp}} - \big(K_i^2\sigma_\mu^2T^2 + K_i\langle\mu\rangle T\big)\Big)\times\big(2K_i\sigma_\mu^2T^2 + \langle\mu\rangle T)\big)\end{aligned} \tag{39}$$

$$
\begin{aligned}
&= -2\Big\{2\sigma_\mu^2 T^2 \sum_{m=l+1}^{N} \mathrm{Cov}_{lm}^{\mathrm{emp}} K_m - 2\sigma_\mu^4 T^4 K_i \sum_{m=i+1}^{N} K_m^2 + 2K_i \mathrm{Cov}_{ii}^{\mathrm{emp}} \sigma_\mu^2 T^2 \\
&+ \mathrm{Cov}_{ii}^{\mathrm{emp}} \langle\mu\rangle T - \big(K_i^2 \sigma_\mu^2 T^2 + K_i \langle\mu\rangle T\big)\big(2K_i \sigma_\mu^2 T^2 + \langle\mu\rangle T\big)\Big\} \qquad (40) \\
&= 0.
\end{aligned}
$$

We have nonlinear equations with respect to $K_i(s)$

$$
\begin{aligned}
&2\sigma_\mu^4(s) \sum_{T=1}^{T_{\max}} T^2 \sum_{m=1}^{N} \mathrm{Cov}_{im}^{\mathrm{emp}}(T, s) K_m(s) \\
&- 2\sigma_\mu^4(s) K_i(s) \sum_{T=1}^{T_{\max}} T^4 \sum_{m=1}^{N} K_m^2(s) \\
&- 3K_i^2(s)\langle\mu\rangle(s)\sigma_\mu^2(s) \sum_{T=1}^{T_{\max}} T^3 - K_i(s)\langle\mu\rangle^2(s) \sum_{T=1}^{T_{\max}} T^2 \\
&+ \langle\mu\rangle(s) \sum_{T=1}^{T_{\max}} \mathrm{Cov}_{ii}^{\mathrm{emp}}(T, s) T = 0. \quad (i = 1, \ldots, N) \qquad (41)
\end{aligned}
$$

By solving the nonlinear equations described as Eq. (41) numerically, we can finally obtain normalize $K_i(s)$.

To examine the method to calibrate parameters, we use actual tick-by-tick data of the foreign exchange market for 31 currency pairs consisting of 22 currencies (ICAP 2013).[1] We obtain real solutions for $K_i(s)$ by using Eq. (41), and the calculated values are shown in Table 2. The calibrated values are consistent with the values computed from relative frequencies of the number of quotations.

As shown in Fig. 7, it is found that the coefficients estimated from the model with computed parameters $K_i(s)$ and those values directly computed from the data showed similar temporal development. The estimated parameters can be used to capture the collective behavior of market participants in the foreign exchange markets.

The proposed method may measure the weights of participants in socioeconomic activities from the number of counts for specific activities.

[1] USD: US Dollar, CHF: Swiss Franc, EUR: European Union Euro, JPY: Japanese Yen, NZD: New Zealand Dollar, AUD: Australian Dollar, GBP: British Pound, CAD: Canadian Dollar, XAU: Gold, XAG: Silver, SEK: Swedish Krona, XPD: Palladium, XPT: Platinum, SGD: Singapore Dollar, HKD: Hong Kong Dollar, NOK: Norway Krona, ZAR: South African Rand, MXN: Mexico Peso, DKK: Danish Krone, CZK: Czech Koruna, TRY: Turkish Lira, and RUB: Russian Ruble.

Table 2 Calibrated values of $K_i(s)$ for 31 currency pairs consisting of 22 currencies and precious metals for the period from 3 to 10 August 2008 at $T_{\max} = 100$. The value is compared with the calibrated value computed from their relative frequencies, $\tilde{K}_i(s) = \sum_{t=0}^{Q_s-1} X_{i,T}(t,s) / \sum_{t=0}^{Q_s-1} \sum_{i=1}^{N} X_{i,T}(t,s)$

Currency pair	Calibration	Comp.
USD/CHF	0.066246	0.061597
EUR/USD	0.060267	0.067412
USD/JPY	0.040629	0.054125
EUR/JPY	0.061965	0.076751
EUR/CHF	0.037817	0.033362
NZD/USD	0.006404	0.008336
AUD/USD	0.036921	0.038402
GBP/USD	0.051995	0.050295
USD/CAD	0.030301	0.025032
AUD/NZD	0.001012	0.001172
EUR/GBP	0.035353	0.040190
XAU/USD	0.080827	0.077078
XAG/USD	0.051762	0.038437
EUR/SEK	0.017530	0.012714
CHF/JPY	0.028117	0.035791
XPD/USD	0.004156	0.002822
XPT/USD	0.005090	0.004679
USD/SGD	0.000019	0.000030
USD/HKD	0.000657	0.001297
EUR/NOK	0.006757	0.005836
USD/ZAR	0.000010	0.000013
USD/MXN	0.032158	0.025681
EUR/DKK	0.000135	0.000092
EUR/CZK	0.000000	0.000002
USD/RUB	0.036570	0.024212
GBP/JPY	0.080507	0.100893
GBP/CHF	0.029609	0.025867
AUD/JPY	0.050981	0.063165
USD/TRY	0.022449	0.014726
NZD/JPY	0.043717	0.052984
CAD/JPY	0.050737	0.057006

3 Conclusion

This chapter investigated the quotation and transaction activities of the foreign exchange markets as an observable example of collective man–machine behavior. We proposed a simple multivariate Poisson model with the common mode to describe both the quotation and transaction activities of the foreign exchange market. We developed a method to compute fluctuation coefficients of the common mode from actual observations. We found that the fluctuation coefficients increase during unstable market situations. It was concluded that the fluctuation coefficients of the common mode for quotation and transaction activities may be used to quantify the situations of the foreign exchange markets from a comprehensive viewpoint.

The proposed method may measure collective human behavior and use the weights of participants in socioeconomic activities from the number of counts for specific activities as indicators to quantify the importance of specific behavior.

Acknowledgements This work was supported by the Grant-in-Aid for Young Scientists (B) by the Japan Society for the Promotion of Science (JSPS) KAKENHI (#23760074). The author further expresses his sincere gratitude to Prof. Dr. Janusz A. Hołyst from Warsaw University of Technology for his stimulating advice.

References

Alfarano S, Lux T, Wagner F (2006) Estimation of a simple agent-based model of financial markets: an application to Australian stock and foreign exchange data. Physica A 370:38–42
Bak P, Paczuski M, Shubik M (1997) Price variations in a stock market with many agents. Physica A 246:430–453
Black F, Scholes M (1973) The pricing of options and corporate liabilities. Polit J Econ 81:637–654
Bollerslev T (1986) Generalized autoregressive conditional heteroskedasticity. J Econ 31:307–327
Bonanno G, Lillo F, Mantegna RN (2000) Dynamics of the number of trades of financial securities. Physica A 280:136–141
Brock WA, LeBaron BD (1996) A dynamic structural model for stock return volatility and trading volume. Rev Econ Stat 78:94–110
Egenter E, Lux T, Stauffer D (1999) Finite-size effects in Monte Carlo simulations of two stock market models. Physica A 268:250–256
Eisler Z, Kertész J (2006) Scaling theory of temporal correlations and size-dependent fluctuations in the traded value of stocks. Phys Rev E 73:046109
Eisler Z, Bartos I, Kertész J (2008) Fluctuation scaling in complex systems: Taylor's law and beyond. Adv Phys 57:89–142
Engle RF (1982) Autoregressive conditional heteroscedasticity with estimates of the variance of United Kingdom inflation. Econometrica 50:987–1007
Engle RF, Russell JR (1998) Autoregressive conditional duration: a new model for irregularly spaced transaction data. Econometrica 66:1127–1162
Fronczak A, Fronczak P et al (2010) Taylor's power law for fluctuation scaling in traffic. Acta Phys Pol B Proc Suppl 3:327–333; Origins of Taylor's power law for fluctuation scaling in complex systems. Phys Rev E 81:066112
González MC, Hidalgo CA, Barabási A-L (2008) Understanding individual human mobility patterns. Nature 453:779–782
Guerrero-Cusumano J-L, The entropy of the multivariate poisson: an approximation. Inf Sci 86:1–17 (1995)
Gworek S, Kwapień J, Drożdż S (2010) Sign and amplitude representation of the Forex networks. Acta Phys Pol A 117:681–687
Haken H (1983) Synergetics: an introduction. Springer, Berlin
Hashimoto Y, Ito T (2010) Effects of Japanese macroeconomic statistic announcements on the dollar/yen exchange rate: high-resolution picture. J Jpn Int Econ 24:334–354
Helbing D (2001) Traffic and related self-driven many-particle systems. Rev Mod Phys 73:1067–1141
Hirabayashi T, Takayasu H, Miura H, Hamada K (1993) The behavior of a threshold model of market price in stock exchange. Fractals 1:29–40

ICAP (2013) The data is purchased from ICAP EBS: http://www.icap.com
Inoue JI, Sazuka N (2010) Queueing theoretical analysis of foreign currency exchange rates. Quant Finan 10:121–130
Jefferies P, Hart ML, Hui PM, Johnson NF (2001) From market games to real-world markets. Eur Phys J B 20:493–501
Johnson NF, Jarvis S, Jonson R, Cheung P, Kwong YR (1998) Volatility and agent adaptability in a self-organizing market. Physica A 258:230–236
Kaltwasser PR (2010) Uncertainty about fundamentals and herding behavior in the FOREX market. Physica A 389:1215–1222
Karlis D, Meligkotsidou L Finite mixtures of multivariate Poisson distributions with application. J Stat Plann Infer 137:1942–1960 (2007)
Kendal WS (2004) Taylor's ecological power law as a consequence of scale invariant exponential dispersion models. Ecol Complex 1:193–209
Kendal W, Jorgensen B (2011) Taylor's power law and fluctuation scaling explained by a central-limit-like convergence. Phys Rev E 83:066115; Tweedie convergence: A mathematical basis for Taylor's power law, 1/f noise, and multifractality. Phys Rev E 84:066120
Kilpatrick AM, Ives AR (2003) Species interactions can explain Taylor's power law for ecological time series. Nature 422:65–68
Kyle AS (1985) Continous auctions and insider trading. Econometrica 53:1315–1335
Liu LZ, Qian XY, Lu HY (2010) Cross-sample entropy of foreign exchange time series. Physica A 389:4785–4792
Lukacs E, Beer S (1977) Characterization of the multivariate poisson distribution. J Multivar anal 7:1–12
Lux T, Marchesi M (1999) Scaling and criticality in a stochastic multi-agent model of a financial market. Nature 397:498–500
Mandelbrot B (1997) Fractals and scaling in finance. Springer, Chicago
Mantegna RN, Stanley HE (1995) Scaling behaviour in the dynamics of an economic index. Nature 376:46–49
Matassini L, Franci F (2001) On financial markets trading. Physica A 289:526–542
Merton RC (1973) Theory of rational option pricing. Bell J Econ Manag Sci 3:141–183
Nelson DB (1990) ARCH models as diffusion approximations. J Econ 45:7–38
Ohnishi T, Mizuno T, Aihara K, Takayasu M, Takayasu H (2004) Statistical properties of the moving average price in dollar-yen exchange rates. Physica A 344:207–210
Palmer RG, Arthur WB, Holland JH, LeBaron B, Tayler P (1994) Artificial economic life: a simple model of a stock market. Physica D 75:264–274
Samaniego H, Sérandour G et al (2011) Analyzing Taylor's scaling law: qualitative differences of social and territorial behavior on colonization/extinction dynamics. Popul Ecol 54:213–223
Sato A-H, Takayasu H (1998) Dynamic numerical models of stock market price: from microscopic determinism to macroscopic randomness. Physica A 250:231–252
Sato A-H, Nishimura M, Hołyst JA (2010) Fluctuation scaling of quotation activities in the foreign exchange market. Physica A 389:2793–2804
Sazuka N, Inoue J, Scalas E (2009) The distribution of first-passage times and durations in FOREX and future markets. Physica A 388:2839–2853
Takayasu H, Miura H, Hirabayashi T, Hamada K (1992) Statistical properties of deterministic threshold elements – the case of market price. Physica A 184:127–134
Takayasu H, Takayasu M, Okazak MP, Marumo K, Shimizu T (2000) Fractal properties in economics. In: Paradigms of complexity. World Scientific, Singapore, pp 243–258
Takayasu M, Takayasu H, Okazaki MP (2002) Transaction interval analysis of high resolution foreign exchange data. In: Takayasu H (ed) Procceedings of empirical science of financial fluctuations – the advent of econophysics Springer, Tokyo
Taylor LR (1961) Aggregation, variance and mean. Nature 189:732–735
Teicher H (1954) On the multivariate Poisson distribution. Skand Akt 37:1–9
Wiener N (1961) Cybernetics, 2nd edn. The M.I.T. Press, Cambridge

On Measuring Extreme Synchrony with Network Entropy of Bipartite Graphs

Aki-Hiro Sato

1 Introduction

Recently a comprehensive measurement of agent behavior can be conducted based on high-resolution data on socioeconomic systems due to the development of information and communication technology. In particular, we want to consider a problem that we need to deal with the collective behavior of millions of nodes contribute to the observable dynamical features of such a complex system (de Menezes and Barabási 2004).

To do so, it is necessary to develop an adequate model both which considers states of agent behavior from the comprehensive point of view and which is as simple as one can estimate its parameters from actual data under assumptions. Specifically in socioeconomic activities, if we regard the situation where people exchange things, money, and information with each other as a network, then it further seems to be fruitful to study such social or economic systems from a network point of view.

To make advances in this direction, we need to treat structure on the basis of information transmission among heterogeneous agents in a given socioeconomic system from a limited amount of available data without precise knowledge on communication networks.

Several networks relating to human activity can be described as bipartite graphs with two kinds of nodes as shown in Fig. 1. For example, it is known that financial markets (financial commodities and participants), blog systems (blogs and

A.-H. Sato (✉)
Yokohama City University, Kanazawa-ku, Yokohama-shi, Kanagawa, Japan

Japan Science and Technology Agency PRESTO, Kawaguchi-shi, Saitama, Japan

Statistical Research and Training Institute, Ministry of Internal Affairs and Communications, Shinjuku-ku, Tokyo, Japan
e-mail: ahsato@yokohama-cu.ac.jp

A.-H. Sato (ed.), *Applications of Data-Centric Science to Social Design*,
Agent-Based Social Systems 14, https://doi.org/10.1007/978-981-10-7194-2_16

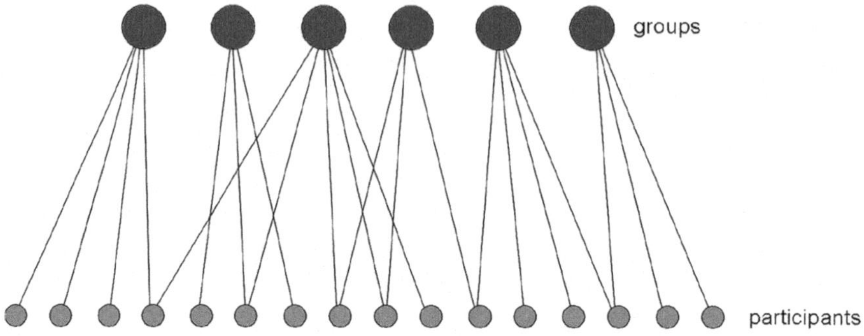

Fig. 1 A conceptual illustration of social systems where M agents (B node) communicate with one another in K groups (A node)

bloggers), and economic systems (firms and goods/consumers) can be described as a directed bipartite graph (Lambiotte et al. 2007; Chmiel et al. 2007; Sato 2007; Sato and Hołyst 2008).

Such a bipartite network is recognized by observers when constituents are transmitted between two kinds of nodes. Links are recognized by observers when a constituent moves from one node to another node. Therefore bipartite network representation also seems to be one of our cognitive categories such as causality, time and space, intensity, quantity, and so on.

This chapter considers a model-based comparative measurement of collective behavior of groups based on their activities (Sato 2017). Specifically, a descriptive multivariate model of a financial market is proposed from a comprehensive point of view. Using comprehensive high-resolution data on the behavior of market participants, correlations of log returns and quotations (transactions) are analyzed on the basis of theoretical insights from the proposed model.

As a subject, we focus on financial markets which are attracting numerous researchers in various fields since they are of complex systems consisting of various types of heterogeneous agents. In the context of finance, the normal mixture hypothesis (or more generally "mixture of distribution hypothesis") has been proposed as an alternative explanation for the description of return distribution of financial assets by several studies (Mandelbrot and Taylor 1967; Clark 1973; Tauchen and Pitts 1983; Richardson and Smith 1994). They have considered trading volume or the number of transactions (or quotations) as a proxy of the latent number of information arrivals. The proposed model may lead to better understanding of dynamics for return-volume relationship including volatility persistence (Lamoureux and Lastrapes 1990; Andersen 1996; Liesenfeld 1998; Watanabe 2000).

Cloud-based services can be used to provide various kinds of real-world applications. In fact, rich data on both mobility and activity of users have been accumulated by the cloud service providers. If secondary usage of the data collected in the cloud-based servers is permitted both legally and adequately and data analytics procedures for anonymized data can be developed, then we will be able to quantify risk and chance of our society based on rich data on human mobility and activity with high resolution.

The chapter aims to propose a method to detect change points of temporally evolutionary bipartite network regarding anonymity of personal information. Moreover, in order to confirm an ability of the proposed method empirically, the proposed method is applied to data of trading activity collected in a cloud-based service of the foreign exchange market and quantified trading activity of the foreign exchange market from a comprehensive point of view.

The network structure of various kinds of physical and social systems has attracted considerable research attention. A many-body system can be described as a network, and the nature of growing networks has been examined well (Albert and Barabási 2002; Miura et al. 2012). Power-law properties can be found in the growing networks, which are called complex networks. These properties are related to the growth of elements and preferential attachment (Albert and Barabási 2002).

A network consists of several nodes and links that connect nodes. In the literature on the physics of socioeconomic systems (Carbone et al. 2007), nodes are assumed to represent agents, goods, and computers, while links express the relationships between nodes (Milaković et al. 2010; Lämmer et al. 2006). The network structure is perceived in many cases through the conveyance of information, knowledge, and energy, among others.

In statistical physics, the number of combinations of possible configurations under given energy constraints is related to "entropy." Entropy is a measure that quantifies the states of thermodynamic systems. In physical systems, entropy naturally increases because of the thermal fluctuations on elements. Boltzmann proposed that entropy S is computed from the possible number of ensembles g by $S = \log g$. For a system that consists of two sub-systems whose respective entropies are S_1 and S_2, the total entropy S is calculated as the sum of one of two subsystems $S_1 + S_2$. This case is attributed to the possible number of ensembles $g_1 g_2$. Entropy in statistical physics is also related to the degree of complexity of a physical system. If the entropy is low (high), then the physical configuration is rarely (often) realized. Energy injection or work in an observed system may be assumed to represent rare situations. Shannon entropy is also used to measure the uncertainty of time series (Carbone and Stanley 2007).

The concept of statistical–physical entropy was applied by Bianconi (2009) to measure network structure. She considered that the complexity of a network is related to the number of possible configurations of nodes and links under some constraints determined by observations. She calculated the network entropy of an arbitrary network in several cases of constraints.

Researchers have used a methodology to characterize network structure with information-theoretic entropy (Dehmer and Mowshowitz 2011; Wilhelm and Hollunder 2007; Rashevsky 1955; Trucco 1956; Mowshowitz 1968; Sato 2009). Several graph invariants such as the number of vertices, vertex degree sequence, and extended degree sequences have been used in the construction of entropy-based measures (Wilhelm and Hollunder 2007; Sato 2009).

2 An Entropy Measure on a Bipartite Network

The number of elements in socioeconomic systems is usually very large, and several restrictions or finiteness of observations can be found. Therefore, we need to develop a method to infer or quantify the affairs of the entire network structure from partial observations. Specifically, many affiliation relationships of socioeconomic systems can be expressed as a bipartite network. Describing the network structure of complex systems that consist of two types of nodes by using the bipartite network is important. A bipartite graph model also can be used as a general model for complex networks (Guillaume and Latapy 2006; Chmiel et al. 2007; Tumminello et al. 2011). Tumminello et al. proposed a statistical method to validate the heterogeneity of bipartite networks (Tumminello et al. 2011).

Suppose a symmetric binary two-mode network can be constructed by linking K groups (A node) and M participants (B node) if the participants belong to groups (Fig. 1). Assume that we can count the number of participants in each group within the time window $[t\delta, (t+1)\delta] \quad (t = 1, 2, 3, \ldots)$, which is defined as $m_i(t) \quad (i = 1, 2, \ldots, K)$.

Let us assume a bipartite graph consisting of A nodes and B nodes, of which the structure at time t is described as an adjacency matrix $C_{ij}(t)$. We also assume that A nodes are observable and B nodes are unobservable. That is, we only know the number of participants (B node) belonging to A nodes $m_i(t)$. We do not know the correct number of B nodes, but we assume that it is M. In this setting, how do we measure the complexity of the bipartite graph from $m_i(t)$ at each observation time t?

The network entropy is defined as a logarithmic form of the number of possible configurations of a network under a constraint (Bianconi 2009). We can introduce the network entropy at time t as a measure to quantify the complexity of a bipartite network structure. The number of possible configurations under the constraint $m_i(t) = \sum_{j=1}^{M} C_{ij}(t)$ may be counted as

$$N(t) = \prod_{i=1}^{K} \binom{M}{m_i(t)} = \prod_{i=1}^{K} \frac{M!}{m_i(t)!(M - m_i(t))!}. \tag{1}$$

Then, the network entropy is defined as $\Sigma(t) = \ln N(t)$. Inserting Eq. (1) into this definition, we have

$$\Sigma(t) = K \sum_{n=1}^{M} \ln n - \sum_{i=1}^{K} \sum_{n=1}^{M-m_i(t)} \ln n - \sum_{i=1}^{K} \sum_{n=1}^{m_i(t)} \ln n. \tag{2}$$

Note that because $0! = 1$, $\sum_{n=1}^{0} \ln n = 0$. Obviously, if $m_i(t) = M$ for any i, then $\Sigma(t) = 0$. If $m_i(t) = 0$ for any i, then $\Sigma(t) = 0$. The lower number of combinations gives a lower value of $\Sigma(t)$. To eliminate a difference in the number of links, we consider the network entropy per link defined as

$$\sigma(t) = \frac{\Sigma(t)}{\sum_{i=1}^{K} m_i(t)}. \tag{3}$$

This quantity shows the degree of complexity of the bipartite network structure. We may capture the temporal development of the network structure from the value of $\sigma(t)$. The network entropy per link $\sigma(t)$ is also an approximation of the ratio of the entropy rate for $m_i(t)$ to its mean so that

$$\sigma[m_1(t), \ldots, m_K(t)] = \frac{\frac{1}{K}\Sigma[m_1(t), \ldots, m_K(t)]}{\frac{1}{K}\sum_{i=1}^{K} m_i(t)} \approx \frac{\Sigma[\mathbf{m}(t)]}{\langle m(t)\rangle}, \tag{4}$$

where the entropy rate and the mean are, respectively, defined as

$$\Sigma[\mathbf{m}(t)] = \lim_{K\to\infty} \frac{1}{K}\Sigma[m_1(t), \ldots, m_K(t)], \tag{5}$$

$$\langle m(t)\rangle = \lim_{K\to\infty} \frac{1}{K}\sum_{i=1}^{K} m_i(t). \tag{6}$$

The ratio of the entropy rate to the mean tells us the uncertainty of the mean from a different point of view from the coefficient of variation ($C.V. =$ standard deviation/mean).

To understand the fundamental properties of Eq. (3), we compute $\sigma(t)$ in simple cases. Consider values of entropy for several cases at $K = 100$ with different M. We assume that the total number of links is fixed at 100, which is the same as the number of A nodes, and we confirm the dependence of $\sigma(t)$ on the degree of monopolization. We assign the same number of links at each A node. That is, we set

$$m_i(t) = \begin{cases} 100/k & (i = 1, \ldots, k) \\ 0 & (i = k+1, \ldots, K) \end{cases}, \tag{7}$$

where k can be set as 1, 2, 4, 5, 10, 20, 50, or 100. In this case, we can calculate $\sigma(t)$ as follows:

$$\begin{aligned} \sigma(t) &= \frac{\sum_{i=1}^{k} \ln \binom{M}{100/k}}{\sum_{i=1}^{k} 100/k} \\ &= \frac{k}{100}\Big(\ln M! - \ln(100/k)! - \ln(M - 100/k)!\Big). \end{aligned} \tag{8}$$

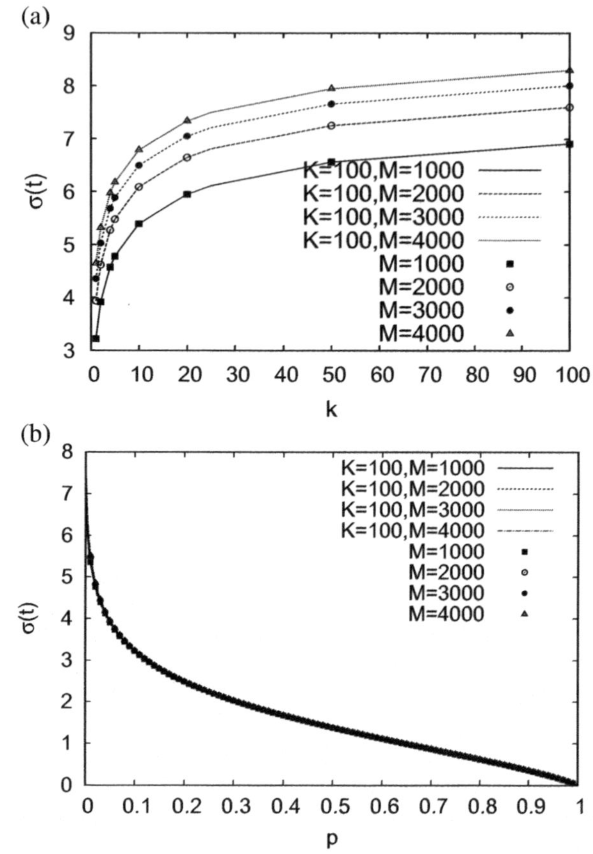

Fig. 2 (**a**) Plots between $\sigma(t)$ and degree of monopolization k. Each curve represents the relation between $\sigma(t)$ and k. Filled squares numerical values for $M = 1000$, unfilled circles for $M = 2000$, filled circles for $M = 3000$, and unfilled triangle for $M = 4000$. (**b**) Plots between $\sigma(t)$ and density of links p. Each curve represents the relation between $\sigma(t)$ and k. Filled squares numerical values for $M = 1000$, unfilled circles for $M = 2000$, filled circles for $M = 3000$, and unfilled triangle for $M = 4000$

Figure 2a shows the relationship between $\sigma(t)$ and the degree of monopolization at $M = 1000$, 2000, 3000, and 4000. The network entropy per link $\sigma(t)$ is small if a small population of nodes occupies a large number of links. The multiplication regime gives a large value of $\sigma(t)$. The value of $\sigma(t)$ is a monotonically increasing function in terms of k. As M increases, the value of $\sigma(t)$ increases. From this instance, we confirmed that $\sigma(t)$ decreases with the degree of monopolization at A nodes.

Next, we confirm the dependency of $\sigma(t)$ on the density of links. We assume that each element of an adjacency matrix $C_{ij}(t)$ is given by an *i.i.d.* Bernoulli random variable with a successful probability of p. Then, $m_i(t) = \sum_{j=1}^{M} C_{ij}(t)$ is sampled from an *i.i.d.* binomial distribution $\mathrm{Bin}(p, M)$. In this case, one can approximate $\sigma(t)$ as

$$\sigma(t) = \frac{\frac{1}{K}\sum_{i=1}^{K} \Sigma[m_i(t)]}{\frac{1}{K}\sum_{i=1}^{K} m_i(t)}$$
$$\approx \frac{\langle \Sigma[m_1(t)] \rangle}{\langle m_1(t) \rangle}$$
$$= \frac{1}{M}\sum_{k=1}^{M} \binom{M}{k} p^{k-1}(1-p)^{M-k} \ln \binom{M}{k}. \quad (9)$$

Figure 2b shows the plots of $\sigma(t)$ versus p obtained from both Monte Carlo simulation with random links drawn from Bernoulli trials and Eq. (9). The number of links at each A node monotonically increases as p increases. $\sigma(t)$ decreases as the density of links decreases. The dependence of the entropy per link on p is independent of M.

3 Empirical Analysis

The application of network analysis to financial time series has been advancing. Several researchers have investigated the network structure of financial markets (Bonanno et al. 2003; Gworek et al. 2010; Podobnik et al. 2009; Iori et al. 2008). Bonanno et al. examined the topological characterization of the correlation-based minimum spanning tree (MST) of real data (Bonanno et al. 2003). Gworek et al. analyzed the exchange rate returns of 38 currencies (including gold) and computed the characteristic path length and average weighted clustering coefficient of the MST topology of the graph extracted from the cross-correlations for several base currencies (Gworek et al. 2010). Podobnik et al. (2009) examined the cross-correlations between volume changes and price changes for the New York Stock Exchange, Standard and Poor's 500 index, and 28 worldwide financial indices. Iori et al. (2008) analyzed the network topology of the Italian segment of the European overnight money market and investigated the evolution of these banks' connectivity structure over the maintenance period. These studies collectively aimed to detect the susceptibility of network structures to macroeconomic situations.

Data collected from the ICAP EBS platform were used. The data period spanned May 28, 2007 to November 30, 2012 (ICAP 2013). The data included records for orders (BID/OFFER) and transactions for currencies and precious metals with a 1-s resolution. The total number of orders included in the data set is 520,973,843, and the total number of transactions is 58,679,809. The data set involved 94 currency pairs consisting of 39 currencies, 11 precious metals, and 2 basket currencies (AUD, NZD, USD, CHF, JPY, EUR, CZK, DKK, GBP, HUF, ISK, NOK, PLN, SEK, SKK, ZAR, CAD, HKD, MXC, MXN, MXT, RUB, SGD, XAG, XAU, XPD, XPT, TRY, THB, RON, BKT, ILS, SAU, DLR, KES, KET, AED, BHD, KWD, SAR, EUQ,

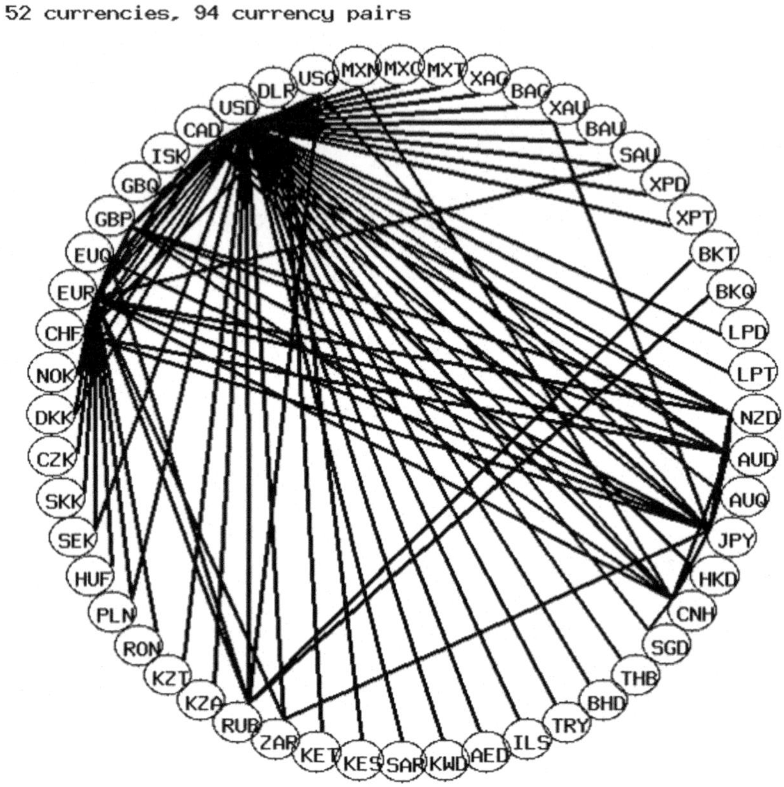

Fig. 3 A network representation of 94 currency pairs included in the data sets. Nodes show currencies and links currency pairs. If there is a link between two currencies, it is shown that the currency pair consisting of them is included in the data set

USQ, CNH, AUQ, GBQ, KZA, KZT, BAG, BAU, BKQ, LPD, and LPT).[1] Figure 3 shows a network representation of 94 currency pairs.

[1] AED, United Arab Emirates dirham; AUD, Australian dollar; AUQ, Australian dollar (small amount); BAG, gold (bank); BAU, silver (bank); BHD, Bahraini dinar; BKT, basket of USD/EUR; BKQ, basket of USD/EUR (small amount); CAD, Canadian dollar; CHF, Swiss franc; CNH, Chinese yuan; CZK, Czech koruna; DKK, Danish krone; EUR, EU euro; EUQ, EU euro (small amount); GBP, British sterling; GBQ, British sterling (small amount); HKD, Hong Kong dollar; HUF, Hungarian forint; ILS, Israeli new shekel; ISK, Iceland krona; JPY, Japanese yen; KES, Kenyan shilling; KET, Kenyan shilling (small amount); KZA, Kazakhstani tenge (small amount); KZT, Kazakhstani tenge; LPD, palladium (London); LPT, platium (London); MXN, Mexican peso; MXQ, Mexican peso (small amount); MXT, Mexican peso (special deals); NOK, Norwegian krone; NZD, New Zealand dollar; PLN, Poland zloty; RON, Romanian leu; RUB, Russian ruble; SAR, Saudi Arabian riyal; SGD, Singapore dollar; SEK, Swedish krona; SKK, Slovak koruna; SAU, silver (small amount); TRY, Turkish lira; THB, Thai baht; USD/DLR, US dollar; USQ, US dollar (small amount); ZAR, South African rand; XAU, gold; SAU, gold (small amount); XAG, silver; XPD, palladium; and XPT, platinum.

3.1 The Total Number

The number of quotations and transactions in each currency pair was extracted from the raw data. Let $m_{X,i}(t) \quad (t = 0, \ldots; i = 1, \ldots, K)$ be the number of quotations $(X = P)$ or transactions $(X = D)$ within every minute $(\delta = 1\,[\mathrm{min}])$ for a currency pair i $(K = 94)$ at time t. Let $c_X(t)$ be denoted as the total number of quotations $(X = P)$ and transactions $(X = D)$, which is defined as

$$c_X(t) = \sum_{i=1}^{K} m_{X,i}(t). \tag{10}$$

Let us consider the maximum value of $c_X(t)$ in each week:

$$w_X(s) = \max_{t \in W(s)} \{c_X(t)\}, \tag{11}$$

where $W(s) \quad (s = 1, \ldots, T)$ represents a set of times included in the s-th week. A total of 288 weeks are included in the data set $(T = 288)$. Figure 4 shows the maximum values $c_X(t)$ for the period from May 28, 2007 to November 30, 2012.

According to the extreme value theorem, the probability density for maximum values can be assumed to be a Gumbel density:

$$P(w_X; \mu_X, \rho_X) = \frac{1}{\rho_X} \exp\Big(-\frac{w_X - \mu_X}{\rho_X} - e^{-\frac{w_X - \mu_X}{\rho_X}}\Big), \tag{12}$$

where μ_X and ρ_X are the location and scale parameters, respectively. Under the assumption of the Gumbel density, these parameters are estimated with the maximum likelihood procedure. The log-likelihood function for T observations $w_X(s') \quad (s' = 1, \ldots, T)$ under Eq. (12) is defined as

$$l(\mu_X, \rho_X) = \sum_{s'=1}^{T} \ln\Big[\frac{1}{\rho_X} \exp\Big(-\frac{w_X(s') - \mu_X}{\rho_X} - e^{-\frac{w_X(s') - \mu_X}{\rho_X}}\Big)\Big]. \tag{13}$$

The maximum likelihood estimators are obtained by maximizing the log-likelihood function. Partially differentiating $l(\mu_X, \rho_X)$ in terms of μ_X and ρ_X and setting them to zero, one has its maximum likelihood estimators as

$$e^{-\frac{\hat{\mu}_X}{\hat{\rho}_X}} = \frac{1}{T} \sum_{s'=1}^{T} e^{-\frac{w_X(s')}{\hat{\rho}_X}}, \tag{14}$$

$$\hat{\rho}_X = \frac{1}{T} \sum_{s'=1}^{T} w_X(s') - \frac{\sum_{s'=1}^{T} e^{-\frac{w_X(s')}{\hat{\rho}_X}} w_X(s')}{\sum_{s'=1}^{T} e^{-\frac{w_X(s')}{\hat{\rho}_X}}}. \tag{15}$$

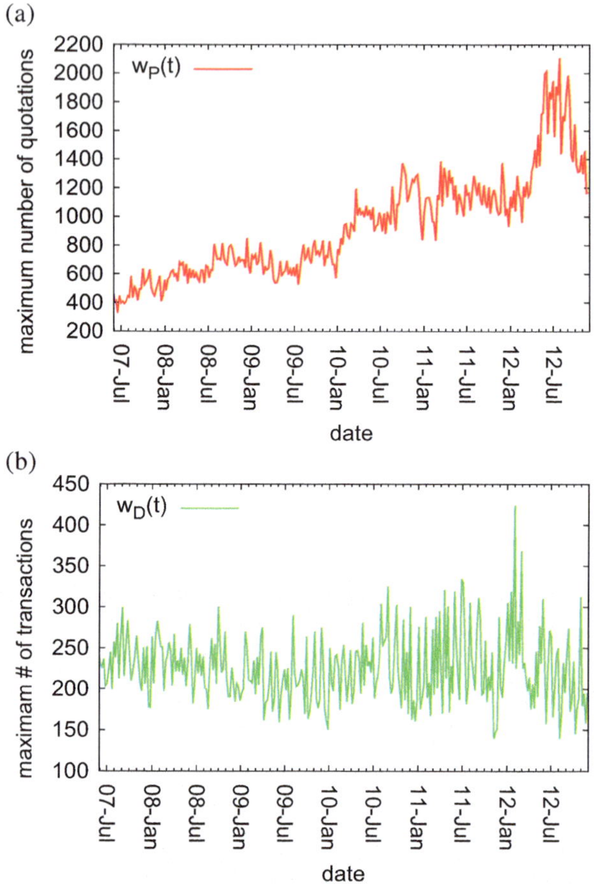

Fig. 4 (**a**) The maximum values of the number of quotations within 1 min in every week. (**b**) The maximum values of the number of transactions within 1 min in every week

The parameters are estimated as $\hat{\mu}_P = 772.179499$, $\hat{\rho}_P = 281.741815$, $\hat{\mu}_D = 206.454884$, and $\hat{\rho}_D = 35.984804$.

The Kolmogorov–Smirnov (KS) test is conducted to determine the statistical significance of the estimated distributions. The KS test is a popular statistical method of assessing the difference between observations and its assumed distribution by p-value, which is a measure of probability where a difference between the two distributions happens by chance. Large p-values imply that the observations are sampled from the assumed distribution in the null hypothesis with high significance. Let $w_X(s) \quad (s = 1, \ldots, T)$ be T observations, and let K_T be a test statistic

$$K_T = \sup_{1 \le s' \le T} \sqrt{T} \Big| F_T(w_X(s')) - F(w_X(s')) \Big|, \tag{16}$$

where $0 \le F(v) \le 1$ is an assumed cumulative distribution in a null hypothesis and $F_T(v)$ an empirical one based on T observations such that $F_T(v) = k/T$, in which k represents the number of observations satisfying $v_X(s) \le v(s = 1, \ldots, T)$. The p-value is computed from the Kolmogorov–Smirnov distribution.

Table 1 The p-values of statistical tests under a stationary assumption of the Gumbel distribution for the maximum values

p-val (P)	KS-val (P)	p-val (D)	KS-val (D)
0.041374	1.392521	0.586818	0.774087

The KS test is conducted under the assumption of the Gumbel distribution for the maximum value corresponding to Eq. (19):

$$F(w_X; \mu_X, \rho_X) = \exp\Big[-\exp(-\frac{w_X - \mu_X}{\rho_X})\Big]. \tag{17}$$

The p-values of the KS test are shown in Table 1. The stationary Gumbel assumption cannot explain the maximum values for quotes with a 5% significance level in the KS test. The stationary Gumbel assumption may not be accepted in the case of the block maximum number of quotes. The dominant reason is the strong nonstationarity of the maximum number of quotes. During the last 5 years, the currencies and pairs quoted in the electronic brokerage market increased. The mean value of the total number constantly increased. In fact, the maximum number of quotations $w_P(t)$ reached the maximum value on 30 July, 2012. The nonstationarity breaks the assumption of the extreme value theorem.

It is confirmed that the stationary Gumbel assumption can be accepted for the block maxima of transactions in each week using the KS test with a 5% significance level. The maximum number of transactions $w_D(t)$ was reached on on January 30, 2012. This period seems to be related to the extreme synchrony.

3.2 Network Entropy Per Link

The proposed method based on statistical–physical entropy is applied to measure the states of the foreign exchange market. The relationship between a bipartite network structure and macroeconomic shocks or crises was investigated, and the occurrence probabilities of extreme synchrony were inferred. We compute a statistical–physical entropy per link from $m_{X,i}(t)(X \in \{P, D\})$ with Eqs. (2) and (3), which are denoted as $\sigma_X(t)$. $\sigma_P(t)$ and $\sigma_D(t)$.

Since small values of $\sigma_X(t)$ correspond to a concentration of links at a few nodes or a dense network structure, let us consider the minimum value of $\sigma_X(t)$ every week:

$$v_X(s) = \min_{t \in W(s)} \{\sigma_X(t)\}, \tag{18}$$

where $W(s) \quad (s = 1, \ldots, T)$ represents a set of times included in the s-th week. A total of 288 weeks are included in the data set ($T = 288$). According to the extreme value theorem, the probability density for minimum values can be assumed to be the Gumbel density:

$$P(v_X;\mu_X,\rho_X)=\frac{1}{\rho_X}\exp\Big(\frac{v_X+\mu_X}{\rho_X}-e^{\frac{v_X+\mu_X}{\rho_X}}\Big), \tag{19}$$

where μ_X and ρ_X are the location and scale parameters, respectively. Under the assumption of the Gumbel density, these parameters are estimated with the maximum likelihood procedure. The log-likelihood function for T observations $v_X(s')\quad (s'=1,\ldots,T)$ under Eq. (19) is defined as

$$l(\mu_X,\rho_X)=\sum_{s'=1}^{T}\ln\Big[\frac{1}{\rho_X}\exp\Big(\frac{v_X(s')+\mu_X}{\rho_X}-e^{\frac{v_X(s')+\mu_X}{\rho_X}}\Big)\Big]. \tag{20}$$

Partially differentiating $l(\mu_X,\rho_X)$ in terms of μ_X and ρ_X and setting them to zero yields its maximum likelihood estimators as

$$e^{-\frac{\hat{\mu}_X}{\hat{\rho}_X}}=\frac{1}{T}\sum_{s'=1}^{T}e^{\frac{v_X(s')}{\hat{\rho}_X}}, \tag{21}$$

$$\hat{\rho}_X=\frac{\sum_{s'=1}^{T}e^{\frac{v_X(s')}{\hat{\rho}_X}}v_X(s')}{\sum_{s'=1}^{T}e^{\frac{v_X(s')}{\hat{\rho}_X}}}-\frac{1}{T}\sum_{s'=1}^{T}v_X(s'). \tag{22}$$

The parameter estimates are computed as $\hat{\mu}_P=-4.865382$, $\hat{\rho}_P=0.110136$, $\hat{\mu}_D=-5.010175$, and $\hat{\rho}_D=0.120809$ with Eqs. (21) and (22).

The KS test is conducted for the Gumbel distribution for the minimum values corresponding to Eq. (19):

$$F(v_X;\mu_X,\rho_X)=1-\exp\Big[-\exp\Big(\frac{v_X+\mu_X}{\rho_X}\Big)\Big]. \tag{23}$$

The p-value of the distribution is shown in Table 2. The stationary Gumbel assumption cannot explain the synchronizations observed in both quotes and transactions completely with a 5% significance level. The stationary Gumbel assumption is rejected because there is a stationary assumption to derive the extreme value distribution. If we can weaken this assumption, then the goodness of fit may be improved.

Table 2 The p-values of statistical tests under a stationary Gumbel ssumption

p-val (P)	KS-val (P)	p-val (D)	KS-val (D)
0.001393	1.906528	0.019241	1.523791

4 Probability of Extreme Synchrony

The literature detecting structural breaks or change points in an economic time series (Goldfeld and Quandt 1973; Preis et al. 2011; Scalas 2007; Cheong et al. 2012) points out that nonstationary time series are constructed from locally stationary segments sampled from different distributions. Goldfeld and Quandt conducted a pioneering work on the separation of stationary segments (Goldfeld and Quandt 1973). Recently, a hierarchical segmentation procedure was also proposed by Choeng et al. under the Gaussian assumption (Cheong et al. 2012). We applied this concept to define the segments for $v_X(s') \quad (s' = 1, \ldots, T)$.

Let us consider the null model L_1, which assumes that all the observations $v_X(s') \quad (s' = 1, \ldots, T)$ are sampled from a stationary Gumbel density parameterized as μ and ρ. An alternative model $L_2(s)$ assumes that the left observations $v_X(s') \quad (s' = 1, \ldots, s)$ are sampled from a stationary Gumbel density parameterized as μ_L and ρ_L and that the right observations $v_X(s') \quad (s' = s+1, \ldots, T)$ are sampled from a stationary Gumbel density parameterized as μ_R and ρ_R.

Denoting likelihood functions as

$$L_1(\mu, \rho) = \prod_{s'=1}^{T} P(v_X(s'); \mu, \rho), \tag{24}$$

$$\begin{aligned} L_2(s; \mu_L, \rho_L, \mu_R, \rho_R) = &\prod_{s'=1}^{s} P(v_X(s'); \mu_L, \rho_L) \\ &\times \prod_{s'=s+1}^{T} P(v_X(s'); \mu_R, \rho_R), \end{aligned} \tag{25}$$

the difference between the log-likelihood functions can be defined as

$$\Delta(s) = \log L_2(s) - \log L_1. \tag{26}$$

$\Delta(s)$ can be approximated as the Shannon entropy $H[p] = -\int_{-\infty}^{\infty} dv \log p(v) p(v)$:

$$\begin{aligned} \Delta(s) \approx\ & T H[P(v_X; \mu, \rho)] \\ & - s H[P(v_X; \mu_L, \rho_L)] - (T - s) H[P(v_X; \mu_R, \rho_R)]. \end{aligned} \tag{27}$$

Since the Shannon entropy of the Gumbel density expressed in Eq. (12) is calculated as

$$H[P(v_X; \mu_X, \rho_X)] = \ln \rho_X - \gamma_E + 1, \tag{28}$$

where γ_E represents Euler's constant, defined as

$$\gamma_E = \int_0^\infty \ln t e^{-t} \mathrm{d}t, \tag{29}$$

we obtain

$$\Delta(s) \approx T \ln \rho - s \ln \rho_L - (T - s) \ln \rho_R. \tag{30}$$

In the context of model selection, several information criteria are proposed. The information criterion provides both goodness of fit of the model to the data and model complexity. For the sake of simplicity, we use the Akaike information criterion (AIC) to determine the adequate model. The AIC for a model with the number of parameters K and the maximum likelihood of L is defined as

$$AIC = -2 \ln L + 2K. \tag{31}$$

We can compute the difference in AIC between model L_2 and model $L_1(s)$ as

$$\begin{aligned} \Delta_{AIC}(s) &= \text{AIC of } L_2(s) - \text{AIC of } L_1 \\ &\approx -2\big(T \ln \hat{\rho} - s \ln \hat{\rho}_L - (T - s) \ln \hat{\rho}_R\big) + 4, \\ &= -2\Delta(s) + 4 \end{aligned} \tag{32}$$

since the number of parameters of L_1 is 2, that of $L_2(s)$ is 4, and the maximum likelihood is obtained by using their maximum likelihood estimators calculated from

$$\hat{\rho} = \frac{\sum_{s'=1}^{T} e^{\frac{v_X(s')}{\hat{\rho}_X}} v_X(s')}{\sum_{s'=1}^{T} e^{\frac{v_X(s')}{\hat{\rho}_X}}} - \frac{1}{T} \sum_{s'=1}^{T} v_X(s') \tag{33}$$

$$\hat{\rho}_L = \frac{\sum_{s'=1}^{s} e^{\frac{v_X(s')}{\hat{\rho}_L}} v_X(s')}{\sum_{s'=1}^{s} e^{\frac{v_X(s')}{\hat{\rho}_L}}} - \frac{1}{s} \sum_{s'=1}^{s} v_X(s') \tag{34}$$

$$\hat{\rho}_R = \frac{\sum_{s'=s+1}^{T} e^{\frac{v_X(s')}{\hat{\rho}_R}} v_X(s')}{\sum_{s'=s+1}^{T} e^{\frac{v_X(s')}{\hat{\rho}_R}}} - \frac{1}{T - s} \sum_{s'=s+1}^{T} v_X(s') \tag{35}$$

Therefore, $P(v_X; \mu_L, \rho_L)$ is maximally different from $P(v_X; \mu_R, \rho_R)$ when $\Delta(s)$ assumes a maximal value. This spectrum has a maximum at some time s^*, which is denoted as

$$\Delta^*_{AIC} = \Delta_{AIC}(s^*) = \max_s \Delta_{AIC}(s). \tag{36}$$

The segmentation can be used recursively to separate the time series into further smaller segments. We do this iteratively until all segment boundaries have converged onto their optimal segment, defined by a stopping (termination) condition.

Several termination conditions were discussed in previous studies (Cheong et al. 2012). Assuming that $\Delta_0 > 0$, we terminate the iteration if Δ^*_{AIC} is less than a typical conservative threshold of $\Delta_0 = 10$, while the procedure is recursively conducted if Δ^*_{AIC} is larger than Δ_0. We checked the robustness of this segmentation procedure for Δ_0. Δ_0 gives a statistical significance level of termination. The value of Δ_0 is related to statistical significance. According to Wilks theorem, $-2\Delta(s)$ follows a chi-squared distribution with a degree of freedom r, where r is given by the difference between the number of parameters assumed in the null hypothesis and one in the alternative hypothesis. In this case, $r = 2$. Hence, the cumulative distribution function of Δ^*_{AIC} may follow

$$\Pr[\Delta^*_{AIC} > x] = 1 - \gamma\left(1, \frac{x-4}{2}\right), \tag{37}$$

where $\gamma(x, a)$ is the regularized incomplete gamma function defined as

$$\gamma(a, x) = \frac{\int_0^x t^{a-1}e^{-t}\mathrm{d}t}{\int_0^\infty t^{a-1}e^{-t}\mathrm{d}t}. \tag{38}$$

Therefore, setting the threshold $\Delta_0 = 10$ implies that the segmentation procedure is tuned as a 4.928% statistical significance level.

Let the number of segments be L_X, the parameter estimates be $\{\mu_{X,j}, \rho_{X,j}\}$ at the j-th segment, and the length of the j-th segment be $\tau_{X,j}$, where $\sum_{j=1}^{R_X} \tau_{X,j} = T$. The cumulative probability distribution for $v_X(s) \quad (s = 1, \ldots, T)$ may be assumed to be a finite mixture of Gumbel distributions:

$$\begin{aligned}\Pr(V_X \leq v_X) &= \int_{-\infty}^{v_X} \sum_{j=1}^{R_X} \frac{\tau_{X,j}}{T} P(v'_X; \mu_{X,j}, \rho_{X,j})\mathrm{d}v'_X \\ &= \sum_{j=1}^{R_X} \frac{\tau_{X,j}}{T}\left\{1 - \exp\left[-e^{\frac{v_X+\mu_{X,j}}{\rho_{X,j}}}\right]\right\},\end{aligned} \tag{39}$$

Tables 3 and 4 show parameter estimates of $v_P(s)$ and $v_D(s)$ using the recursive segmentation procedure. Figure 5 shows the temporal development of $v_P(s)$ and $v_D(s)$ from May 28, 2007 to November 30, 2012. $R_P = 6$ and $R_D = 6$ are obtained from $v_X(s)$ using the proposed segmentation procedure. During the observation period, the global financial system suffered from the following significant macroeconomic shocks and crises: (I) the BNP Paribas shock (August 2007), (II) the Bear Stearns shock (February 2008), (III) the Lehman shock (September 2008 to March 2009), (IV) the European sovereign debt crisis (April to May 2010), (V) the East

Table 3 Parameter estimates obtained from the weekly minimum values of network entropy for quotations with the recursive segmentation procedure

j	Start date	End date	τ/T	$\hat{\mu}_{X,j}$	$\hat{\rho}_{X,j}$
1	May 28, 2007	Oct. 15, 2007	0.072917	−4.965486	0.078571
2	Oct. 22, 2007	Aug. 17, 2009	0.333333	−4.829061	0.080558
3	Aug. 24, 2009	Oct. 21, 2011	0.274306	−4.880156	0.099532
4	Feb. 28, 2011	Jul. 30, 2012	0.260417	−4.798364	0.058265
5	Aug. 6, 2012	Sep. 10, 2012	0.020833	−4.856415	0.121566
6	Sep. 17, 2012	Nov. 26, 2012	0.038194	−5.031930	0.071871

Table 4 Parameter estimates obtained from the weekly minimum values of network entropy for transactions with the recursive segmentation procedure

j	Start date	End date	τ/T	$\hat{\mu}_{X,j}$	$\hat{\rho}_{X,j}$
1	May 28, 2007	Mar. 16, 2009	0.329861	−4.903660	0.072684
2	Mar. 23, 2009	Jun. 14, 2010	0.225694	−4.990569	0.088420
3	Jun. 21, 2010	Dec. 5, 2011	0.267361	−5.065305	0.087721
4	Dec. 12, 2011	Mar. 12, 2012	0.048611	−4.826384	0.174328
5	Mar. 19, 2012	Apr. 30, 2012	0.024306	−5.108679	0.011079
6	May 7, 2012	Nov. 26, 2012	0.104167	−5.131156	0.079630

Japan tsunami (March 2011), (VI) the United States debt-ceiling crisis (May 2011), and (VII) the Bank of Japan's 10 trillion JPY gift on Valentine's Day (February 2012).

Before entering these global affairs, both $v_P(s)$ and $v_D(s)$ took large values. Note that, during the (I) Paribas shock, the (II) Bear Stearns and the (III) Lehman shock $v_P(s)$ and $v_D(s)$ took smaller values than they did during the previous term. This implies that a global shock may drive many participants and that these participants may trade the same currencies at the same time. The smallest values $v_P(s)$ and $v_D(t)$ correspond to the days of the (II) Bear Stearns shock, the (III) Lehman shock, and the (VI) Euro crisis. These days are generally related to the start or the end of macroeconomic shocks or crises. The period from December 2011 to March 2012 shows that the values of $v_D(s)$ are smaller than they were during other periods. This result implies that, during period, singular patterns appeared in the transactions.

Figure 6 shows both the empirical and estimated cumulative distribution functions of $v_P(s)$ and $v_D(s)$. The estimated cumulative distributions are drawn from Eq. (39) with parameter estimates. The KS test verifies this mixing assumption. The distribution estimated by the finite mixture of Gumbel distributions for quotes is well fitted, as shown in Table 5. From the p-values, the mixture of Gumbel distributions for quotations accepts the null hypothesis that $v_P(s)$ is sampled from the mixing distribution with a 5% significance level. The mixture of Gumbel distributions for transactions also accepts the null hypothesis that $v_D(s)$ is sampled from the mixing distribution with a 5% significance level. Extrapolation of cumulative distribution function also provides a guideline of the future probability of extreme events. The finite mixture Gumbel distributions with parameter estimates may be used as an inference of probable extreme synchrony.

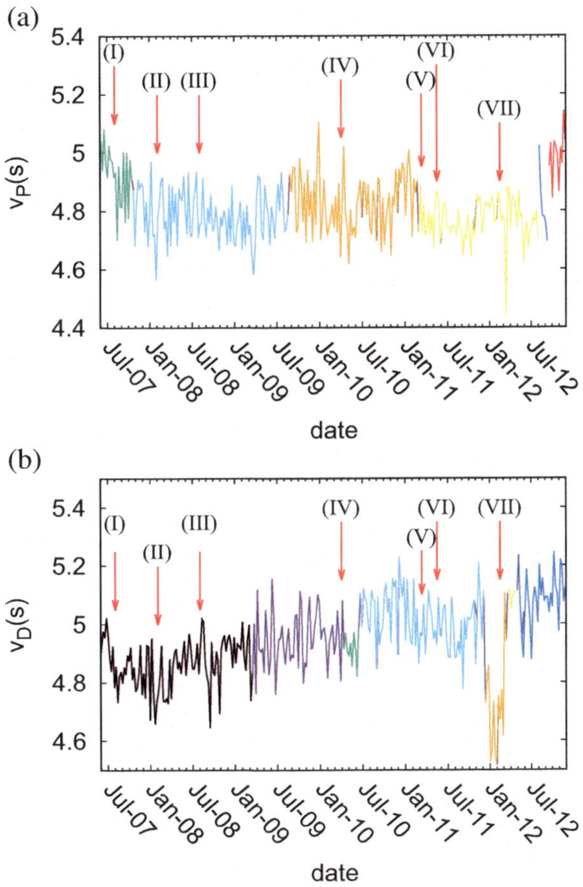

Fig. 5 Temporal development of (**a**) $v_P(s)$ and (**b**) $v_D(s)$ from May 28, 2007 to November 30, 2012. Each color corresponds to a segment which is shown in Tables 3 and 4

5 Conclusion

A method based on the concept of "entropy" in statistical physics was proposed to quantify states of a bipartite network under constraints. The statistical–physical network entropy of a bipartite network was derived under the constraints for the number of links at each group node. Both numerical and theoretical calculations for a binary bipartite graph with random links showed that the network entropy per link can capture both the density and the concentration of links in the bipartite network. The proposed method was applied to measure the structure of bipartite networks consisting of currency pairs and participants in the foreign exchange market.

An empirical investigation of the total number of quotes and transactions was conducted. The nonstationarity of the number of quotes and transactions strongly affected the extreme value distributions. The empirical investigation confirmed that the entropy per link decreased before and after the latest global shocks that have

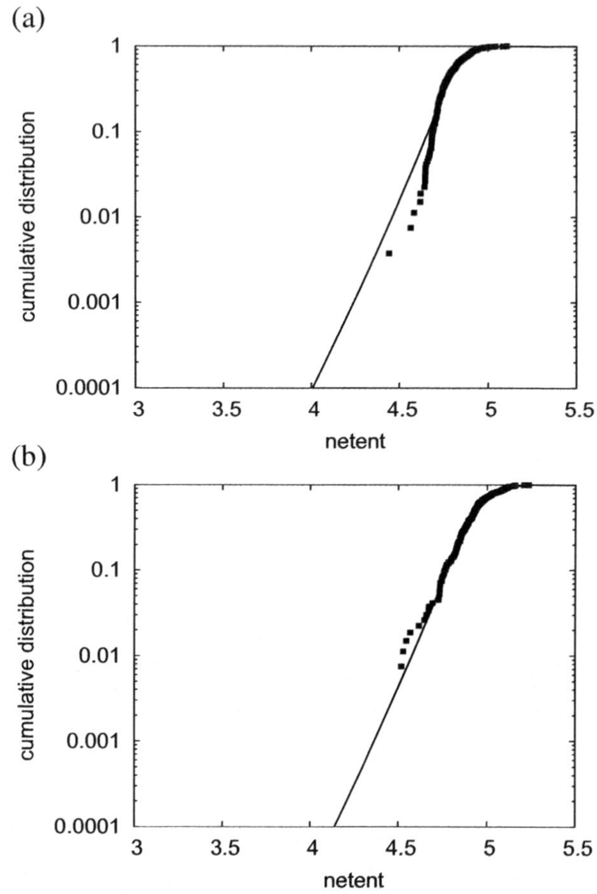

Fig. 6 Cumulative distribution functions for the minimum values of the entropy per link in each week (**a**) $v_P(s)$ and (**b**) $v_D(s)$. Filled squares represent the empirical distribution of $v_P(s)$, and unfilled circles represent the empirical distribution of $v_D(s)$. A solid curve represents the estimated distribution of $v_P(s)$, and a dashed curve represents the estimated distribution of $v_D(s)$

Table 5 The p-values of statistical tests under the assumption of a finite mixture of Gumbel distributions

p-val (P)	KS-val (P)	p-val (D)	KS-val (D)
0.183793	1.092317	0.829013	0.625372

influenced the world economy. A method was proposed to determine segments with recursive segmentation based on the Akaike information criterion between Gumbel distributions with different parameters. Under the assumption of a finite mixture of Gumbel distributions, the estimated distributions were verified by the Kolmogorov–Smirnov test. The finite mixture of Gumbel distributions can estimate the occurrence probabilities of extreme synchrony of a nonstationary system extracted as a bipartite network. The extrapolation of the extreme synchrony can be done based on the estimated mixture of Gumbel distributions.

Acknowledgements This work is supported by Japan Science and Technology Agency (JST) PRESTO Grant Number JPMJPR1504, Japan. The work was also financially supported by the Grant-in-Aid for Young Scientists (B) by the Japan Society for the Promotion of Science (JSPS) KAKENHI (#23760074).

References

Albert R, Barabási A-L (2002) Statistical mechanics of complex networks. Rev Mod Phys 74:47–97

Andersen TG (1996) Return volatility and trading volume: an information flow interpretation of stochastic volatility. J Financ 51:169–204

Bianconi G (2009) Entropy of network ensembles. Phys Rev E 79:036114; Anand K, Bianconi G (2009) Entropy measures for networks: toward an information theory of complex topologies. Phys Rev E 80:045102

Bonanno G, Caldarelli G, Lillo F, Mantegna RN (2003) Topology of correlation-based minimal spanning trees in real and model markets. Phys Rev E 68:046130

Carbone A, Stanley HE (2007) Scaling properties and entropy of long-range correlated time series. Physica A 304:21–24

Carbone A, Kaniadakis G, Scarfone AM (2007) Eur Phys J B 57:121

Cheong SA, Fornia RP, Lee GHT, Kok JL, Yim WS, Xu DY, Zhang Y (2012) The Japanese economy in crises: a time series segmentation study. Econ E-J 2012-5. http://www.economics-ejournal.org

Chmiel AM, Sienkiewicz J, Suchecki K, Hołyst JA (2007) Networks of companies and branches in Poland. Physica A 383:134–138

Clark P (1973) A subordinated stochastic process model with finite variance for speculative prices. Econometrica 41:135–155

Dehmer M, Mowshowitz A (2011) A history of graph entropy measures. Inf Sci 181:57–78

de Menezes MA, Barabási A-L, Fluctuations in network dynamics. Phys Rev Lett 92 (2004) 028701.

Goldfeld SM, Quandt RE (1973) A Markov model for switching regressions. J Econometrics 1:3–15

Guillaume J-L, Latapy M (2006) Bipartite graphs as models of complex networks. Physica A 371:795–813

Gworek S, Kwapień J, Drożdż S (2010) Sign and amplitude representation of the forex networks. Acta Phys Pol A 117:681–687

ICAP (2013). The data is purchased from ICAP EBS: http://www.icap.com

Iori G, Masi GD, Precup OV, Gabbi G, Caldarelli G (2008) A network analysis of the Italian overnight money market. J Econ Dyn Control 32:259–278

Lambiotte R, Ausloos M, Thelwall M (2007) Word statistics in Blogs and RSS feeds: towards empirical universal evidence. J Informetrics 1:277–286

Lämmer S, Gehlsen B, Helbing D (2006) Scaling laws in the spatial structure of urban road networks. Physica A 363:89–95

Lamoureux CG, Lastrapes WD (1990) Heteroskedasticity in stock return data: volume versus GARCH effects. J Financ 45:221–229

Liesenfeld R (1998) Dynamic bivariate mixture models: modeling the behavior of prices and trading volume. J Bus Econ Stat 16:101–109

Mandelbrot BB, Taylor H (1967) On the distribution of stock price differences. Oper Res 15:1057–1062

Milaković M, Alfrano S, Lux T (2010) The small core of the German corporate board network. Comput Math Organ Theory 16:201–215

Miura W, Takayasu H, Takayasu M (2012) Effect of coagulation of nodes in an evolving complex network. Phys Rev Lett 108:168701
Mowshowitz A (1968) Entropy and the complexity of graphs: I. An index of the relative complexity of a graph. Bull Math Biophys 30:175–204
Podobnik B, Horvatic D, Petersen AM, Stanley HE (2009) Cross-correlations between volume change and price change. Proc Natl Acad Sci U S A 106:22079–22084
Preis T, Schneider JJ, Stanley HE (2011) Switching processes in financial markets. Proc Natl Acad Sci U S A 108:7674–7678
Rashevsky N (1955) Life, information theory, and topology. Bull Math Biophys 17:229–235
Richardson M, Smith T (1994) A direct test of the mixture of distributions hypothesis: measuring the daily flow of information. J Financ Quant Anal 29:101–116
Sato A-H (2007) Frequency analysis of tick quotes on the foreign exchange market and agent-based modeling: a spectral distance approach. Physica A 382:258–270
Sato A-H (2010) Comprehensive analysis of information transmission among agents: similarity and heterogeneity of collective behavior. In: Chen S-H et al (eds) Agent-based in economic and social systems VI: post-proceedings of the AESCS international workshop 2009, agent-based social systems, vol 8. Springer, Tokyo, pp 1–17
Sato A-H (2017) Inference of extreme synchrony with an entropy measure on a bipartite network. In: 2017 IEEE 41st annual computer software and applications conference (COMPSAC), pp. 766–771
Sato A-H, Hołyst JA (2008) Characteristic periodicities of collective behavior at the foreign exchange market. Eur Phys J B 62:373–380
Scalas E (2007) Mixtures of compound Poisson processes as models of tick-by-tick financial data. Chaos, Solitons Fractals 34:33–40
Tauchen T, Pitts M (1983) The price variability-volume relationship on speculative markets. Econometrica 51:485–505
Trucco E (1956) A note on the information content of graphs. Bull Math Biophys 18:129–135
Tumminello M, Miccichè S, Lillo F, Pillo J, Mantegna RN (2011) Statistically validated networks in bipartite complex systems. PLoS One 6:e17994
Watanabe T (2000) A nonlinear filtering approach to stochastic volatility models with an application to daily stock returns. J Bus Econ Stat 18:199–210
Wilhelm T, Hollunder J (2007) Information theoretic description of networks. Physica A 385:385–396